NOTICE TO THE READER

COVER PHOTOS: Circuit board photo © Manfred Kage/Peter Arnold, Inc.
  Photo of technician with oven and LTS-2020 courtesy of Analog Devices, Norwood, MA
  Photo of autoranging multimeter courtesy of Simpson Electric Company, Elgin, IL

Delmar Staff

Associate Editor: Cameron O. Anderson
Editing Manager: Gerry East
Project Editor: Christine E. Worden
Production Coordinator: Judith Block
Design Coordinator: Susan C. Mathews

For information address Delmar Publishers Inc.,
2 Computer Drive West, Box 15-015
Albany, New York 12212

Library of Congress Cataloging in Publication Data

Bignell, James.
  Digital electronics/James Bignell, Robert Donovan.—2nd ed.
     p.   cm.
  Includes index.
  ISBN 0–8273–3134–7. — ISBN 0–8273–3135–5 (instructors guide)
  1. Digital electronics.   I. Donovan, Robert.   II. Title.
TK7868.D5B54   1989
621.3815—dc19   88–39507
                                    CIP

DIGITAL
ELECTRONICS
Second Edition

James Bignell
Robert Donovan

 Delmar Publishers Inc.®

Contents

Preface

Digital Electronics was designed to be used as the primary text for a systematic and sequential course in digital theory. The content of each chapter builds on the knowledge learned in the previous chapters. The authors have tried to write the material in clear and succinct form. The text contains many drawings. Each chapter includes lists of objectives for the reader. There are exercises and labs at the end of each chapter. The 15 chapters can be covered in a typical semester course at the rate of one chapter per week.

The labs are designed to reinforce the material presented in the chapter and are closely correlated to the chapter material. In the labs the theory comes alive and practical hands-on skills are learned. Many of the labs have a section which covers troubleshooting the circuits in the lab. By wiring the circuits and making them work in the lab, you develop the kind of analytical troubleshooting techniques needed to repair digital circuits. Pinouts for the ICs used in the labs are listed in Appendix C but this is no substitute for a good TTL and CMOS manual. The skill developed by using a data manual for solving circuit problems and obtaining needed pinouts for ICs is invaluable. It is recommended that you have a TTL and CMOS manual from one of the major IC manufacturers. The text can be studied without performing the labs but the hands-on experience increases the knowledge acquired to a much higher level.

A new chapter (16) has been added which combines the very important memories chapter (formerly Appendix D) with valuable content on the microcomputer. This final chapter allows for a smooth transition into the more advanced area of microprocessors.

A brief definition of many of the terms and phrases used in this text are listed in the Glossary. Refer to it often if you are unsure of the terminology being used. You should have a good working knowledge of ac and dc electronics and a beginning knowledge of semiconductors such as diodes and transistors. Mathematics is not stressed but you should be able to handle basic college algebra. You should also have developed skills for using meters, oscilloscopes and power supplies.

Acknowledgments

The authors would like to thank the reviewers for their encouragement and help in developing this text. These reviewers include Russ Bonine of San Diego City College; George Crossman of Broward Community College; Michael Cushing of North Seattle Community College; Ray Dong of Foothill College; Fred Driscoll of Wentworth Institute; Thomas Grady of Arizona State University; Richard Hardman of SUNY Alfred Ag. and Tech. College; Stan Hochman of North Seattle Community College; Bill Kramer of Santa Rosa Junior College; Vincent Loizzo of DeVry Institute of Technology; Frank Morrison of

Electronics Institute of Technology; Duane Overby of West Valley College; S. Neil Towey of DeVry Institute of Technology; Edward Tuba of Heald College; Harry Waller of San Antonio College; Gus Rummel of Central Texas College; Jim Knight of ECPI; Steve Lenoir of RETS Electronic Institute; and Al Brunkow of SECC.

We would like to thank National Semiconductor Corporation for permission to use the following figures: Figures 2-5, 2-12, 2-21, 2-27, 2-34, 4-4, 4-9, 4-21, 4-23, 5-13, 5-17, 6-1, 6-5, 6-7, 6-10, 6-11, 6-14, 6-29, 6-30, 6-31, 6-32, 10-11, 12-7, 14-9, 14-10, 14-16, 15-3, 15-4, 15-5, 15-6, 15-7, and for Appendix C.

CHAPTER 1

Number Systems

1.1 BINARY NUMBER SYSTEM

Digital electronics makes extensive use of the binary number system. Binary is useful in electronics because it uses only two digits, 1 and 0. Binary digits are used to represent the two voltage levels used in digital electronics, HIGH or LOW. In most digital systems, a high voltage level is represented by 1; a low voltage level or zero volts is represented by 0. A switch, a light, or a transistor can be ON and represented by 1 or OFF and represented by 0. A decimal number like 32 must be converted into binary and represented by ones and zeros before it can be manipulated by a digital computer.

Since we use the decimal number system daily, we are most familiar with it. First, we will examine a characteristic of the decimal number system and then compare the binary system to it. In the decimal system, we work with ten different digits, zero through nine. These ten digits make the decimal system a base-ten system. In the binary system, we work with two different digits, 0 and 1. These two digits make the binary system a base-two system.

To count in the decimal system, start in the first column or decimal place with 0 and count up to 9. Once the first place is "full," reset the first column to 0 and add 1 to the next column to the left. After 9 comes 10. Now the first column can be "filled" again. After 10 comes 11, 12, 13, etc. When the first column is full again, reset to 0 and add 1 to the next column to the left. After 19 comes 20. When both columns are full, reset both to zeros and add 1 to the next column on the left. After 99 comes 100.

To count in binary, start in the first column or binary place with 0 and count up to 1. The first column is full. Reset and add 1 to the next binary place on the left. After 0 comes 1, 10. Now the first column can be filled again. After 10 comes 11. Both columns are full. Reset both and add one to the next binary place to the left. After 11 comes 100. Now the first column can be filled again. After 100 comes 101, 110, 111, 1000, 1001, 1010, 1011, 1100, 1101, and so on. Counting in binary we have

0	1000	10000	11000
1	1001	10001	11001
10	1010	10010	11010
11	1011	10011	11011
100	1100	10100	11100
101	1101	10101	11101
110	1110	10110	11110
111	1111	10111	11111

Try writing the binary numbers from 11111 to 1000000.

The word *bit* is a contraction of the words *binary digit*. Each place in a binary number is called a bit. The binary number 10110 is a 5-bit binary number. The first place on the right is called the least significant bit (LSB) and the left-most place is called the most significant bit (MSB).

Most Significant Bit (MSB)

10110 is a 5-bit binary number

Least Significant Bit (LSB)

Using three bits we can count in binary to 111 or 7. Including 000, we have eight different combinations. In general, with N bits we can count up to $2^N - 1$ for a total of 2^N different numbers.

Example: How high can you count using a 4-bit number?

Solution:

With $N = 4$, we can count up to $2^4 - 1 = 15$.

Example: How many different numbers can be represented with 6 bits?

Solution:

With $N = 6$, there are 2^N combinations, $2^6 = 64$.

1.2 BINARY TO DECIMAL CONVERSION

In the decimal system, the first decimal place to the left of the decimal point is called the one's or unit's place. Each column to the left increases by a factor of ten (base-ten system). Moving to the left from the decimal the values can be expressed in terms of the base-ten as 10^0, 10^1, 10^2, 10^3, and so on. The decimal number 3954 means $(3 \times 1000) + (9 \times 100) + (5 \times 10) + (4 \times 1)$, for a total value of three thousand, nine hundred fifty-four.

In the binary system the first binary place to the left of the binary point is still the one's or unit's place. Each column to the left increases by a factor of 2 (base-two system). Moving to the left from the binary point the column values are 1, 2, 4, 8, 16, 32, 64, 128, 256, 512, 1024, and so on. These values can be represented in terms of the base two as 2^0, 2^1, 2^2, 2^3, 2^4, 2^5, 2^6, 2^7, 2^8, 2^9, 2^{10}, and so on. The binary number 10110 means $(1 \times 16) + (0 \times 8) + (1 \times 4) + (1 \times 2) + (0 \times 1)$, for a total value of 22. The binary number 10110 is the same as the decimal number 22. A binary number is often distinguished from a decimal number by writing the base as a subscript. Thus

$$10110_2 = 22_{10}$$

To convert a binary number to a decimal number, list the value of each place; then total the values that are represented by ones.

Example: Convert 1000111_2 to a decimal number.

Solution:

List the value of each place.

1	0	0	0	1	1	1
64	32	16	8	4	2	1

Total the values that are represented by ones.
$64 + 4 + 2 + 1 = 71$

$1000111_2 = 71_{10}$

Example: Convert 101011_2 to a decimal number.

Solution:

1	0	1	0	1	1
32	16	8	4	2	1

$32 + 8 + 2 + 1 = 43$

$101011_2 = 43_{10}$

Example: Convert 11001100_2 to a decimal number.

Solution:

1	1	0	0	1	1	0	0
128	64	32	16	8	4	2	1

$$128 + 64 + 8 + 4 = 204$$

$$11001100_2 = 204_{10}$$

1.3   DECIMAL TO BINARY CONVERSION

Two methods are presented for converting decimal numbers to binary numbers.

Method 1

Label the binary places until you reach the place with a value which exceeds the decimal number to be converted. For example, to convert 23_{10} to a binary number:

32	16	8	4	2	1

There are no 32's in 23, but there is a 16. Place a 1 in the 16's column, and subtract 16 from 23 to see how much is left to convert.

	1				
32	16	8	4	2	1

$$23 - 16 = 7$$

There are no eights in 7, but there is a 4. Place a 0 in the 8's column and a 1 in the 4's place and subtract 4 from 7 to see what remains.

	1	0	1		
32	16	8	4	2	1

$$7 - 4 = 3$$

There is a 2 in 3. Place a 1 in the 2's column and subtract 2 from 3 to see what remains.

	1	0	1	1	
32	16	8	4	2	1

$$3 - 2 = 1$$

Place a 1 in the 1's column and subtract 1 from 1 to see what remains.

	1	0	1	1	1
32	16	8	4	2	1

$$1 - 1 = 0 \qquad \text{The process is finished.}$$

$$23_{10} = 10111_2$$

Example: Convert 45_{10} to a binary number.

Solution:

1	0	1	1	0	1
64	32	16	8	4	2	1

$$45 - 32 = 13$$
$$13 - 8 = 5$$
$$5 - 4 = 1$$
$$1 - 1 = 0$$

$$45_{10} = 101101_2$$

Example: Convert 132_{10} to a binary number.

Solution:

1	0	0	0	0	1	0	0
256	128	64	32	16	8	4	2	1

$$132 - 128 = 4$$
$$4 - 4 = 0$$

$$132_{10} = 10000100_2$$

Method 2

Divide the decimal number to be converted successively by 2, ignoring the remainders, until you have a quotient of 0. The remainders will be used later to determine the answer. For example, to convert 101_{10} to a binary number:

$$101 \div 2 = 50 \text{ remainder } 1 \qquad \text{LSB}$$
$$50 \div 2 = 25 \text{ remainder } 0$$
$$25 \div 2 = 12 \text{ remainder } 1$$
$$12 \div 2 = 6 \text{ remainder } 0$$
$$6 \div 2 = 3 \text{ remainder } 0$$
$$3 \div 2 = 1 \text{ remainder } 1$$
$$1 \div 2 = 0 \text{ remainder } 1 \qquad \text{MSB}$$

Read the remainders from bottom to top to form the answer.

$$1100101$$

Therefore,

$$101_{10} = 1100101_2$$

Example: Convert 291_{10} to a binary number.

Solution:

$291 \div 2 = 145$ remainder 1   LSB
$145 \div 2 = 72$ remainder 1
$72 \div 2 = 36$ remainder 0
$36 \div 2 = 18$ remainder 0
$18 \div 2 = 9$ remainder 0
$9 \div 2 = 4$ remainder 1
$4 \div 2 = 2$ remainder 0
$2 \div 2 = 1$ remainder 0
$1 \div 2 = 0$ remainder 1   MSB

$291_{10} = 100100011_2$

Example: Convert 1024_{10} to a binary number.

Solution:

$1024 \div 2 = 512$ remainder 0   LSB
$512 \div 2 = 256$ remainder 0
$256 \div 2 = 128$ remainder 0
$128 \div 2 = 64$ remainder 0
$64 \div 2 = 32$ remainder 0
$32 \div 2 = 16$ remainder 0
$16 \div 2 = 8$ remainder 0
$8 \div 2 = 4$ remainder 0
$4 \div 2 = 2$ remainder 0
$2 \div 2 = 1$ remainder 0
$1 \div 2 = 0$ remainder 1   MSB

$1024_{10} = 10000000000_2$

Although binary numbers are ideal for digital machines, they are cumbersome for humans to manipulate. It is very difficult to copy a string of 8-bit binary numbers without losing or transposing a 1 or 0. Octal and hexadecimal number systems are used as an aid in handling binary numbers. First, we will examine the characteristics of octal numbers and use them to represent binary numbers. Then we will examine hexadecimal numbers and use them to represent binary numbers.

1.4   OCTAL NUMBER SYSTEM

Octal is a base-eight number system. There are eight different digits to work with, zero through seven. To count in octal, start in the first column to the left of the octal point and count from zero to seven. The first column is full, so reset to zero and add one to the next column. After 7 comes 10. Now fill the first column again. After 10 comes 11, 12, 13, 14, 15, 16, 17. The first column is again full, so reset and add one to the next column to the left. After 17 comes 20, 21, 22 and so on. When the first two columns are full, reset both and add one to the next column. After 77 comes 100, 101, 102 and so on. After 757 comes 760, 761, 762 and so on.

Example: Count in octal from 666_8 to 720_8.

Solution:

666	675	704	713
667	676	705	714
670	677	706	715
671	700	707	716
672	701	710	717
673	702	711	720
674	703	712	

In the octal system the first place to the left of the octal point is the one's or unit's place. Each column to the left increases the factor of 8 (base-eight system). Moving to the left from the octal point the values of the columns are 1, 8, 64, 512, 4096, and so on. These values can be expressed in terms of the base number 8 as 8^0, 8^1, 8^2, 8^3, 8^4, and so on. The octal number 6405_8 means $(6 \times 512) + (4 \times 64) + (0 \times 8) + (5 \times 1)$ for a total value of 3333_{10}. An octal number is distinguished from a decimal number by writing the base as a subscript.

$$6405_8 = 3333_{10}$$

Comparing decimal, binary and octal we have

Decimal	Binary	Octal
0	000	0
1	001	1
2	010	2
3	011	3
4	100	4
5	101	5
6	110	6
7	111	7
8	1000	10
9	1001	11
10	1010	12
11	1011	13
12	1100	14

Notice that three binary bits correspond perfectly with one octal digit. That is, it takes exactly three bits to count from zero to seven.

1.5   BINARY TO OCTAL CONVERSION

The fact that three binary bits represent eight different digits yields an easy method for converting from binary to octal. Starting at the binary point and moving to the left, mark off groups of three. Then using weights of 1, 2, and 4, convert each group into the corresponding octal digit.

Example: Convert 10111101_2 to an octal number.

Solution:

10	111	101
2	7	5

Note that the most significant group only has two bits.

$$10111101_2 = 275_8$$

An eight-bit binary word can be represented by three octal digits, which are much easier to handle.

Example: Convert 10101010_2 to an octal number.

Solution:

10	101	010
2	5	2

$$10101010_2 = 252_8$$

Example: Convert $11010100110111101001000_2$ to an octal number.

Solution:

11	010	100	110	111	101	001	000
3	2	4	6	7	5	1	0

$$11010100110111101001000_2 = 32467510_8$$

1.6   OCTAL TO BINARY CONVERSION

To convert from octal back to binary is just as easy. For each octal digit write the three corresponding binary bits. For example, to convert 3062_8 to a binary number:

3	0	6	2
11	000	110	010

The number $3062_8 = 11000110010_2$. Note that the 2 is written 010 with a leading 0 added to complete the three bits and that 0 is written 000 to hold three places. On the most significant digit, leading zeros can be suppressed. The 3 was written as 11 and not 011.

Example: Convert 377_8 to a binary number.

Solution:

$$
\begin{array}{ccc}
3 & 7 & 7 \\
11 & 111 & 111
\end{array}
$$

$377_8 = 11111111_2$

Example: Convert 647015_8 to a binary number.

Solution:

$$
\begin{array}{cccccc}
6 & 4 & 7 & 0 & 1 & 5 \\
110 & 100 & 111 & 000 & 001 & 101
\end{array}
$$

$647015_8 = 110100111000001101_2$

1.7 HEXADECIMAL NUMBER SYSTEM

An alternate method for handling binary numbers is to use the hexadecimal number system. Hexadecimal is a base-sixteen number system which means that a choice of 16 digits is available for each column. Those 16 are 0, 1, 2, 3, 4, 5, 6, 7, 8, 9, A, B, C, D, E, and F. To count in the hexadecimal system, start in the first column to the left of the hexadecimal point and count from 0 to F. Once the first column is full, reset and add one to the second column. After 18, 19, 1A, 1B, 1C, 1D, 1E, 1F comes 20, 21, and so on. After 9FFF comes A000, and so on.

Example: Count in the hexadecimal number system from AE9 to B00.

Solution:

AE9	AEF	AF5	AFB
AEA	AF0	AF6	AFC
AEB	AF1	AF7	AFD
AEC	AF2	AF8	AFE
AED	AF3	AF9	AFF
AEE	AF4	AFA	B00

The first column to the left of the hexadecimal point is the one's or unit's place. Moving to the left, each column increases by a factor of 16, giving values of 1, 16, 256, 4096, 65536, 1048576, and so on. The hexadecimal number $A6F0_{16}$ means $(10 \times 4096) + (6 \times 256) + (15 \times 16) + (0 \times 1)$ for a total value of 42736. A hexadecimal number is distinguished from a decimal number by writing the base number as a subscript.

$$A6F0_{16} = 42736_{10}$$

Comparing decimal, binary and hexadecimal numbers, we have

Decimal	Binary	Hexadecimal
0	0000	0
1	0001	1
2	0010	2
3	0011	3
4	0100	4
5	0101	5
6	0110	6
7	0111	7
8	1000	8
9	1001	9
10	1010	A
11	1011	B
12	1100	C
13	1101	D
14	1110	E
15	1111	F
16	10000	10

1.8 BINARY TO HEXADECIMAL CONVERSION

Notice that four binary bits correspond perfectly with one hexadecimal digit. That is, it takes exactly four binary bits to count from 0 to F. To represent binary numbers as hexadecimal numbers, mark off groups of four, starting at the binary point and moving to the left. Then convert each group into the corresponding hexadecimal digit. Until you learn the binary to hexadecimal conversions, refer to the chart in section 1.7 or, better yet, make your own chart in the margin of your paper. With use, the conversions become automatic.

Example: Convert 10111001_2 to a hexadecimal number.

Solution:

$$1011 \qquad 1001$$
$$\text{B} \qquad\quad 9$$

$$10111001_2 = \text{B9}_{16}$$

An 8-bit binary word can be represented quite nicely with two hexadecimal digits.

Example: Convert 01011110_2 to a hexadecimal number.

Solution:

$$0101 \qquad 1110$$
$$5 \qquad\quad \text{E}$$

$$01011110_2 = 5\text{E}_{16}$$

Example: Convert 11110000001110_2 to a hexadecimal number.

Solution:

11	1100	0000	1110
3	C	0	E

$11110000001110_2 = 3C0E_{16}$

1.9   HEXADECIMAL TO BINARY CONVERSION

Conversion from hexadecimal back to binary is just as easy. For each hexadecimal digit write the corresponding four binary bits. Refer to the chart in section 1.7 until you learn the conversions.

Example: Convert $C3A6_{16}$ to a binary number.

Solution:

C	3	A	6
1100	0011	1010	0110

$C3A6_{16} = 1100001110100110_2$

Note that 3 is written 0011 to complete the four bits required and 6 is written 0110. The leading zeros must be added to work with groups of four bits. On the most significant digit, leading zeros can be suppressed.

Example: Convert $48BA_{16}$ to a binary number.

Solution:

4	8	B	A
100	1000	1011	1010

$48BA_{16} = 100100010111010_2$

Example: Convert $1FC02_{16}$ to a binary number.

Solution:

1	F	C	0	2
1	1111	1100	0000	0010

$1FC02_{16} = 11111110000000010_2$

1.10   BINARY CODED DECIMAL (BCD)

Some binary machines represent decimal numbers in codes other than straight binary. One such code is *Binary Coded Decimal* (BCD). In BCD each decimal digit is represented with four binary bits, according to the weighted 1, 2, 4, 8 system that you have already learned.

Example: Convert 3906_{10} to BCD.

Solution:

3	9	0	6
11	1001	0000	0110

$3906_{10} = 1110010000110_{BCD}$

Note that leading zeros are added to ensure that each digit is represented by four bits. The leading zeros can be suppressed on the most significant digit.

Example: Convert 5437_{10} to BCD.

Solution:

5	4	3	7
101	0100	0011	0111

$5437_{10} = 101010000110111_{BCD}$

Converting BCD back to decimal is just as easy. Starting at the BCD point and moving to the left, mark off groups of four. Convert each group of four bits to the corresponding decimal digit.

Example: Convert 11010010011_{BCD} to a decimal number.

Solution:

110	1001	0011
6	9	3

$11010010011_{BCD} = 693_{10}$

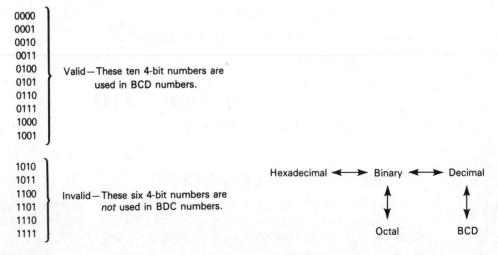

```
0000
0001
0010
0011
0100    Valid—These ten 4-bit numbers are
0101         used in BCD numbers.
0110
0111
1000
1001
```

```
1010
1011
1100    Invalid—These six 4-bit numbers are
1101         not used in BDC numbers.
1110
1111
```

Hexadecimal ⟷ Binary ⟷ Decimal

Octal   BCD

FIGURE 1-1   Valid and invalid BCD numbers   **FIGURE 1-2   Conversion flowchart**

Using four bits, we can count from 1 to 15. The six numbers over 9 are not valid BCD numbers because these numbers do not convert to a single decimal digit. Care must be taken that these numbers are not used in the BCD system. For example, 1010 is not a legitimate BCD number because 1010 does not convert to a single decimal digit. Figure 1-1 lists the ten valid BCD numbers and the six that are invalid and must be avoided.

The flowchart in Figure 1-2 summarizes the conversions that have been discussed so far. From any of the five number systems you can convert to any or all of the others. There is no line directly from octal to BCD because that direct conversion was not presented. To convert from octal to BCD, convert to binary, then decimal, then BCD.

Example: Convert 157_8 to BCD.

Solution:

Convert octal to binary.
$157_8 = 1101111_2$

Convert binary to decimal.
$1101111_2 = 111_{10}$

Convert decimal to BCD.
$111_{10} = 100010001_{BCD}$

Example: Convert 362_8 to hexadecimal, binary, decimal, and BCD.

Solution:

Convert octal to binary.
$362_8 = 11110010_2$

Convert binary to hexadecimal and binary to decimal.
$11110010_2 = F2_{16}$
$11110010_2 = 242_{10}$

Convert decimal to BCD.
$242_{10} = 1001000010_{BCD}$

1.11   BINARY ADDITION

The table in Figure 1-3 summarizes the results that can occur when adding two binary bits, *A* and *B*. The outputs are listed as a sum and a carry. The carry indicates whether a 1 has to be added to the next column to the left. The first three lines are exactly what you would expect. In the last line, $1 + 1 = 2$, and 2 in binary is 10_2. Therefore, the sum is 0 and the carry is 1.

The table in Figure 1-4 covers the situation in which a carry from a previous column, the carry-in, is added to *A* and *B*. The outputs are sum and carry-out. In the top four lines of the table in Figure 1-4, the carry-in is zero and the results are

$$\begin{array}{c} A \\ + \ B \\ \hline \text{Carry} \quad \text{Sum} \end{array}$$

Inputs		Outputs	
A	B	Sum	Carry
0	0	0	0
0	1	1	0
1	0	1	0
1	1	0	1

FIGURE 1-3   Binary addition

$$\begin{array}{c} \text{Carry-in} \\ A \\ + \ B \\ \hline \text{Carry-out} \quad \text{Sum} \end{array}$$

Inputs			Outputs	
Carry-in	A	B	Sum	Carry-out
0	0	0	0	0
0	0	1	1	0
0	1	0	1	0
0	1	1	0	1
1	0	0	1	0
1	0	1	0	1
1	1	0	0	1
1	1	1	1	1

FIGURE 1-4   Binary addition with carry-in

the same as those shown in Figure 1-3. In the last four lines, the carry-in is 1. The addition required on lines 5, 6, and 7, is covered in Figure 1-4. On line 8, $1 + 1 + 1 = 3$, and 3 in binary is 11_2. Therefore, the sum is 1 and the carry-out is 1.

Example: Add 11110_2 and 1100_2.

Solution:

11	*Carries*	Check.
11110_2		30_{10}
$+ \ 1100_2$		$+ 12_{10}$
101010_2		42_{10}

Example: Add 1011_2, 101_2, and 1001_2.

Solution:

1111	*Carries*	Check.
1011_2		11_{10}
101_2		5_{10}
$+ \ 1001_2$		$+ \ 9_{10}$
11001_2		25_{10}

1.12 BINARY SUBTRACTION

The table in Figure 1-5 summarizes the results that can occur when subtracting two binary bits, A and B. The outputs are listed as a difference and a borrow. The borrow indicates whether a 2 has to be borrowed from the column on the left to complete the subtraction. The second line is the most difficult to understand. To subtract 1 from 0, we must borrow from the next column on the left, which makes the problem:

$$10_2 - 1_2 \text{ or } 2 - 1$$

which is equal to 1. We borrowed 1 and had a difference of 1.

Borrow

$\begin{array}{r} A \\ -B \\ \hline \end{array}$

Difference

Inputs		Outputs	
A	B	Difference	Borrow
0	0	0	0
0	1	1	1
1	0	1	0
1	1	0	0

FIGURE 1-5   Binary subtraction

Example: Subtract 1001_2 from 10011_2.

Solution:

$$\begin{array}{r} 10011_2 \\ -\ \ 1001_2 \\ \hline 1010_2 \end{array} \qquad \text{Check.} \qquad \begin{array}{r} 19_{10} \\ -\ \ 9_{10} \\ \hline 10_{10} \end{array}$$

Example: Subtract 1010_2 from 10001_2.

Solution:

$$\begin{array}{r} 10001_2 \\ -\ \ 1010_2 \\ \hline 00111_2 \end{array} \qquad \text{Check.} \qquad \begin{array}{r} 17_{10} \\ -10_{10} \\ \hline 7_{10} \end{array}$$

Example: Subtract 1001_2 from 1110_2.

Solution:

$$\begin{array}{r} 1110_2 \\ -1001_2 \\ \hline 0101_2 \end{array} \qquad \text{Check.} \qquad \begin{array}{r} 14_{10} \\ -\ \ 9_{10} \\ \hline 5_{10} \end{array}$$

This method of binary subtraction parallels the longhand method used in decimal arithmetic. It is possible to program a machine to subtract this way, but most computers utilize a complement method of subtraction that converts the problem into addition.

The concept of overflow will be used in each complement system presented. In addition problems, overflow occurs when the sum of the most significant column (left-most column) yields a carry. For example,

$$
\begin{array}{r}
872 \\
+345 \\
\hline
1\ 217
\end{array}
\qquad\qquad
\begin{array}{r}
7326 \\
+0074 \\
\hline
0\ 7400
\end{array}
$$

Overflow ⟋   No Overflow ⟋

In the second example, leading zeros are added to block in the two numbers.

The binary 1's complement method for subtraction is analogous to the 9's complement method in decimal arithmetic. The decimal 9's complement will be presented first; then the binary 1's method will be compared to it.

1.13 DECIMAL 9'S COMPLEMENT FOR SUBTRACTION

A digit plus its 9's complement equals nine. The 9's complements are listed.

Digit	9's Complement
0	9
1	8
2	7
3	6
4	5
5	4
6	3
7	2
8	1
9	0

The idea can be extended to multiple digits. The 9's complement of 362 is 637.

Example: Find the 9's complement of 9204.

Solution:

Number.   9 2 0 4
9's complement.   0 7 9 5

Note that the sum of these two numbers is 9999.

To subtract using the 9's complement:

1. Take the 9's complement of the subtrahend (bottom number).
2. Add the 9's complement to the minuend (top number).
3. Overflow indicates that the answer is positive. Add the overflow to the least significant digit. This operation is called end-around carry (EAC).

4. If there is no overflow the answer is negative. Take the 9's complement of the result to obtain the true magnitude of the answer.

Example: Subtract 153 from 342 using the 9's complement.

Solution:

$$
\begin{array}{r}
342 \\
-153 \\
\end{array}
\longrightarrow
\begin{array}{r}
342 \\
+846 \\
\hline
1\ \ 188
\end{array}
\longrightarrow
\begin{array}{r}
188 \\
+\ \ \ 1 \\
\hline
189
\end{array}
$$

EAC

Overflow

The answer is positive 189. The overflow was added to the least significant digit (end-around carry) and the answer is positive.

Example: Subtract 594 from 793 using the 9's complement.

Solution:

$$
\begin{array}{r}
793 \\
-594 \\
\end{array}
\longrightarrow
\begin{array}{r}
793 \\
+405 \\
\hline
1\ \ 198
\end{array}
\longrightarrow
\begin{array}{r}
198 \\
+\ \ \ 1 \\
\hline
199
\end{array}
$$

EAC

Overflow

The answer is positive 199.

The same process is followed when the subtrahend is larger than the minuend.

Example: Subtract. 4783 − 9496

Solution:

$$
\begin{array}{r}
4783 \\
-9496 \\
\end{array}
\longrightarrow
\begin{array}{r}
4783 \\
+0503 \\
\hline
0\ \ 5286
\end{array}
\longrightarrow -4713
$$

No Overflow

No overflow indicates that the answer is negative and the true magnitude is the 9's complement of 5286 or 4713. The answer is −4713.

Example: Subtract. 706 − 915

Solution:

$$
\begin{array}{r}
706 \\
-915 \\
\end{array}
\longrightarrow
\begin{array}{r}
706 \\
+084 \\
\hline
0\ \ 790
\end{array}
\longrightarrow -209
$$

No Overflow

No overflow indicates answer is negative and the true magnitude is the 9's complement of 790 or 209. The answer is −209.

Example: Subtract. 9236 − 81

Solution:

$$
\begin{array}{r}
9236 \\
- \quad 81 \\
\hline
\end{array}
\longrightarrow
\begin{array}{r}
9236 \\
+9918 \\
\hline
1\ 9154
\end{array}
\longrightarrow
\begin{array}{r}
9154 \quad \text{EAC} \\
+ \quad 1 \\
\hline
9155
\end{array}
$$

Overflow

Leading zeros are converted to nines. Overflow means the answer is positive 9155.

1.14   BINARY 1'S COMPLEMENT FOR SUBTRACTION

To take the 1's complement of a binary number, simply change each bit. The 1's complement of 1 is 0 and vice versa. The 1's complement of 1001010 is 0110101. To subtract using 1's complement:

1. Take the 1's complement of the subtrahend (bottom number).
2. Add the 1's complement to the minuend (top number).
3. Overflow indicates that the answer is positive. Add the overflow to the least significant bit. This operation is called end-around carry (EAC).
4. If there is no overflow then the answer is negative. Take the 1's complement of the original addition to obtain the true magnitude of the answer.

Example: Subtract. $11001_2 - 10001_2$

Solution:

$$
\begin{array}{r}
11001 \\
- 10001 \\
\hline
\end{array}
\longrightarrow
\begin{array}{r}
11001 \\
+01110 \\
\hline
1\ 00111
\end{array}
\longrightarrow
\begin{array}{r}
00111 \quad \text{EAC} \\
+ \quad 1 \\
\hline
1000
\end{array}
$$

Overflow

The answer is +1000.

Check.
$25_{10} - 17_{10} = 8_{10}$

Example: Subtract. $1011_2 - 101_2$

Solution:

$$
\begin{array}{r}
1011 \\
- \quad 101 \\
\hline
\end{array}
\longrightarrow
\begin{array}{r}
1011 \\
+ 1010 \\
\hline
1\ 0101
\end{array}
\longrightarrow
\begin{array}{r}
0101 \quad \text{EAC} \\
+ \quad 1 \\
\hline
0110
\end{array}
$$

Overflow

Note that the leading 0 becomes a 1. The answer is $+110$.

Check.
$$11_{10} - 5_{10} = 6_{10}$$

The same process is used when the subtrahend is larger than the minuend.

Example: Subtract. $101_2 - 11000_2$

Solution:

$$
\begin{array}{r}
101 \\
-11000 \\
\end{array}
\longrightarrow
\begin{array}{r}
101 \\
+00111 \\
\hline
01100
\end{array}
$$

No Overflow $\longleftarrow$

The answer is negative. The true magnitude is the 1's complement of 01100 or 10011. The answer is -10011.

Check.
$$5_{10} - 24_{10} = -19_{10}$$

Example: Subtract. $10000_2 - 11101_2$

Solution:

$$
\begin{array}{r}
10000 \\
-11101 \\
\end{array}
\longrightarrow
\begin{array}{r}
10000 \\
+00010 \\
\hline
10010
\end{array}
$$

No Overflow $\longleftarrow$

The answer is negative. The true magnitude is the 1's complement of 10010 or 01101. The answer is -01101.

Check.
$$16_{10} - 29_{10} = -13_{10}$$

1.15   DECIMAL 10'S COMPLEMENT FOR SUBTRACTION

The method of using the 10's complement for subtraction is a variation of the method which uses the 9's complement. It is analogous to the often used 2's complement method in binary subtraction. The 10's method will be discussed first and then the 2's complement method will be compared to it. To form the 10's complement of a number, first take the 9's complement and add one. The 10's complement of 283 is $716 + 1 = 717$. To subtract using the 10's complement:

1. Take the 10's complement of the subtrahend (bottom number).
2. Add it to the minuend (top number).
3. Overflow indicates that the answer is positive. Ignore the overflow. (Do not do an end-around carry.)
4. No overflow indicates that the answer is negative. Take the 10's complement of the original addition to obtain the true magnitude of the answer.

Example: Subtract. $683 - 129$

Solution:

$$
\begin{array}{r}
683 \\
-129 \\
\end{array}
\longrightarrow
\begin{array}{r}
683 \\
+871 \\
\hline
1\ 554 \\
\end{array}
$$

Overflow ⟶

9's complement = 870
+ 1

10's complement = 871

The answer is positive 554.

Example: Subtract. $700 - 86$

Solution:

$$
\begin{array}{r}
700 \\
-\ \ 86 \\
\end{array}
\longrightarrow
\begin{array}{r}
700 \\
+914 \\
\hline
1\ 614 \\
\end{array}
$$

Overflow ⟶

9's complement = 913
+ 1

10's complement = 914

The answer is positive 614.

The same process is followed when the subtrahend is larger than the minuend.

Example: Subtract. 532 − 913

Solution:

$$\begin{array}{r} 532 \\ -913 \end{array} \longrightarrow \begin{array}{r} 532 \\ +087 \\ \hline 619 \end{array}$$

No Overflow ⌐

9's complement =   086
                  +   1
10's complement =   087

The answer is negative. Take the 10's complement for the true magnitude of the answer.

9's complement =   380
                  +   1
10's complement =   381

The answer is −381.

Example: Subtract. 32 − 9813

Solution:

$$\begin{array}{r} 32 \\ -9813 \end{array} \longrightarrow \begin{array}{r} 32 \\ +0187 \\ \hline 0219 \end{array}$$

No Overflow ⌐

9's Complement =   0186
                  +   1
10's complement =   0187

The answer is negative. Take the 10's complement for the true magnitude of the answer.

9's complement =   9780
                  +   1
10's complement =   9781

The answer is −9781.

1.16 BINARY 2'S COMPLEMENT FOR SUBTRACTION

To form the 2's complement of a number, first take the 1's complement and then add 1. The 2's complement of 10110 is 01001 + 1 = 01010. A shorter method is to start at the least significant bit and, moving to the left, leave each bit the same until the first 1 is passed. Then change each bit thereafter.

Example: Find the 2's complement of 101101000.

Solution:

Change each bit to the left of the first 1.
Number   = 101101000
2's complement = 010011000

Example: Find the 2's complement of 1011011.

Solution:

Method 1.
1's complement   0100100
Add 1   +   1
2's complement   0100101

Method 2.
Change each bit to the left of the first 1.
1011011
0100101

Example: Find the 2's complement of 101000000.

Solution:

Method 1.
1's complement   010111111
Add 1   +   1
2's complement   011000000

Method 2.
Change each bit to the left of the first 1.
101000000
011000000

To subtract using the 2's complement:
1. Take the 2's complement of the subtrahend (bottom number).
2. Add it to the minuend (top number).
3. Overflow indicates that the answer is positive. Ignore the overflow (no end-around carry).
4. No overflow indicates that the answer is negative. Take the 2's complement of the original addition to obtain the true magnitude of the answer.

Example: Subtract. $1011_2 - 100_2$

Solution:

$$
\begin{array}{r}
1011 \\
-\;100 \\
\hline
\end{array}
\longrightarrow
\begin{array}{r}
1011 \\
+1100 \\
\hline
1\;0111
\end{array}
$$

Overflow ⌐

1's complement =   1011
                   +   1
2's complement =   1100

The answer is positive 111.

Check.
$11_{10} - 4_{10} = 7_{10}$

Example: Subtract. $10011_2 - 10010_2$

Solution:

$$
\begin{array}{r}
10011 \\
-10010 \\
\hline
\end{array}
\longrightarrow
\begin{array}{r}
10011 \\
+01110 \\
\hline
1\;00001
\end{array}
$$

Overflow ⌐

The answer is positive 1.

Check.
$19_{10} - 18_{10} = 1_{10}$

The process is the same when the subtrahend is larger than the minuend.

Example: Subtract. $10010_2 - 11000_2$

Solution:

$$
\begin{array}{r}
10010 \\
-11000 \\
\hline
\end{array}
\longrightarrow
\begin{array}{r}
10010 \\
+01000 \\
\hline
11010
\end{array}
$$

No Overflow ⌐

The answer is negative. The true magnitude is the 2's complement of 11010 or 110. The answer is -110.

Check.
$18_{10} - 24_{10} = -6_{10}$

Example: Subtract. $1001_2 - 10101_2$

Solution:

$$
\begin{array}{r}
1001 \\
-10101
\end{array}
\longrightarrow
\begin{array}{r}
01001 \\
+01011 \\
\hline
10100
\end{array}
$$

No Overflow ⏋

The answer is negative. The true magnitude is the 2's complement of 10100 or 1100. The answer is -1100.

Two advantages of subtraction by a complement system are

1. The procedure is the same whether the subtrahend is larger or smaller than the minuend. This saves the extra time or circuitry for a digital machine to decide if one number is larger or smaller than another.
2. The subtraction problem is converted to an addition problem. The same circuitry could be used for both processes.

1.17   SIGNED 2'S COMPLEMENT NUMBERS

In some systems, the most significant bit is used to indicate the sign of the number and the remaining bits are used to indicate the magnitude. A zero in the sign bit usually indicates that the number is positive and the remaining bits express the number in true magnitude form.

Example: Convert 00101101 in a signed 2's complement system to a decimal number.

Solution:

0	0	1	0	1	1	0	1
	64	32	16	8	4	2	1

Sign Bit ⏋

The true magnitude is $32 + 8 + 4 + 1 = 45$. The number is positive. The answer is 45_{10}.

Example: What is the highest positive number that can be represented in an 8-bit signed 2's complement system?

Solution:

The highest positive number that can be represented in an 8-bit signed 2's complement system is 01111111.

0	1	1	1	1	1	1	1
	64	32	16	8	4	2	1

Sign Bit ⏋

The true magnitude is $64 + 32 + 16 + 8 + 4 + 2 + 1 = 127$. The number is positive. The positive number 127_{10} is the highest decimal number that can be represented in this system.

A 1 in the sign bit usually indicates that the number is negative. The remaining bits express the number in 2's complement form.

Example: Convert 10010011 in a signed 2's complement system to a decimal number.

Solution:

$$10010011$$

Sign Bit ⟋   Magnitude Bits
2's Complement of True Magnitude

To find the true magnitude, take the 2's complement of the complete number, including the sign bit.

2's complement = 10010011
True magnitude = 01101101

0	1	1	0	1	1	0	1
128	64	32	16	8	4	2	1

$64 + 32 + 8 + 4 + 1 = 109$
The number is negative. The answer is -109_{10}.

Example: Convert 11110000 in a signed 2's complement system to a decimal number.

Solution:

$$11110000$$

Sign Bit ⟋   Magnitude Bits
2's Complement of True Magnitude

2's complement = 11110000
True magnitude = 00010000

0	0	0	1	0	0	0	0
128	64	32	16	8	4	2	1

The number is negative. The answer is -16_{10}.

Example: What is the most negative number that can be represented in an 8-bit signed 2's complement system?

Solution:

The most negative number that can be represented in an 8-bit signed 2's complement system is 10000000.

$$\text{Sign Bit} \nearrow \overbrace{10000000}$$

Magnitude Bits
2's Complement of True Magnitude

2's complement = 10000000
True magnitude = 10000000

1	0	0	0	0	0	0	0
128	64	32	16	8	4	2	1

The negative number -128_{10} is the most negative number that can be represented in this system.

In an 8-bit signed 2's complement system, the numbers can range from -128_{10} to $+127_{10}$.

Example: Express -78_{10} as an 8-bit signed 2's complement number.

Solution:

$78_{10} =$

0	1	0	0	1	1	1	0
128	64	32	16	8	4	2	1

True magnitude = 01001110
2's complement = 10110010
-78_{10} = 10110010 (signed 2's complement)

Check.

$$\text{Sign Bit} \nearrow \overbrace{10110010}$$

Magnitude Bits
2's Complement of True Magnitude

2's complement = 10110010
True magnitude = 01001110

0	1	0	0	1	1	1	0
128	64	32	16	8	4	2	1

The number is negative. The 8-bit signed 2's complement number 10110010 is equal to $-(64 + 8 + 4 + 2) = -78_{10}$.

Numbers in signed 2's complement form can be added using straight binary addition and subtracted by taking the 2's complement of the subtrahend and adding. The sign bit will indicate the sign of the answer. Positive answers will be in true magnitude form. Negative answers will be in 2's complement form.

Example: Add these 8-bit signed 2's complement numbers.
01011001 + 10101101

Solution:

```
   0 1 0 1 1 0 0 1      (+89)
 + 1 0 1 0 1 1 0 1      (-83)
 1 0 0 0 0 0 1 1 0      (+ 6)
```

↗
⌐ Ignore Overflow

Example: Add these 8-bit signed 2's complement numbers. Express the answer in decimal form.
11011001 + 10101101

Solution:

```
   1 1 0 1 1 0 0 1      (-  39)
 + 1 0 1 0 1 1 0 1      (-  83)
 1 1 0 0 0 0 1 1 0      (- 122)
```

Ignore Overflow ⟶

Sign Bit ⟶   Magnitude Bits
2's Complement of True Magnitude

2's complement   = 10000110
True magnitude = 01111010

0	1	1	1	1	0	1	0
128	64	32	16	8	4	2	1

The number is negative. The answer is
$-(64 + 32 + 16 + 8 + 2) = -122$

Check.
$-39 + (-83) = -122$

Example: Subtract these 8-bit signed 2's complement numbers. Express the answer in decimal form.
01011011 - 11100101

Solution:

To subtract, take the 2's complement of the subtrahend and add.

```
   0 1 0 1 1 0 1 1           0 1 0 1 1 0 1 1
 - 1 1 1 0 0 1 0 1    ⟶    + 0 0 0 1 1 0 1 1
                            0 1 1 1 0 1 1 0
```

No Overflow ⌐

Sign Bit ⌐

1	1	1	0	1	1	0
64	32	16	8	4	2	1

The answer is positive. The answer is
64 + 32 + 16 + 4 + 2 = 118

Check.
91 − (−27) = 118

Example: Subtract these 8-bit signed 2's complement numbers. Express the answer in decimal form.
10001010 − 11111100

Solution:

To subtract, take the 2's complement of the subtrahend and add.

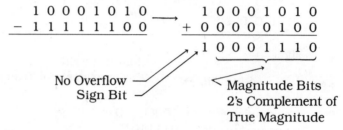

2's complement = 10001110
True magnitude = 01110010

0	1	1	1	0	0	1	0
128	64	32	16	8	4	2	1

The number is negative. The answer is
−(64 + 32 + 16 + 2) = −114

Check.
−118 − (−4) = −114

In each of the examples of signed 2's complement mathematics presented so far, the result has been correct. To ensure that the result is correct, the carry from column 7 into the sign bit and the overflow must be monitored and the following rules observed.

1. If there is a carry from column 7 into the sign bit and an overflow, the answer is correct.
2. If there is no carry from column 7 into the sign bit and no overflow, the answer is correct.
3. If there is no carry from column 7 into the sign bit and overflow occurs or vice versa, the answer is not correct.

Systems that use signed 2's complement mathematics must monitor the carry from column 7 into the sign bit and the overflow to signal whether an error has occurred.
In the following two examples, the results are not correct.

Example: Subtract these 8-bit signed 2's complement numbers. Express the answer in decimal form.
$10000101 - 01111111$

Solution:

To subtract, take the 2's complement of the subtrahend and add.

$$
\begin{array}{r}
1\ 0\ 0\ 0\ 0\ 1\ 0\ 1 \\
-\ 0\ 1\ 1\ 1\ 1\ 1\ 1\ 1 \\
\end{array}
\longrightarrow
\begin{array}{r}
1\ 0\ 0\ 0\ 0\ 1\ 0\ 1 \\
+\ 1\ 0\ 0\ 0\ 0\ 0\ 0\ 1 \\
\hline
1\ 0\ 0\ 0\ 0\ 0\ 1\ 1\ 0 \\
\end{array}
\begin{array}{r}
(-123) \\
-(+127) \\
\hline
(-250)
\end{array}
$$

Ignore Overflow $\longrightarrow$

The result indicates that the answer is $+6$, but it should be -250. There was no carry into the sign bit, but there was overflow. Since these differ the result is incorrect. The error has occurred because the true answer is too large to be expressed in an 8-bit signed 2's complement system.

Example: Add these 8-bit signed 2's complement numbers. Express the answer in decimal form.
$01111110 + 00111101$

Solution:

$$
\begin{array}{rl}
0\ 1\ 1\ 1\ 1\ 1\ 1\ 0 & =\ +126 \\
+\ 0\ 0\ 1\ 1\ 1\ 1\ 0\ 1 & =\ +\ 61 \\
\hline
0\ 1\ 0\ 1\ 1\ 1\ 0\ 1\ 1 & \neq\ +187 \\
\end{array}
$$

No Overflow $\longrightarrow$

The result indicates that the answer is negative. A carry from bit 7 into the sign bit has changed the sign bit to a 1. No overflow occurred. Since they differ the result is incorrect. The error occurred because $+187$ is too large to be represented in this 8-bit signed 2's complement system.

Exercise 1

1. Write the binary numbers from 100_2 to 1000_2.
2. Write the binary numbers from 1011_2 to 10101_2.
3. Write the octal numbers from 66_8 to 110_8.
4. Write the octal numbers from 767_8 to 1010_8.
5. Write the hexadecimal numbers from DD_{16} to 101_{16}.
6. Write the hexadecimal numbers from EFD_{16} to $F10_{16}$.
7. Write the BCD numbers from 10001001_{BCD} to 100000001_{BCD}.
8. Write the BCD numbers from 1101000_{BCD} to 10010000_{BCD}.

9. a. How high can you count with a 4-bit binary number?
   b. How many different numbers are represented?
10. a. How high can you count with an 8-bit binary number?
    b. How many different numbers are represented?
11. a. How high can you count with a 16-bit binary number?
    b. How many different numbers are represented?
12. a. How many different digits are used in the octal number system?
    b. How many different digits are used in the hexadecimal number system?
    c. How many different digits are used in the BCD number system?
13. Complete the chart.

Octal	Hexadecimal	Binary	Decimal	BCD
36				
	A9			
		10010		
			99	
				1100111

14. Complete the chart.

Octal	Hexadecimal	Binary	Decimal	BCD
54				
	3C			
		1011100		
			100	
				10000001

15. Add in binary.

a. $\begin{array}{r} 1001_2 \\ +\,1101_2 \\ \hline \end{array}$ b. $\begin{array}{r} 1_2 \\ 1011_2 \\ +\,1001_2 \\ \hline \end{array}$ c. $\begin{array}{r} 10010_2 \\ 1100_2 \\ +\,11101_2 \\ \hline \end{array}$

16. Subtract in binary.

    a. 1001_2 b. 10101_2 c. 1101_2 d. 10010100_2
        $-\ \ \ 110_2$ $-\ 1110_2$ -100100_2 $-\ 1010010_2$

17. Subtract in binary.

    a. 1100_2 b. 11010_2 c. 1101_2 d. 101_2
        $-\ \ \ 101_2$ $-\ 1011_2$ -100111_2 -10010_2

18. Subtract using 1's complement.

    a. 1010_2 b. 10001_2
        -1000_2 -11101_2

19. Subtract using 1's complement.

    a. 1101_2 b. 1001_2
        $-\ \ \ 100_2$ -1100_2

20. Subtract using 2's complement.

    a. 11010_2 b. 10010_2
        $-\ 1100_2$ -11110_2

21. Subtract using 2's complement.

    a. 100101_2 b. 10101_2
        $-\ 1001_2$ -11000_2

22. Express the following decimal numbers in 8-bit signed 2's complement form.

    a. -38 c. -12 e. -100
    b. $+57$ d. $+12$ f. $+60$

23. Express the following decimal numbers in 8-bit signed 2's complement form.

    a. -50 c. -2 e. -120
    b. $+43$ d. $+8$ f. $+83$

24. Add these 8-bit signed 2's complement numbers. Use the carry from column 7 and the overflow to tell whether the answer is correct.

    a. 00011110 + 00111000   c. 11100011 + 10000001
    b. 01011101 + 00111100   d. 00110011 + 11001100

25. Add these 8-bit signed 2's complement numbers. Use the carry from column 7 and the overflow to tell whether the answer is correct.

    a. 00111101 + 11010110   c. 10011100 + 10011011
    b. 01100111 + 11001001   d. 01111111 + 01111111

26. Use the 2's complement method to subtract these 8-bit signed 2's complement numbers. Use the carry from column 7 and the overflow to indicate whether the answer is correct.
    a. $00101010 - 01101101$
    b. $01111111 - 10000000$
    c. $10001111 - 10100000$
    d. $10000000 - 10000000$

27. Use the 2's complement method to subtract these 8-bit signed 2's complement numbers. Use the carry from column 7 and the overflow to indicate whether the answer is correct.
    a. $00111001 - 11000110$
    b. $10101010 - 10011010$
    c. $11111111 - 10101101$
    d. $10001011 - 01110101$

28. Why do computers make extensive use of the binary number system?

29. Discuss the sign bit in a signed 2's complement number.

30. Name two advantages of a complement system of subtraction over the longhand method.

31. Discuss the use of binary and hexadecimal number systems in digital work.

LAB

7483 4-Bit Adder

OBJECTIVES

After completing this lab, you should be able to:

■ connect a 7483 IC for proper operation.

■ use a 7483 as an adder.

■ use a 7483 as a 1's complement subtractor.

■ use a 7483 as a 2's complement subtractor.

■ cascade two 7483s to operate as an 8-bit adder.

COMPONENTS NEEDED

2	7483 ICs
9	LEDs
9	330-Ω resistors

PREPARATION

A 7483 4-bit adder will be used in this lab to add binary numbers. The 7483 integrated circuit will be studied in detail in Chapter 5, but you can use it here to confirm what you have learned in Chapter 1. The 7483 is a TTL integrated circuit. Detailed specifications of the TTL family of integrated circuits will be presented in Chapter 6. The facts you need to know about TTL to perform this lab are listed.

1. The power supply connections are called V_{CC} and ground. V_{CC} is always connected to +5 V and ground is connected to 0 V.
2. A legitimate 0 on an output pin can range from 0 V to 0.4 V. A legitimate 1 on an output pin can range from 2.4 V to 5 V.
3. Input pins that are left unconnected are usually taken by the IC to be at a 1 level. In this lab, input signals will be connected to +5 V for a 1 and to ground for a 0.

You will wire this lab on the protoboard of your TTL/CMOS trainer. The protoboard has buses, across the top and bottom, that you can connect to +5 V and ground. Each bus is in two halves. You must install jumper wires across with hook-up wire to have a continuous bus.

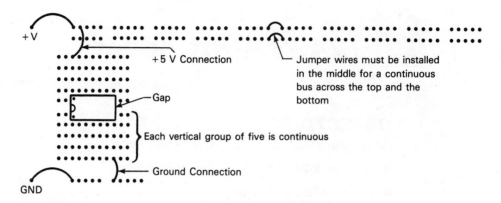

+5 V Connection

Jumper wires must be installed in the middle for a continuous bus across the top and the bottom

Gap

Each vertical group of five is continuous

Ground Connection

ICs are plugged into the protoboard so that they span the gap. You then have access to each pin via the vertical groups of 5 continuous connections. Short wires are connected to the power supply connection of each IC to be used. The switches at the bottom of the trainer can be used to supply ones and zeros for inputs. Up supplies a one; down supplies a zero. Other inputs must be manually connected to +5 V or 0 V with a hookup wire. Wire outputs to the 4 light emitting diodes (LEDs) at the top of the trainer. Any other outputs must be connected to added LEDs through a current limiting resistor of approximately 330 Ω as shown.

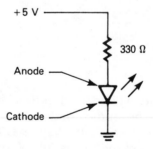

Four ways to tell the anode lead from the cathode lead are listed.

    1. The cathode lead has the flag.

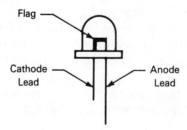

2. The anode lead is usually longer than the cathode lead.
3. The flat edge of the package is on the cathode side of the LED.

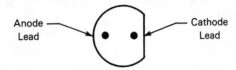

4. Connect the LED in a test circuit. If it does not light, turn the LED around. When lit, the lead connected to ground is the cathode lead. Do not forget the current limiting resistor or the LED will become a dark emitting diode.

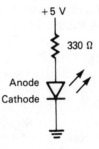

The power supply has a fixed 5-V output for work in TTL or a variable supply for work in CMOS. Be sure that the supply is set to TTL. If your trainer or dc power supply does not have a fixed 5-V output, connect a voltmeter across the output of a variable supply and adjust it to 5 V. A higher supply voltage can destroy a TTL integrated circuit. Switch off supply voltages before inserting or removing integrated circuits.

Integrated circuit pins are arranged in a definite pattern. One end of the top of the IC has a notch or indentation. Starting from the notch the pins are numbered counterclockwise. The IC shown is a 16-pin Dual In-line Package or DIP. The 14-, 20-, 24-, and 40-pin ICs are also common.

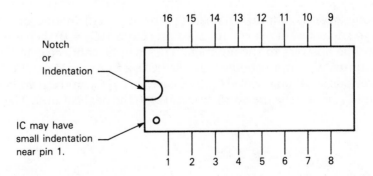

ICs are identified by a number code stamped on the top. The first two letters indicate the device technology. The next two numbers denote the family of ICs such as TTL or CMOS. If letters follow, they indicate the subfamily of the IC. The next numbers indicate the function of the IC, and the last letters indicate the package style. For example:

D M	7 4	L S	2 8 3	N
Digital Monolithic	TTL Commercial	Low Power Schottky	4-Bit Binary Adder	Plastic Package

Do not confuse the IC number with the date code that is also often stamped on the IC. The number 7436 indicates that the IC was manufactured in 1974 during the 36th week.

Integrated circuits are divided into categories according to their complexity. Small-scale integration (SSI) circuits such as gates and flip-flops have circuitry that is equivalent to less than 12 gates. Medium-scale integration (MSI) circuits, such as decoders, counters, multiplexers, and the adder IC that you will use in this lab, have circuitry that is equivalent to 12 or more gates but less than 100. Large-scale integration (LSI) circuits contain the equivalent of 100 or more gates.

In a TTL spec book, find the IC to obtain the pinout (where the power supply, input signals, and output signals are connected). You may have to find the MSI section of your spec book. Plug your IC into the protoboard and make the power connections V_{CC} (+5 V) and ground (0 V). Get in the habit of making those connections first. Many hours are lost troubleshooting the circuit when the only problem is faulty power supply. This IC adds two 4-bit numbers, $A_4A_3A_2A_1$ and $B_4B_3B_2B_1$, plus a carry C_0 from some previous addition.

$$\begin{array}{c} C_0 \\ A_4 \ A_3 \ A_2 \ A_1 \\ + \ B_4 \ B_3 \ B_2 \ B_1 \\ \hline \end{array}$$ You are going to supply those inputs, +5 V for a 1, 0 V for a zero.

Don't forget C_0. Inputs left floating (unconnected) are interpreted by the IC as ones. If you do not want a carry into the first column ($C_0 = 0$) then ground it. The outputs are the answers to the addition. They will be connected to LEDs to monitor the outputs. A 1 output should light the LED; a zero should not. Note that a carry from a previous addition is called C_0. The carries, C_1, C_2, and C_3, are handled inside the IC and C_4 represents the overflow or carry to the next column. The 7483 can only add.

To subtract $B_4B_3B_2B_1$ from $A_4A_3A_2A_1$ you must take the 1's or 2's complement of $B_4B_3B_2B_1$ and input it into the IC. It cannot do it for you. Labs are hard work. Do not get frustrated. Avoid the trap of rushing through a lab to get home early and missing the point. You must pay close attention to details, but don't miss the overall concepts involved.

$$
\begin{array}{ccccc}
 & & & & C_0 \\
 & A_4 & A_3 & A_2 & A_1 \\
 & B_4 & B_3 & B_2 & B_1 \\
\hline
C_4 & \Sigma_4 & \Sigma_3 & \Sigma_2 & \Sigma_1
\end{array}
$$

If your circuit does not work properly consider these points:

1. V_{CC} and ground. Use a voltmeter to check for $+5$ V on V_{CC} and 0 V on ground directly on the pins of the IC. If it measures otherwise, trace the wiring back to find the fault.
2. Inputs. Use a voltmeter to check that each input is at the level you expected. Check directly on the pins of the IC itself. Correct any discrepancies. Since these inputs are supplied directly from the switches or power supply buses, a 1 should be close to $+5$ V and 0 close to ground.
3. Outputs. Use a voltmeter to check the outputs directly on the pins of the IC (2.4 V to 5 V for a 1, 0 V to 0.4 V for a 0). If steps 1 and 2 check and step 3 does not, then either the IC is bad or something connected to the output is loading it down. A common beginning mistake is to tie the outputs, especially C_4, to ground or V_{CC}. Disconnect the wires from an output pin that does not check and see if the proper value is restored.
4. Pinout. Are you using the right pinout for your IC? Consult your spec book.
5. Think and act. You cannot correct a circuit by staring at it. Use your voltmeter. Get involved. Discuss it with your lab partner.
6. Hookup wires are sometimes shoved too far into the protoboards so that the insulation prevents electrical connections. You should be able to track down such a situation with your voltmeter.
7. If you don't understand after really trying, ask!

PROCEDURE

1. Connect the IC as shown.

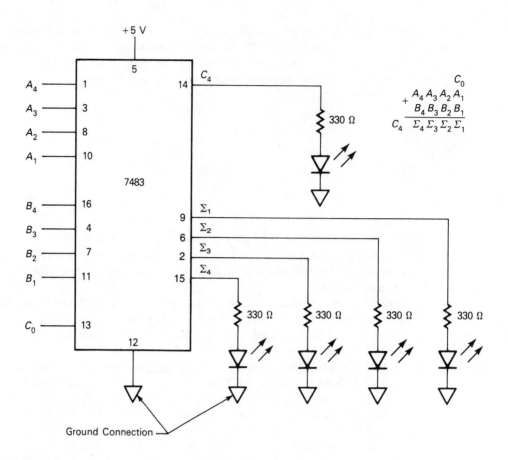

2. Let $A = 10_{10}$ and $B = 6_{10}$ and $C_0 = 0$.
3. Verify the output.
4. Let $A = 7_{10}$ and $B = 4_{10}$ and $C_0 = 1$.
5. Verify the output.
6. Try several other combinations.
7. Input the 1's complement of B to work the following problems ($C_0 = 0$). Use the form $A - B = C$.
    a. $14_{10} - 7_{10}$
    b. $10_{10} - 6_{10}$
    c. $7_{10} - 8_{10}$
8. Try several other examples of subtraction using the 1's complement.
9. Input the 2's complement of B to work the following problems. Use the form $A - B = C$.
    a. $14_{10} - 8_{10}$
    b. $6_{10} - 12_{10}$
    c. $7_{10} - 8_{10}$

10. Interconnect two 7483 ICs as shown to form an 8-bit full adder.

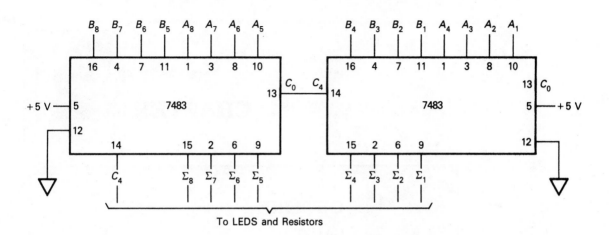

To LEDS and Resistors

11. Work the following problems. Verify your answers using the 7483s.
    a. $150_{10} + 201_{10} =$ ———————————— $(C_0 = 0)$.
    b. $255_{10} + \quad 1_{10} =$ ———————————— $(C_0 = 1)$.
    c. $128_{10} - \quad 31_{10} =$ ———————————— (use 1's complement).
    d. $500_{10} - \quad 63_{10} =$ ———————————— (use 2's complement).

CHAPTER 2

OUTLINE

Gates and Inverters

OBJECTIVES

After completing this chapter, you should be able to:

■ draw symbols for the inverter and each of the basic gates.

■ write truth tables for the inverter and each of the basic gates.

■ write Boolean expressions for the output of the inverter and each of the basic gates.

■ describe how to enable each of the two-input gates.

■ describe how to inhibit each of the two-input gates.

■ use a NAND or NOR gate as an inverter.

■ expand an AND, NAND, OR, and NOR.

2.1   GATES

Gates are circuits that are used to combine digital logic levels (ones and zeros) in specific ways. A system called Boolean algebra is used to express the output in terms of the inputs. The basic gates are the AND, NAND, OR, and NOR.

2.2   INVERTERS

Another basic circuit is the inverter. The *inverter* is a single-input gate whose output is the complement of the input. It inverts the signal on the input. If A is 0 then X is 1 and if A is 1 then X is 0. The symbol for the inverter is shown in Figure 2-1. The operation of the inverter can be summarized in a truth table by listing all possible inputs and the corresponding outputs, Figure 2-2.

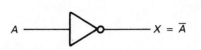

FIGURE 2-1   Inverter

Input	Output
A	X
0	1
1	0

FIGURE 2-2   Inverter truth table

The small circle on the output is called a bubble. The input is not bubbled. Read 0 for a bubbled input or output, and 1 for a nonbubbled input or output. The symbol is read, "1 in, 0 out." The bubble on the output indicates that the output is active low, and the absence of a bubble on the input indicates that the input is active high. The input is "looking for" a 1 level to produce a 0, active low, output. The Boolean expression for the output is $\overline{A}$, which is read "A complement" or "A not."

An alternate symbol, Figure 2-3, called an inverted logic symbol or functional logic symbol has a bubble on the input but none on the output. This symbol is read "0 in, 1 out." The result is the same either way. Both symbols are used on

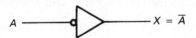

FIGURE 2-3   Inverted logic symbol for inverter

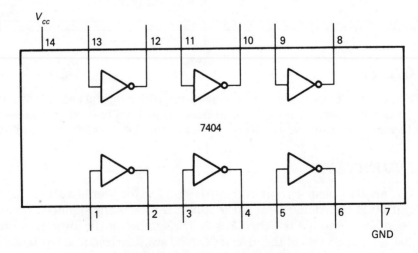

FIGURE 2-4   Equivalent inverter symbols

FIGURE 2-5   Hex inverter pinout

schematics and both should be learned. Equivalent inverter symbols are shown in Figure 2-4.

Inverters are available in 14-pin DIP packages in both TTL and CMOS. In the TTL family, the 7404 is a hex inverter. *Hex* signifies that six inverters are contained on the same IC. Each is independent from the others and each can be used in a different part of the circuit. The supply voltage, V_{CC}, is $+5$ V and is applied to pin 14 with pin 7 connected to ground, Figure 2-5.

In the CMOS family, the 4069 is a general purpose hex inverter. The IC functions the same as the TTL inverter and has the same pinout. The supply voltage, V_{DD}, can range between $+3$ V and $+15$ V. Pin 7, V_{SS}, is connected to ground. Some of the available inverter ICs are listed in Table 2-1.

TABLE 2-1   Inverter ICs

NUMBER	FAMILY	DESCRIPTION
7404	TTL	Hex inverter
74C04	CMOS	Hex inverter
4069	CMOS	Hex inverter

Note: The 74xx series of ICs are TTL; 74Cxx are CMOS, with the same pinout as the 74xx IC of the same number; 40xx ICs are CMOS.

In addition to the conventional logic symbol shown in Figure 2-5, the IEC (International Electrotechnical Commission) and the IEEE (Institute of Electrical and Electronics Engineers) have developed a system of logic symbols that shows the relationship of each input to each output, without showing the internal circuitry.

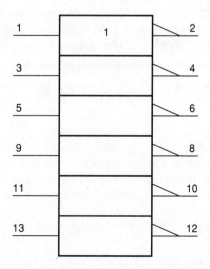

FIGURE 2-6   IEC symbol—7406 hex inverter

The IEC symbol for a 7406 hex inverter is shown in Figure 2-6. Since each inverter functions independently of the others, each is drawn in its own rectangle. The "1" in the top rectangle indicates that one input must be active to produce the output. The triangle on the right is equivalent to the bubble in the conventional symbol. An active high input produces an active low output.

2.3 OR GATES

The *OR gate* is a circuit that produces a 1 on its output when any of its inputs are 1. Figure 2-7 shows the symbol for a two-input OR gate with inputs A and B and output X.

The Boolean expression for the output is $A + B$, which is read "A OR B." The output X is 1 when A is 1 or B is 1 or both. The truth table in Figure 2-8 summarizes the operation of the OR gate. All possible input combinations are listed by counting in binary from 00 to 11.

The symbol shown in Figure 2-7 represents an OR function. Since there are no bubbles shown on the inputs or outputs, the symbol is read "1 OR 1 in, 1 out." This statement summarizes the last three lines of the truth table in Figure 2-8. The first line of the truth table shows the only condition in which the output is 0. This is called the unique state of the gate.

Example: Predict the output of each gate.

Solution:

In the first gate the inputs are different (line 2 or 3 of the truth table), and the output is 1. In the second gate, both inputs are 0 (line 1 of the truth table), and the output must be 0.

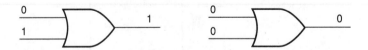

Figure 2-9 shows an alternate symbol for the two-input OR gate, called the inverted logic symbol. The shape of the symbol represents the AND function. Both the inputs and outputs are bubbled. This represents the first line of the truth table, whereas the symbol in Figure 2-7 represents the last three lines. The alternate symbol can be read "0 AND 0 in, 0 out."

The Boolean expression for the output of the gate in Figure 2-9 is developed as follows:

1. Since A is bubbled, write A complement, $\overline{A}$.
2. Since the B is bubbled, write B complement, $\overline{B}$.

3. Since the shape of the gate is AND, which is written as a multiplication sign (or omitted), write $\overline{A} \cdot \overline{B}$ or $\overline{A}\overline{B}$.
4. To find X, complement the whole expression $\overline{\overline{A} \cdot \overline{B}}$.

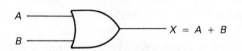

$X = A + B$

FIGURE 2-7   The two-input OR gate

Inputs		Output
B	A	X
0	0	0
0	1	1
1	0	1
1	1	1

FIGURE 2-8   Truth table for two-input OR gate

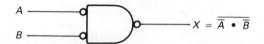

$X = \overline{A} \cdot \overline{B}$

FIGURE 2-9   Two-input OR inverted logic symbol

Since the symbols shown in Figures 2-7 and 2-9 are equivalent, then the outputs are equivalent and $A + B = \overline{\overline{A} \cdot \overline{B}}$. Both symbols are used in schematics and both should be learned. Equivalent OR gate symbols are shown in Figure 2-10.

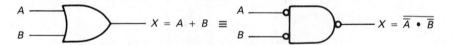

$X = A + B \equiv$ $X = \overline{\overline{A} \cdot \overline{B}}$

FIGURE 2-10   Equivalent OR gate symbols

Example: Predict the output of each gate.

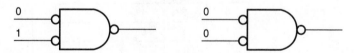

Solution:

This alternate symbol states that a 0 AND a 0 will produce a 0 out. In the first gate the inputs are not both 0s and the output is a 1. In the second gate the inputs are both 0s, and the output is 0.

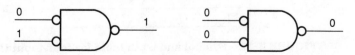

The inverted logic symbols appear on schematics because of the nature of the signals that the gates combine.

Some signals are normally HIGH and go to LOW when they are active. These are called active LOW signals. Others are normally LOW and go HIGH when they are active. They are called active HIGH signals. A gate that is combining active HIGH signals is usually drawn without bubbles on the inputs. A gate that is combining active LOW inputs is sometimes drawn in the inverted logic form with bubbles on its inputs.

A good example of the use of an inverted logic symbol occurs when a Z-80 microprocessor needs to store a word in memory. It issues two control signals, $\overline{\text{MEMORY REQUEST}}$ and $\overline{\text{WRITE}}$, both active LOW. The complement over their names indicates that they are active LOW. These two signals need to be combined to

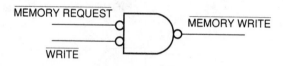

FIGURE 2-11

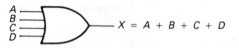

$$X = A + B + C + D$$

Inputs				Output
D	C	B	A	X
0	0	0	0	0
0	0	0	1	1
0	0	1	0	1
0	0	1	1	1
0	1	0	0	1
0	1	0	1	1
0	1	1	0	1
0	1	1	1	1
1	0	0	0	1
1	0	0	1	1
1	0	1	0	1
1	0	1	1	1
1	1	0	0	1
1	1	0	1	1
1	1	1	0	1
1	1	1	1	1

FIGURE 2-12   Symbol and truth table for four-input OR gate

produce a new active LOW signal called $\overline{\text{MEMORY WRITE}}$. $\overline{\text{MEMORY WRITE}}$ should go LOW when both inputs are LOW. The inverted logic OR symbol fits this situation perfectly. $\overline{\text{MEMORY WRITE}}$ will be LOW when $\overline{\text{MEMORY REQUEST}}$ and $\overline{\text{WRITE}}$ are LOW.

A variety of forms of OR gates are available in TTL and CMOS. The 7432 is a quad (meaning four) two-input TTL OR gate IC. The four are independent. Each can be used in a different part of a circuit without feedback. Power is supplied to the IC through a V_{CC} (+5 V) and ground connection. The 4072 is a dual (meaning two gates) four-input CMOS OR gate IC. The symbol and truth table for a four-input OR gate are shown in Figure 2-12.

The pinouts for the 7432 and 4072 are shown in Figure 2-13.

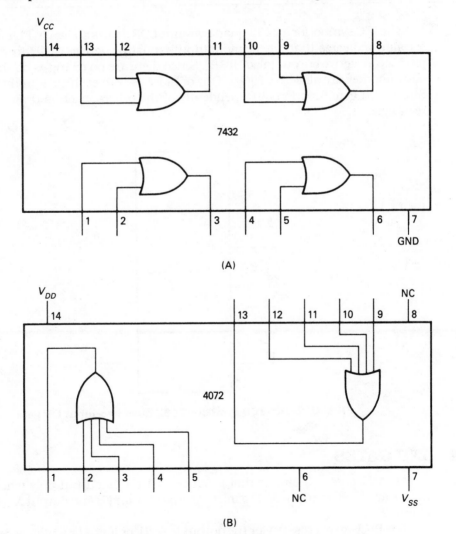

FIGURE 2-13 OR gate pinouts

Some of the available OR gate ICs are listed in Table 2-2.

TABLE 2-2 OR Gates

NUMBER	FAMILY	DESCRIPTION
7432	TTL	Quad 2-input OR
74C32	CMOS	Quad 2-input OR
4071	CMOS	Quad 2-input OR
4072	CMOS	Dual 4-input OR

The IEC symbol for a 7432 quad two-input OR gate is shown in Figure 2-14. The ≥ 1 sign indicates that one or more inputs must be active (HIGH in this case) to produce an active output (also HIGH). Since there are no triangles on the inputs or outputs, they are all active HIGH. One or more 1s into each gate produces a 1 out. Since each gate functions independently of the others, each is drawn in its own rectangle.

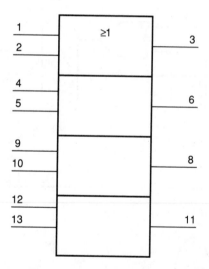

FIGURE 2-14 IEC symbol—7432 quad two-input OR gate

2.4 AND GATES

An *AND gate* is a circuit that produces a 1 on its output only when all of its inputs are 1. A two-input AND gate, with inputs A and B and output X, is shown in Figure 2-15.

The Boolean expression for the output is $A \cdot B$, or just AB, which is read "A AND B." The output X is 1 only when both A and B are ones. The truth table, Figure 2-16, summarizes the operation of the gate. All possible input combinations are listed by counting in binary from 00 to 11.

The AND symbol describes the operation of the gate. Since there are no bubbles on the input or the output, the gate is read "1 AND 1 in, 1 out." This statement describes the last line of the truth table, and the only situation in which the output is 1. This is the unique state of the AND gate.

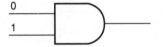

$X = A \cdot B$

Inputs		Output
A	B	X
0	0	0
0	1	0
1	0	0
1	1	1

FIGURE 2-15 Two-input AND gate

FIGURE 2-16 Two-input AND truth table

Example: Predict the output of each gate.

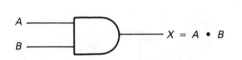

Solution:

In the first gate the inputs are different (line 2 or 3 of the truth table), and the output is 0. In the second gate, both inputs are 1 (last line of the truth table), and the output is 1.

   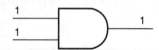

The top three lines of the AND gate truth table show that if a 0 is present on A or B (or both) then the output will be 0. This is summarized as "0 OR 0 in, 0 out." The symbol that represents this statement is shown in Figure 2-17. Both inputs and outputs are bubbled in this inverted logic symbol.

The Boolean expression for the inverted logic symbol is developed as follows:

1. Since the A input is bubbled, write $\overline{A}$.
2. Since the B input is bubbled, write $\overline{B}$.

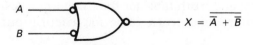

$X = \overline{A} + \overline{B}$

FIGURE 2-17 Inverted logic symbol for two-input AND gate

3. The shape of the gate is OR, which is written $+$.
4. Write $\overline{A} + \overline{B}$. Since the output is bubbled, $\overline{A} + \overline{B}$ is the Boolean expression for $\overline{X}$. $\overline{X} = \overline{A} + \overline{B}$.
5. To find X, complement the whole expression $\overline{\overline{A} + \overline{B}}$.

Since the symbols shown in Figures 2-15 and 2-17 are equivalent, their outputs are equivalent and $A \cdot B = \overline{\overline{A} + \overline{B}}$. Both symbols are used on schematic diagrams and both should be learned. Equivalent AND gate symbols are shown in Figure 2-18.

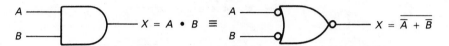

FIGURE 2-18   Equivalent AND gate symbols

Example: Predict the output of each gate.

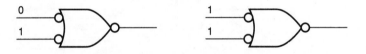

Solution:

This alternate symbol states that any 0 in yields a 0 out. In gate 1, there is a 0 input, and the output must be 0. In the second gate, there are no zeros in, and the output must be a 1.

A variety of forms of AND gates are available in TTL and CMOS. The 7408 IC is a quad (meaning four) two-input AND IC. There are four independent two-input gates. The 7411 is a triple (meaning three gates) three-input TTL AND gate IC, and the 4082 is a dual (meaning two gates on one IC) four-input CMOS AND gate IC.

The symbol and truth table for the three-input AND gate are shown in Figure 2-19. The symbol and truth table for a four-input AND gate are shown in Figure 2-20.

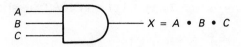

Inputs			Output
C	B	A	X
0	0	0	0
0	0	1	0
0	1	0	0
0	1	1	0
1	0	0	0
1	0	1	0
1	1	0	0
1	1	1	1

FIGURE 2-19 Symbol and truth table for three-input AND gate

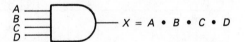

Inputs				Output
D	C	B	A	X
0	0	0	0	0
0	0	0	1	0
0	0	1	0	0
0	0	1	1	0
0	1	0	0	0
0	1	0	1	0
0	1	1	0	0
0	1	1	1	0
1	0	0	0	0
1	0	0	1	0
1	0	1	0	0
1	0	1	1	0
1	1	0	0	0
1	1	0	1	0
1	1	1	0	0
1	1	1	1	1

FIGURE 2-20 Symbol and truth table for four-input AND gate

The pinouts for the 7408, 7411 and 4082 are shown in Figure 2-21.

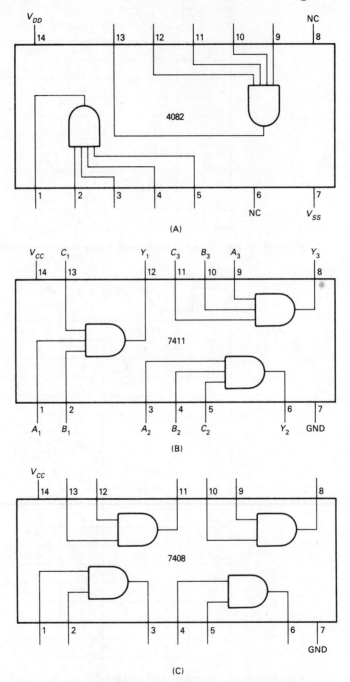

(A)

(B)

(C)

FIGURE 2-21   AND gate pinouts

Some of the available AND gates are listed in Table 2-3.

TABLE 2-3   AND Gates

NUMBER	FAMILY	DESCRIPTION
7408	TTL	Quad 2-input AND
74C08	CMOS	Quad 2-input AND
4081	CMOS	Quad 2-input AND
7411	TTL	Triple 3-input AND
7421	TTL	Dual 4-input AND
4082	CMOS	Dual 4-input AND

The IEC chose the ampersand (&) sign to represent the AND function. Figure 2-22A shows the IEC symbol for a 7408 quad two-input AND gate integrated circuit. The symbol for a 4082 dual four-input AND gate integrated circuit is shown in Figure 2-22B.

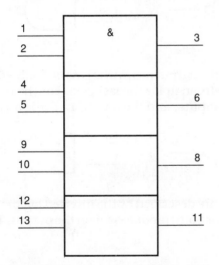

FIGURE 2-22A   IEC symbol—7408 quad two-input AND gate

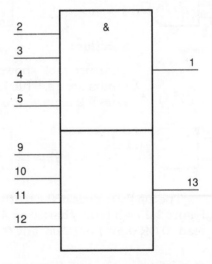

FIGURE 2-22B   IEC symbol—4082 quad four-input AND gate

2.5   NAND GATES

A *NAND gate* is a circuit that produces a 0 on its output only when all of its inputs are ones. NAND is the contraction of the words "not" and "and." Its symbol is the AND symbol with an inverted (bubbled) output, Figure 2-23.

The truth table for a NAND gate is shown in Figure 2-24. Notice that its output is the complement of the AND gate output.

The symbol describes the operation of the gate. Since the inputs are not bubbled and the output is, the symbol is read "1 AND 1 in, 0 out." This describes line four of the truth table which is the unique state of this gate (the only situation that produces a 0).

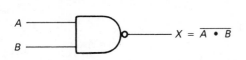

FIGURE 2-23   **Two-input NAND gate**

Inputs		Output
B	B	X
0	0	1
0	1	1
1	0	1
1	1	0

FIGURE 2-24   **Truth table for two-input NAND gate**

Example: Predict the output of each gate.

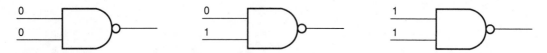

Solution:

The truth table shows that the output of a NAND gate is 0 only when all inputs are 1s. This is the situation for the last gate. For the first two gates a 0 is present on the inputs, and the output must be a 1.

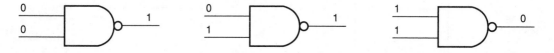

The top three lines of the truth table are described by the inverted logic symbol, Figure 2-25, and state that a 0 on A or B (or both) produces a 1 on the output. This is read "0 OR 0 in, 1 out," or "any 0 in, 1 out."

$$X = \overline{A} + \overline{B}$$

FIGURE 2-25   **Inverted logic symbol for two-input NAND gate**

$$X = \overline{A \cdot B} \equiv X = \overline{A} + \overline{B}$$

FIGURE 2-26   **Equivalent NAND gate symbols**

The Boolean expression for the inverted logic symbol is developed as follows:

1. Since A is bubbled, write A complement, $\overline{A}$.
2. Since B is bubbled, write B complement, $\overline{B}$.
3. Since the shape of the gate is OR, write $\overline{A} + \overline{B}$.

The Boolean expression for the output is $\overline{A} + \overline{B}$, which is read "A complement OR B complement."

Both symbols represent the NAND gate, both are used in schematics, and both should be learned. Since the symbols shown in Figures 2-23 and 2-25 are equivalent, their outputs are equivalent and $\overline{A \cdot B} = \overline{A} + \overline{B}$. Equivalent NAND gate symbols are shown in Figure 2-26.

Example: Predict the output of each gate.

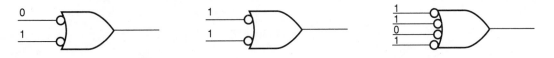

Solution:

The alternate symbol for the NAND gate states any 0 in, 1 out. The first and last gates of this example have 0s in and their outputs are 1s. The middle gate has no 0s in, and its output must be 0.

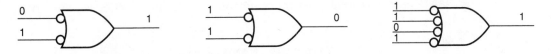

The pinouts for some common NAND gates are shown in Figure 2-27.

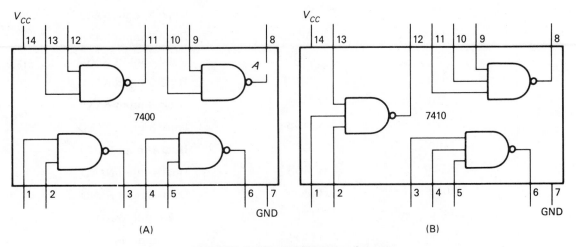

FIGURE 2-27 NAND gate pinouts

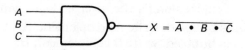

	Inputs		Output
C	B	A	X
0	0	0	1
0	0	1	1
0	1	0	1
0	1	1	1
1	0	0	1
1	0	1	1
1	1	0	1
1	1	1	0

FIGURE 2-28   Symbol and truth table for three-input NAND gate

The 7400 is a quad two-input TTL NAND gate IC, and the 7410 is a triple three-input NAND gate IC. Their pinouts are shown in Figure 2-27. The 74C30 is an eight-input CMOS NAND gate. Figure 2-28 shows the symbol and the truth table for a three-input NAND gate.

The NAND gate is available in many forms in TTL and CMOS, as shown in Table 2-4.

TABLE 2-4   NAND Gates

<u>NUMBER</u>	<u>FAMILY</u>	<u>DESCRIPTION</u>
7400	TTL	Quad 2-input NAND
74C00	CMOS	Quad 2-input NAND
4011	CMOS	Quad 2-input NAND
7410	TTL	Triple 3-input NAND
74C10	CMOS	Triple 3-input NAND
4023	CMOS	Triple 3-input NAND
7420	TTL	Dual 4-input NAND
74C20	CMOS	Dual 4-input NAND
4012	CMOS	Dual 4-input NAND
7430	TTL	8-input NAND
74C30	CMOS	8-input NAND

The IEC symbol for a 7400 quad two-input NAND gate is shown in Figure 2-29. The triangle on the output of each gate indicates an active LOW output. The symbol shows that inputs 1 and 2 are ANDed together to produce an active LOW output on pin 3.

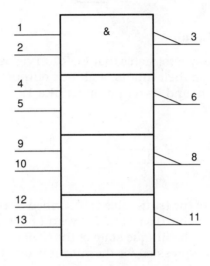

FIGURE 2-29   IEC symbol—7400 quad two-input NAND gate

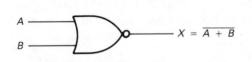

$X = \overline{A + B}$

FIGURE 2-30   Two-input NOR gate

Inputs		Output
B	A	X
0	0	1
0	1	0
1	0	0
1	1	0

FIGURE 2-31   Two-input NOR truth table

2.6   NOR GATES

A *NOR gate* is a circuit that produces a 0 on its output when one or more of its inputs are 1. NOR is the contraction of the words "not" and "or." Its symbol is the OR symbol with an inverted or bubbled output, Figure 2-30.

The truth table for a NOR gate is shown in Figure 2-31. Notice that its output is the complement of the OR gate output.

The symbol describes the operation of the gate. Since the inputs are not bubbled and the output is, the symbol is read "1 OR 1 in, 0 out." This describes the last three lines of the truth table in Figure 2-31.

Example: Predict the output of each gate

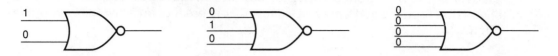

Solution:

The symbol states that 1 OR 1 in yields a 0 out. The first two gates have a 1 on their inputs, and their outputs must be 0. The last gate has no 1s in, and its output must be a 1.

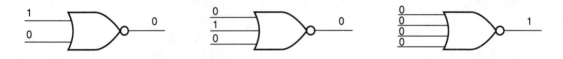

The top line of the truth table is described by the inverted logic symbol, Figure 2-32, which states that 0 on A and B gives a 1 out on X. The symbol is read "0 AND 0 in, 1 out." This is the unique state of the NOR gate.

The Boolean expression for the output is developed as follows:

1. Since A is bubbled, write A complement, $\overline{A}$.
2. Since B is bubbled, write B complement, $\overline{B}$.
3. Since the shape of the gate is AND, write $\overline{A} \cdot \overline{B}$.

Both symbols represent the NOR gate, both are used in schematics, and both should be learned. Since the symbols shown in Figures 2-30 and 2-32 are equivalent their outputs are equivalent, and $\overline{A + B} = \overline{A} \cdot \overline{B}$. Equivalent NOR gate symbols are shown in Figure 2-33. The pinouts for some common NOR gates are shown in Figure 2-34.

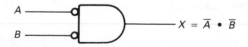

FIGURE 2-32   Inverted logic symbol for two-input NOR gate

$$X = \overline{A} + B \quad \equiv \quad X = \overline{A} \cdot \overline{B}$$

FIGURE 2-33   Equivalent NOR gate symbols

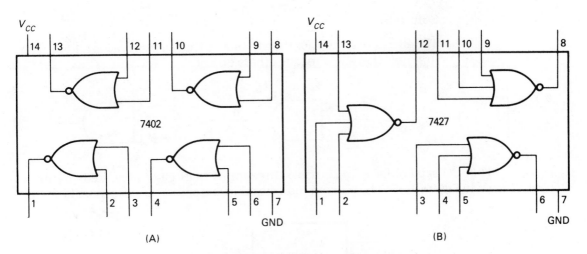

(A)   (B)

FIGURE 2-34   NOR gate pinouts

TABLE 2-5   NOR Gates

NUMBER	FAMILY	DESCRIPTION
7402	TTL	Quad 2-input NOR
7425	TTL	Dual 4-input NOR
7427	TTL	Triple 3-input NOR
74C02	CMOS	Quad 2-input NOR
4000	CMOS	Dual 3-input NOR and inverter
4001	CMOS	Quad 2-input NOR
4002	CMOS	Dual 4-input NOR
4025	CMOS	Triple 3-input NOR

Some of the available NOR gate ICs are listed in Table 2-5.

Example: Predict the outputs of each gate.

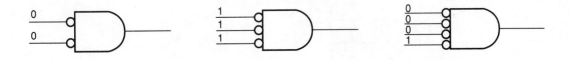

Solution:

The alternate symbol for the NOR states that all 0s in yields a 1 out. The first gate has all 0s in; the others do not.

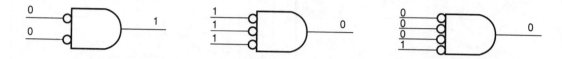

The IEC symbol for a 7427 triple three-input NOR gate integrated circuit is shown in Figure 2-35. The ≥1 sign indicates that one or more inputs must be active

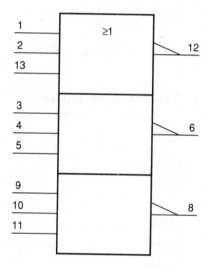

FIGURE 2-35   IEC symbol—7427 triple three-input NOR gate

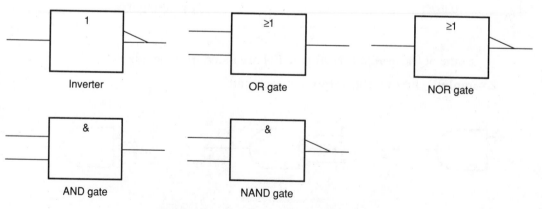

FIGURE 2-36   IEC symbol—basic gates

to cause an active output. Since the outputs are active LOW (triangles), if one or more inputs are HIGH the corresponding output goes LOW.

Figure 2-36 summarizes the IEC symbols discussed in this chapter.

2.7 DATA CONTROL ENABLE/INHIBIT

One of the common uses of gates is to control or gate the flow of data from the input to the output. One input in that mode of operation is used as the control and the other presents the data to be passed to the output. If the data is allowed to pass through, the gate is said to be *enabled.* If the data is not allowed to pass through, the gate is said to be *inhibited.*

2.8 AND GATE ENABLE/INHIBIT

If the voltage on the control input is 0 (top two lines of the truth table in Figure 2-37), the output of the gate is 0 regardless of the data present on the data input. The data does not pass through the gate, and the gate is said to be inhibited. The output is "locked up" in the 0 state.

		Inputs		Output	
		Control	Data	X	
Inhibit		0	0	0	Output locked at 0
		0	1	0	
Enable		1	0	0	Data passes through
		1	1	1	

FIGURE 2-37 AND enable/inhibit

If the voltage on the control is 1 (bottom two lines of the truth table in Figure 2-37), then whatever is present on the data input appears on the output and the gate is said to be enabled. The data "passes through" the gate.

Example: Predict the output of each AND gate.

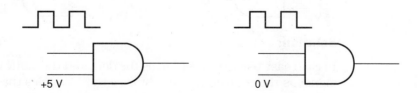

Solution:

In each case use the waveform as the data and the static (unchanging) signal as the control input. In the first case, the 1 enables the gate and data passes through unaltered. In the second case, the 0 inhibits the gate and the output is locked at 0. The data input is ignored.

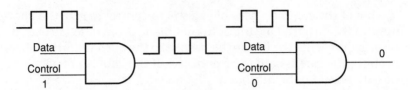

2.9 NAND GATE ENABLE/INHIBIT

If the signal on the control line is 0 (top two lines of the truth table in Figure 2-38), the signal on the data input is ignored and the output is "locked up" in the one state. The gate is said to be inhibited even though the output is 1.

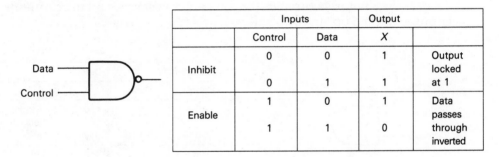

		Inputs		Output	
		Control	Data	X	
	Inhibit	0	0	1	Output locked at 1
		0	1	1	
	Enable	1	0	1	Data passes through inverted
		1	1	0	

FIGURE 2-38   NAND enable/inhibit

If the signal on the control line is 1 (bottom two lines of the truth table in Figure 2-38), the signal on the data input is passed through the gate but is inverted in the process. The gate is said to be enabled.

Example: Predict the output of each NAND gate.

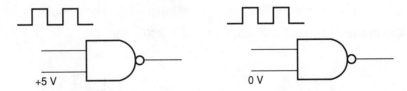

Solution:

In each case use the waveform as the data and the static (unchanging) signal as the control. In the first case the 1 enables the gate, and the data passes through inverted. In the second case the 0 inhibits the gate, and the output is locked at 1. The data input is ignored.

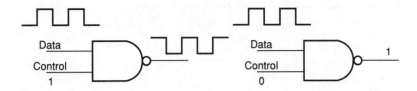

2.10   OR GATE ENABLE/INHIBIT

If the signal on the control line is 0 (top two lines of the truth table in Figure 2-39), the signal on the data input passes through to the output and the gate is enabled.

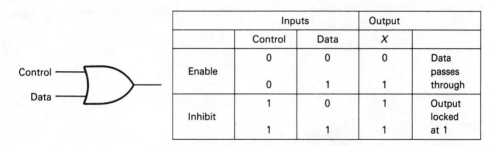

		Inputs		Output	
		Control	Data	X	
	Enable	0	0	0	Data passes through
		0	1	1	
	Inhibit	1	0	1	Output locked at 1
		1	1	1	

FIGURE 2-39   OR enable/inhibit

If the signal on the control line is 1 (bottom two lines of the truth table in Figure 2-39), the signal on the data input is ignored and the output is "locked up" in the 1 state. The gate is said to be inhibited.

Example: Predict the output of each OR gate.

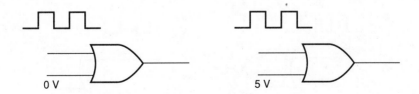

Solution:

In each case use the waveform as the data and the static signal as the control. In the first case the 0 enables the gate, and data passes through unaltered. In the second case the 1 inhibits the gate, and the output is locked at 1. The data input is ignored.

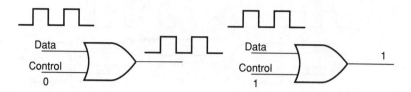

2.11 NOR GATE ENABLE/INHIBIT

If the voltage on the control input is 0 (top two lines of the truth table in Figure 2-40), whatever is present on the data input appears at the output inverted. The gate is enabled.

If the voltage on the control is 1 (bottom two lines of the truth table in Figure 2-40), the output of the gate is 0 regardless of the data present at the data input. The gate is said to be inhibited.

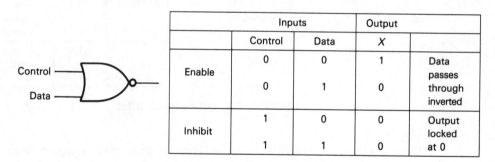

		Inputs		Output	
		Control	Data	X	
Enable		0	0	1	Data passes through inverted
		0	1	0	
Inhibit		1	0	0	Output locked at 0
		1	1	0	

FIGURE 2-40   NOR enable/inhibit

Example: Predict the output of each gate.

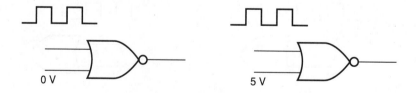

Solution:

In each case use the waveform as the data and the static signal as the control. In the first case the 0 enables the gate, and the data passes through inverted. In the second case the 1 inhibits the gate, and the output is locked at 0. The data input is ignored.

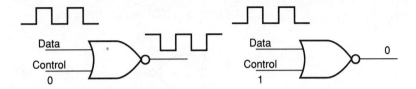

2.12 SUMMARY ENABLE/INHIBIT

Each gate is enabled or inhibited in its own fashion. There is no need to memorize the function of each gate. By examining the truth table of each gate, its method of operation can be determined. The operation of each is summarized in Table 2-6.

TABLE 2-6   Enable/Inhibit

GATE	CONTROL INPUT	GATE CONDITION	OUTPUT
AND	0	Inhibit	0
	1	Enable	Data
NAND	0	Inhibit	1
	1	Enable	$\overline{\text{Data}}$
OR	0	Enable	Data
	1	Inhibit	1
NOR	0	Enable	$\overline{\text{Data}}$
	1	Inhibit	0

Note: Data—Data passes through unaltered.
$\overline{\text{Data}}$—Data passes through inverted.

2.13 NAND AS AN INVERTER

Suppose we apply the same signal to both inputs of a two-input NAND. Then either both inputs are 0 or both inputs are 1. If A is 0 then the output is 1. If A is 1 then the output is 0. The output is always the complement of the input. Figure 2-41 shows a NAND as an inverter.

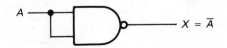

FIGURE 2-41   NAND—as an inverter

2.14 NOR AS AN INVERTER

If we apply the same signal to both inputs of a two-input NOR, then either both inputs are 0 or both inputs are 1. In either case the output is always the complement of the input. Figure 2-42 shows a NOR as an inverter.

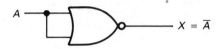

FIGURE 2-42   NOR—as an inverter

Why would we use a two-input NAND or NOR for an inverter? Sometimes there is an extra NAND or NOR gate on an IC that has been used in a circuit. It is better to use the extra gate as an inverter than to add an inverter IC. The extra IC would take up space on the circuit board (real estate), consume extra power, generate extra heat, and add extra expense.

2.15 EXPANDING AN AND GATE

A three-input AND gate can be created from two two-input gates as shown in Figure 2-43. The output is the same as if we had fed A, B, and C into a three-input AND, $X = A \cdot B \cdot C$. You can expand an AND with another AND.

2.16 EXPANDING A NAND GATE

The output from a three-input NAND would be $\overline{A \cdot B \cdot C}$. At first you might think that you could expand a NAND gate with another NAND. But watch what happens in Figure 2-44. $\overline{\overline{A \cdot B} \cdot C}$ does not give the desired output, $\overline{A \cdot B \cdot C}$.

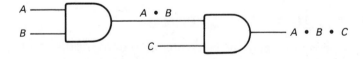

FIGURE 2-43   Expanding an AND

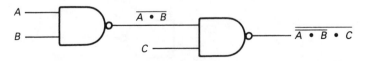

FIGURE 2-44   Attempt to expand a NAND with a NAND

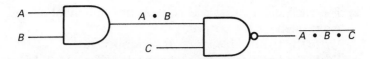

FIGURE 2-45   Expanding a NAND

Consider the circuit in Figure 2-45. This yields the desired output. You can expand a NAND with an AND, but you cannot expand a NAND with another NAND.

2.17   EXPANDING AN OR GATE

A three-input OR gate can be created from the two-input gates as shown in Figure 2-46. You can expand an OR with another OR.

2.18   EXPANDING A NOR GATE

The output from a three-input NOR gate is $\overline{A + B + C}$. As with the NAND, you cannot expand a NOR with another NOR. A NOR can be expanded as shown in Figure 2-47. This yields the desired output. You can expand a NOR with an OR, but you cannot expand a NOR with another NOR.

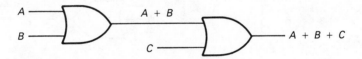

FIGURE 2-46   Expanding an OR

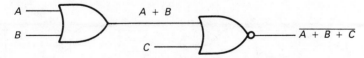

FIGURE 2-47   Expanding a NOR

Exercise 2

1. Draw the symbol for each of the following gates, label the inputs and write the Boolean expression for the output.
   a. Inverter
   b. OR
   c. NOR
   d. AND
   e. NAND

2. For each gate, draw the equivalent logic symbol and write the Boolean expression for the output.
3. Write the truth table for each two-input gate.
4. Write the truth table for three-input AND, NAND, OR, and NOR gates.
5. For a two-input AND gate
   a. Draw the symbol and write the Boolean expression for the output.
   b. Draw the inverted logic symbol and write the Boolean expression for the output.
   c. Write the truth table and indicate the line that represents the unique state.
6. Repeat problem 5 for a two-input NAND.
7. Repeat problem 5 for a two-input OR.
8. Repeat problem 5 for a two-input NOR.
9. Predict the output of each gate.

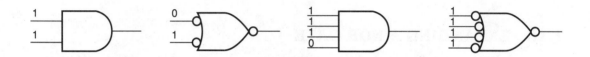

10. Predict the output of each gate.

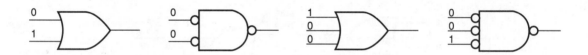

11. Predict the output of each gate.

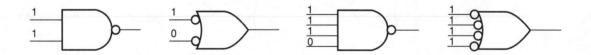

12. Predict the output of each gate.

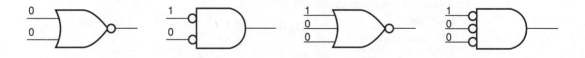

13. Predict the output of each gate.

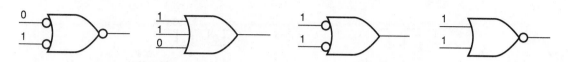

14. Predict the output of each gate.

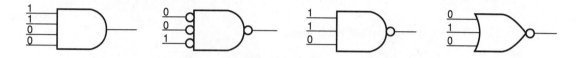

15. A 0 on the control input of a NOR gate (enables, inhibits) the gate.
16. When a NAND gate is enabled, the data passes through (unaltered, inverted).
17. How do you inhibit an OR gate?
18. How do you enable an AND gate?
19. When an AND gate is inhibited the output is a (0,1).
20. When a NOR gate is inhibited, what is the state of the output?
21. When a NOR gate is enabled, the data passes through (unaltered, inverted).
22. How do you inhibit a NOR gate?
23. How do you enable a NAND gate?
24. When a NAND gate is inhibited, what is the state of the output?
25. When an OR gate is inhibited, what is the state of the output?
26. When an OR gate is enabled, the data passes through (unaltered, inverted).
27. When an AND is enabled, data passes through (inverted, unaltered).
28. Predict the output of each gate.

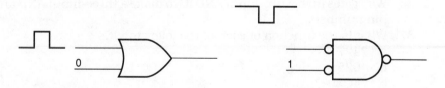

29. Predict the output of each gate.

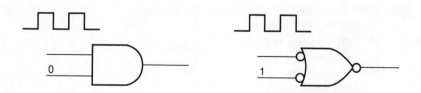

30. Predict the output of each gate.

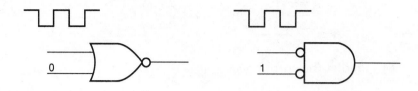

31. Predict the output of each gate.

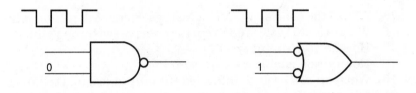

32. Wire two gates from a two-input NAND IC to make an AND gate and show the pin numbers.
33. Wire a two-input NOR gate to act as an inverter and show the pin numbers.
34. Wire two gates from a two-input NOR IC to make an OR gate and show the pin numbers.
35. Wire gates from a two-input NAND IC to make a three-input NAND gate (expand a NAND) and show the pin numbers.
36. Wire gates from a two-input AND IC to make a three-input AND and show the pin numbers.
37. What is the function of each of the following ICs?
    a. 7427
    b. 4025
    c. 74C20
    d. 7410
    e. 4081
    f. 4069
38. How does the IEC specify the following:
    a. Active LOW input
    b. Active HIGH input
    c. Inverter function
    d. OR function
    e. AND function
    f. NOR function
    g. NAND function

39. Draw the IEC symbol for each of the following integrated circuits:
    a. Eight-input NAND gate—7430
    b. Dual four-input AND gate—4082
    c. Triple three-input NAND gate—7410
    d. Dual four-input OR gate—4072
    e. Triple three-input NOR gate—4025

LAB 2 Gates

OBJECTIVES

After completing this lab, you should be able to:

■ determine the truth table of a gate.

■ use each gate in the enable/inhibit mode.

■ use a NAND as an inverter.

■ use a NOR as an inverter.

■ expand a NAND.

■ expand a NOR.

COMPONENTS NEEDED

1	7400 IC
1	7402 IC
1	7404 IC
1	7408 IC
1	7411 IC
1	7432 IC
1	4001 IC
1	4011 IC
1	4069 IC
1	4071 IC
1	4081 IC
1	LED
1	330-Ω resistor

PREPARATION

In Lab 2 both CMOS and TTL gates will be used. On TTL ICs, an input left floating (disconnected) is usually taken as a 1. This is not true in CMOS. Due to the extremely high input impedance of a CMOS gate, unused inputs can drift between a 1 level and a 0 level. Tie all unused inputs on the gates to be used to the power supply voltage or ground. Supply $+5$ V to V_{DD} and connect V_{SS} to ground.

For CMOS operating at $+5$ V:

1. A legitimate 1 input can range from 3.5 V to 5 V. A legitimate 0 input can range from 0 V to 1.5 V.
2. With inputs maintained within these ranges the outputs should remain within 0.05 V of the supply levels. A 0 output should not rise above 0.05 V and a 1 output should not fall below 4.95 V.

For TTL ICs:

1. A legitimate 1 input can range from 2.0 V to 5 V, and a legitimate 0 input can range from 0 V to 0.8 V.
2. A legitimate 1 output can range from 2.4 V to 5 V and a legitimate 0 output can range from 0 V to 0.4 V.

Care must be taken in the handling of CMOS ICs since they can be destroyed from excessive static build-up between pins. These guidelines should be followed:

1. Store CMOS ICs in antistatic tubes or in black conductive foam.
   Never push CMOS into styrofoam. They can be wrapped in aluminum foil.
2. In low humidity environments where static build-up is a problem, avoid touching the pins of a CMOS IC when they are removed from storage unless precautions have been taken to bleed off the static charge. Conductive wrist straps connected through a resistor to ground is one method used.
3. Apply dc voltage to the CMOS circuit before signals are applied.
4. Remove signal sources before the dc supply is switched off.
5. Switch off supply voltages before inserting or removing CMOS devices from a circuit.

In Part I of this lab, static voltage levels (unchanging ones or zeros) will be used to verify the truth table of a variety of gates.

In Part II of this lab, the square wave generator on the trainer will be used to generate the signal for the data input of the gate. An oscilloscope will be used to compare the data in signal and the output signal. Use both channels of your scope, synchronize on the input signal, and use dc coupling so that the 1 and 0 levels can be detected.

Part III of this lab calls for logic diagrams to be drawn for various configurations. A logic diagram shows the symbol for the gates used, the pin numbers, and the IC numbers.

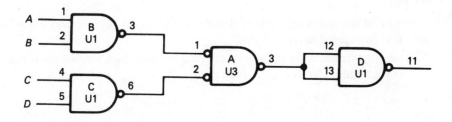

If an accurate logic diagram is drawn before connecting the circuit, it can be used as a guide in troubleshooting. Since the gates are independent and can be used in different parts of a circuit, they are often identified with a letter, A, B, C, etc., and IC number, U1, U2, etc. The three NAND gates are from the same IC U1 and the OR gate is from IC U3.

PROCEDURE

Part I

To determine the truth table of a gate:

A. Connect the inputs of the gate to the switches provided on the trainer or hook them directly to the +5 V supply or ground as required.

B. Connect the outputs to the LED monitors on the trainer or hook them to an LED and current limiting resistor of approximately 330 Ω.

C. Write the input portion of the truth table by counting in binary with one bit for each input. For example,

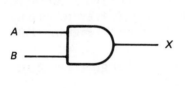

Inputs		Output
A	B	X
0	0	
0	1	
1	0	
1	1	

D. Determine the output of the truth table by supplying the inputs called for in each line of the truth table. Use this procedure to write the truth table for a gate in a 7408, 7411, 7432, 7404, 7400, 7402, 4001, 4069, 4071, 4081, 4011.

Part II

For one of the gates on each of the ICs 7400, 4001, 4071, and 7408, verify the enable/inhibit operation as follows:

A. Use the square wave generator on your trainer as a source for the data input on your gate. Use the 10 kHz setting.

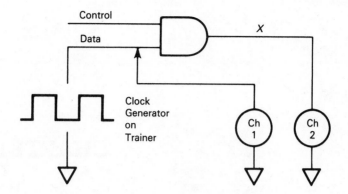

B. Monitor the data input and output with channel 1 and 2 of your scope. Use dc coupling so that 1 and 0 levels can be detected. Sync on the data input signal.
C. Supply a 1 or 0 to the control input to enable or inhibit the gate. Observe the relationships between input and output on the scope.
D. Summarize the operation with a truth table and written description of the operation.

Part III

In each of the following, draw a logic diagram, including pin numbers and IC numbers, of the circuit. Then connect the circuit and verify the operations.

A. Use a NAND as an inverter.
B. Use a NOR as an inverter.
C. Expand a two-input NAND into a three-input NAND.
D. Expand a two-input NOR into a three-input NOR.
E. Using one two-input NAND gate chip, wire a three-input NAND gate.

CHAPTER 3

Waveforms and Boolean Algebra

OBJECTIVES

After completing this chapter, you should be able to:

- predict the output waveforms for each of the gates, given input waveforms.

- develop the Boolean expression for the output of a combinational logic circuit.

- combine signals from a shift counter and predict the outputs.

- select proper shift counter signals and gates to produce required outputs.

- design and construct a logic circuit to implement a given truth table.

3.1 WAVEFORM ANALYSIS

In Chapter 2 you learned the truth tables for the basic gates. Once you know the truth table of a gate it becomes an easy chore to predict the output waveforms from given inputs. First determine the unique state of the gate and the inputs that produce that output. Then find all times at which those inputs occur. Graph the unique output for those times and its complement at all other times.

AND Gate

The unique state of the AND is all 1s in, 1 out.

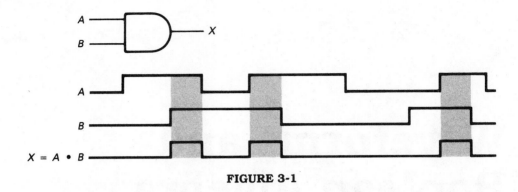

FIGURE 3-1

Example: Refer to Figure 3-1. If A and B are as shown, predict the output X.

Solution:

Watch for areas where both A and B are HIGH. The output is HIGH during those times and LOW at all other times. The waveform for X is shown in Figure 3-1.

NAND Gate

The unique state of the NAND is all 1s in, 0 out.

Example: Refer to Figure 3-2. If A, B, and C are as shown, predict the output X.

Solution:

The areas shaded are those in which all three inputs are HIGH. At those times the output goes LOW. At all other times the output is HIGH. The waveform for X is shown in Figure 3-2.

OR Gate

The unique state of an OR gate is all 0s in, 0 out.

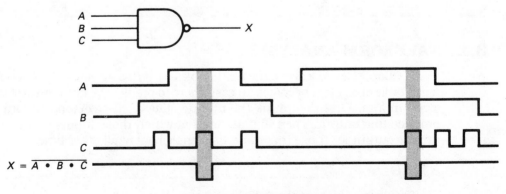

FIGURE 3-2

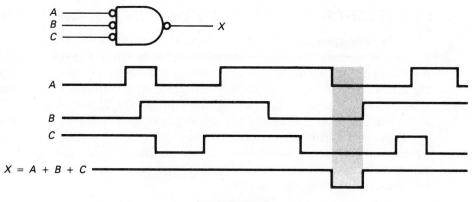

FIGURE 3-3

Example: Refer to Figure 3-3. If A, B, and C are as shown, predict the output X.

Solution:

There is only one period of time at which all three inputs are LOW. At that time, the output goes LOW. At all other times the output is HIGH. The waveform for X is shown in Figure 3-3.

NOR Gate

The unique state of the NOR is all 0s in, 1 out.

Example: Refer to Figure 3-4. If A, B, and C vary as shown, predict the output X.

Solution:

The areas shaded are those in which all three inputs are LOW. At those times the output goes HIGH. At all other times the output is LOW. The waveform for X is shown in Figure 3-4.

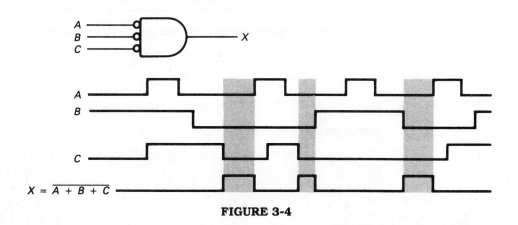

FIGURE 3-4

3.2 DELAYED-CLOCK AND SHIFT-COUNTER WAVEFORMS

In later chapters, we will build circuits that produce a delayed or nonoverlapping clock system. The output waveforms are shown in Figure 3-5 as *CP* for clock pulse and *CP′* for the delayed or nonoverlapping clock. Also shown in Figure 3-5 are the output waveforms, *A*, *Ā*, *B*, *B̄*, *C*, *C̄* from a shift counter. The outputs from the shift counter change on the trailing edge (HIGH to LOW transition) of *CP*. Time 1 is at the trailing edge of clock pulse 1. Time 2 is at the trailing edge of pulse 2, and so on. These waveforms are continuous. After *CP* reaches 6 it starts again at 1. This circuit will also be constructed in a later chapter. From these waveforms a great variety of control signals can be generated. Refer to the graphs in Figure 3-5 often as you master the following examples.

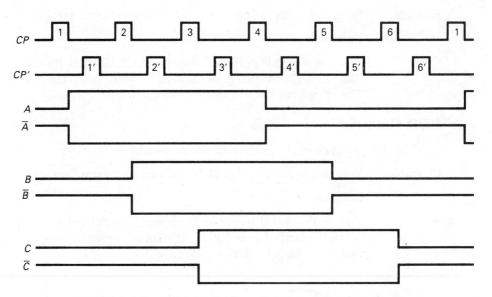

FIGURE 3-5   Delayed clock and shift counter waveforms

Example: Suppose we AND *A* and *C* together. What does the output look like?

Solution:

The unique state of an AND gate is all 1s in, 1 out. *A* goes HIGH at 1 and enables the AND gate. When *C* goes HIGH at 3, the output goes HIGH. The output stays HIGH until 4, when *A* goes LOW and inhibits the output. In other words, A and C are both HIGH from 3 to 4, and the output can be represented as shown in Figure 3-6.

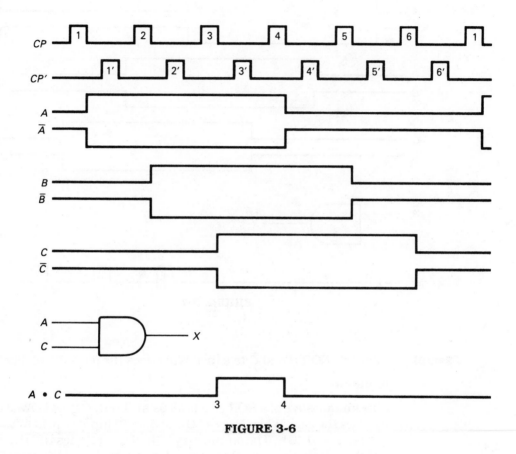

FIGURE 3-6

Example: Suppose we AND $\bar{A}$ and CP' together. What does the output look like?

Solution:

$\bar{A}$ goes HIGH at time 4, which is after the 3' pulse and before the 4' pulse, and goes back LOW at time 1, which is after the 6' pulse. $\bar{A}$ enables the gate during the 4', 5', 6' pulses, and they appear at the output X. See Figure 3-7.

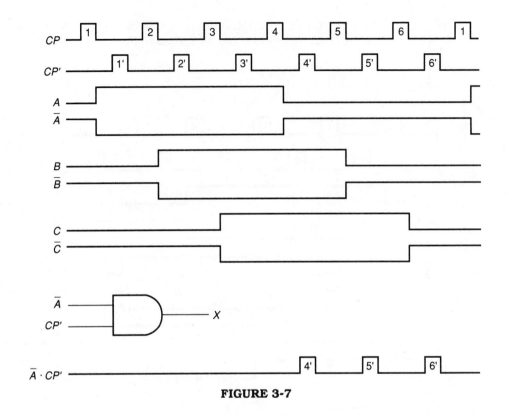

FIGURE 3-7

Example: Now let's NOR B and $\overline{C}$ together. What does the output look like?

Solution:

The unique state of a NOR gate is all 0s in, 1 out. $\overline{C}$ goes LOW at 3 and enables the gate. When B goes LOW at 5, both inputs are LOW and the output goes HIGH. The output stays HIGH until $\overline{C}$ goes HIGH at 6 and inhibits the output. In other words, B and $\overline{C}$ are both LOW from 5 to 6. The output can be represented as shown in Figure 3-8.

Example: Suppose that we NOR $\overline{C}$ and CP' together. Predict the output.

Solution:

Find the times that $\overline{C}$ and CP' are 0s. $\overline{C}$ is LOW from time 3 to time 6. CP' is LOW during that time except during the 3', 4', and 5' pulses. The output will be high from 3 to 6 except during the 3', 4', and 5' pulses. See Figure 3-9. Another way to analyze this problem is that $\overline{C}$

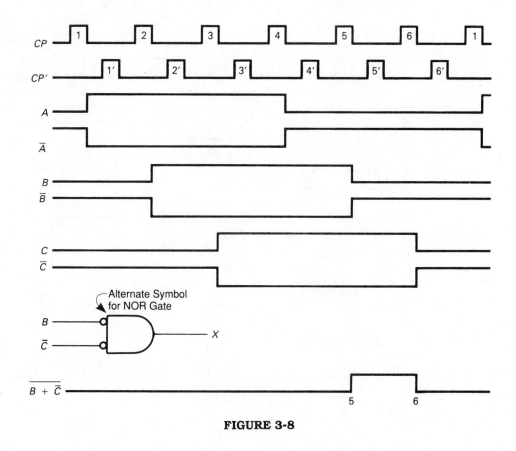

FIGURE 3-8

enables the NOR gate with a 0 from time 3 to time 6 and during that time CP' passes through inverted. While $\overline{C}$ is HIGH, the gate is inhibited and the output is LOW.

Example: Suppose that we NAND three signals together: CP', $\overline{A}$, and $\overline{B}$.

Solution:

The unique state of a NAND is all 1s in, 0 out. First, let's find when $\overline{A}$ and $\overline{B}$ are both 1s. $\overline{A}$ goes HIGH at 4, $\overline{B}$ follows at 5. Both are HIGH from 5 until 1 when A goes LOW again. Between times 5 and 1, CP' goes HIGH in pulses $5'$ and $6'$. During those times all three inputs are HIGH and the output is LOW. The output can be represented as shown in Figure 3-10.

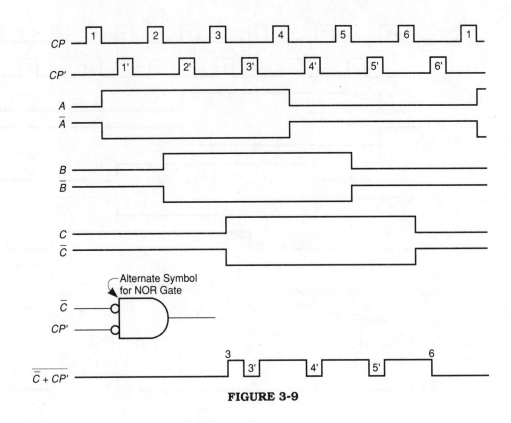

FIGURE 3-9

Example: Suppose we NAND $\overline{C}$ and CP' together. Predict the output.

Solution:

$\overline{C}$ goes HIGH at 6 and enables the NAND gate until time 3. During that time CP' passes through inverted. See Figure 3-11. While $\overline{C}$ is LOW, the NAND is inhibited and the output is locked HIGH.

Example: Let's OR three signals together: CP', A, and C.

Solution:

The unique state of an OR gate is all 0s in, 0 out. A and C are both 0s from 6 to 1. During that time, CP' is 0 except during the 6 pulse. The output can be represented as shown in Figure 3-12.

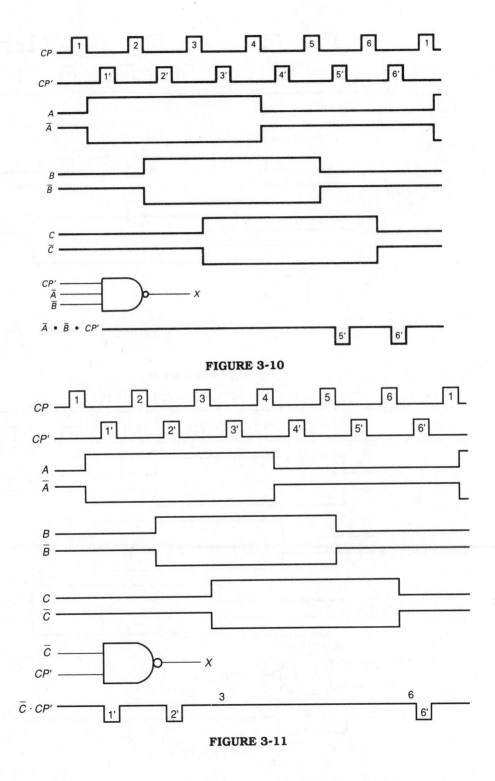

FIGURE 3-10

FIGURE 3-11

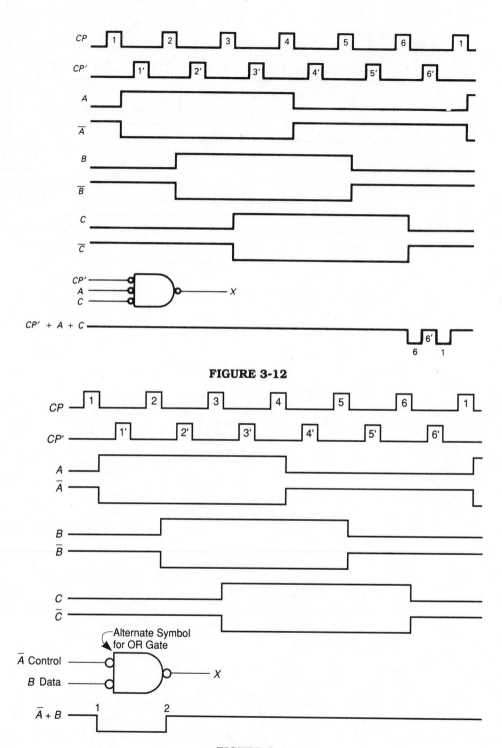

FIGURE 3-12

FIGURE 3-13

Example: OR $\overline{A}$ with B. Predict the output.

Solution:

A 0 enables the OR gate and allows the data to pass unaltered. With $\overline{A}$ as the control and B as the data, the gate is enabled from time 1 to time 4. During that time B passes through unaltered. The rest of the time the output is HIGH. The output is shown in Figure 3-13.

Example: Find a combination that yields only the 6′ pulse.

Solution:

An AND gate yields a 1 out for all 1s in. To get the 6′ pulse out, one input must be CP'. The other two inputs must enable and inhibit the three-input gate at the proper times. For the 6′ pulse to be enabled, one of the inputs must go HIGH at 6 to enable the gate. The other must already be HIGH at 6 to go back LOW at 1 to keep any other pulses from appearing at the output. $\overline{C}$ goes HIGH at 6, $\overline{A}$ goes back LOW at 1. The inputs are $\overline{A}$, $\overline{C}$, and CP'. See Figure 3-14.

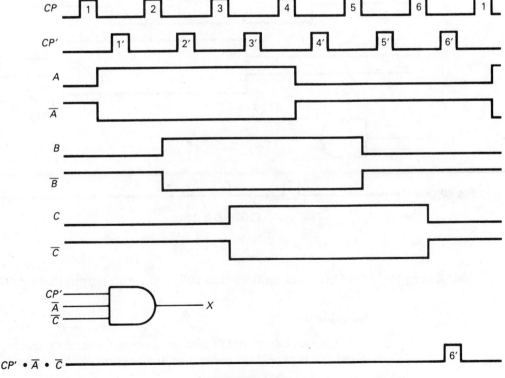

FIGURE 3-14

Example: Find a combination that will produce an output of 2′ and 3′.

Solution:

Using a 3-input AND gate, B goes HIGH at 2 to enable the gate and allow 2′ to appear at the output. A goes LOW at 4 after 3′ has appeared at the output and inhibits the gate so that 4′ cannot pass. The other input must be $CP′$. See Figure 3-15.

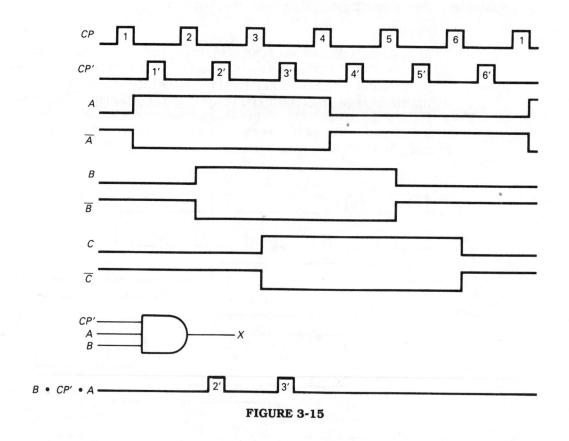

FIGURE 3-15

Example: Find two combinations that will produce an output that is LOW from 3 to 5.

Solution 1:

The unique state of an OR gate is all 0s in, 0 out. Find one signal that goes LOW at time 3 to enable the OR and another signal that is already

LOW and goes HIGH at time 5 to inhibit the gate. $\overline{C}$ goes LOW at 3, and $\overline{B}$ goes back high at 5. $\overline{B} + \overline{C}$ solves the problem. See Figure 3-16.

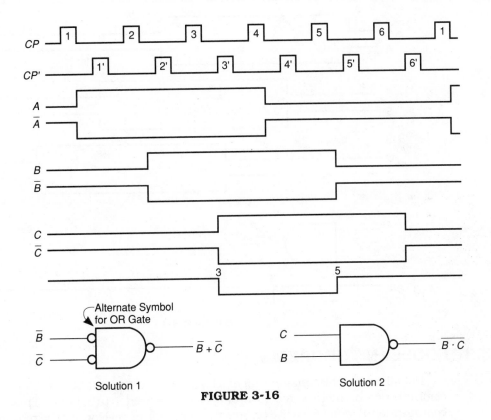

FIGURE 3-16

Solution 2:

The unique state of a NAND gate is all 1s in, 0 out. Find one signal that goes HIGH at time 3 to enable the NAND and another that is already HIGH and goes LOW at time 5 to inhibit the gate. C goes HIGH at 3, and B goes LOW at 5. $\overline{B \cdot C}$ will solve the problem. See Figure 3-16.

Example: Find a combination that will yield LOW going pulses at $1'$ and $2'$.

Solution:

All 1s into a NAND produce a 0 out. Find a signal that goes HIGH before $1'$ and one that is already HIGH and goes LOW after $2'$. A and $\overline{C}$ satisfy these conditions. Put A, $\overline{C}$, and C'_p into a NAND gate. See Figure 3-17.

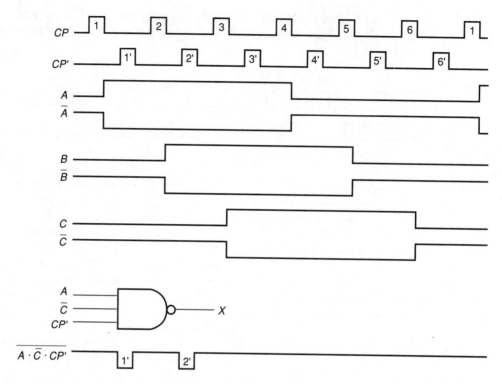

FIGURE 3-17

3.3 BOOLEAN THEOREMS

Not all truth tables match that of a basic gate. At times it becomes necessary to combine two or more gates to produce the desired truth table. This is called *combinational logic.* We will examine two methods of developing the required logic diagram. The first method requires Boolean algebra and De Morgan's theorems to reduce the expressions produced to lowest terms (minimal expressions). The second method is a variation of the first and uses a tool called the *Karnaugh map.* Boolean algebra is based on the following set of eleven fundamental theorems and the two De Morgan theorems. Each is discussed or proved.

1. $\overline{\overline{A}} = A$ (Refer to Figure 3-18)
   A is either 0 or 1.
   Case I: If $A = 0$, then $\overline{A} = 1$ and $\overline{\overline{A}} = 0$.
   Case II: If $A = 1$, then $\overline{A} = 0$ and $\overline{\overline{A}} = 1$.
   In either case $A = \overline{\overline{A}}$.
2. $A \cdot 0 = 0$ (Refer to Figure 3-19)
   The 0 inhibits the AND gate and the output will always be 0.
3. $A + 0 = A$ (Refer to Figure 3-20)
   The 0 input enables the gate.

Case I: If $A = 1$, the output is 1.
Case II: If $A = 0$, the output is 0.
The output is always the same as A.
4. $A \cdot 1 = A$ (Refer to Figure 3-21)
The 1 input enables the gate.
Case I: If $A = 1$, the output is one.
Case II: If $A = 0$, the output is zero.
The output is always the same as A.
5. $A + 1 = 1$ (Refer to Figure 3-22)
The 1 input inhibits the gate and "locks up" the output at 1. The output does not respond to changes in A.
6. $A + A = A$ (Refer to Figure 3-23)
Case I: If $A = 0$, then $0 + 0 = 0$.
Case II: If $A = 1$, then $1 + 1 = 1$.
In either case the output follows A.
7. $A \cdot A = A$ (Refer to Figure 3-24)
Case I: If $A = 0$, then two 0s into an AND yields 0 out.
Case II: If $A = 1$, then two 1s in yields 1 out.
In either case, the output is the same as the inputs.

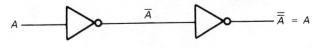

FIGURE 3-18

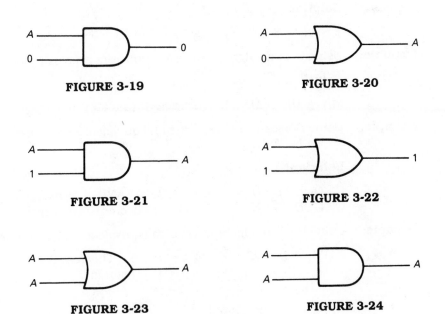

FIGURE 3-19

FIGURE 3-20

FIGURE 3-21

FIGURE 3-22

FIGURE 3-23

FIGURE 3-24

8. $A + \overline{A} = 1$ (Refer to Figure 3-25)
   Case I: If $A = 1$, then the output is 1.
   Case II: If $A = 0$, then $\overline{A} = 1$ and the output is 1.
   The output is always 1.
9. $A \cdot \overline{A} = 0$ (Refer to Figure 3-26)
   Case I: If $A = 1$, then $\overline{A} = 0$ and the output is 0.
   Case II: If $A = 0$, then the output is 0.
   In either case, the output is 0.

Example: Using Boolean algebra, write an equivalent expression for $Y \cdot Z \cdot 0$.

Solution:

$$Y \cdot Z \cdot 0 = 0 \quad \text{by Theorem 2}$$

Example: Using Boolean algebra, write an equivalent expression for $Y + \overline{Z} + 0$.

Solution:

$$Y + \overline{Z} + 0 = Y + \overline{Z} \quad \text{by Theorem 3}$$

Example: Using Boolean algebra, write an equivalent expression for $\overline{D} \cdot E \cdot 1$.

Solution:

$$\overline{D} \cdot E \cdot 1 = \overline{D} \cdot E \quad \text{by Theorem 4}$$

Example: Using Boolean algebra, write an equivalent expression for

$$\overline{E} + H + N + 1.$$

Solution:

$$\overline{E} + H + N + 1 = 1 \quad \text{by Theorem 5}$$

Example: Using Boolean algebra, write an equivalent expression for $M \cdot \overline{N} + M \cdot \overline{N}$.

Solution:

$$M\overline{N} + M\overline{N} = M\overline{N} \quad \text{by Theorem 6}$$

Example: Using Boolean algebra, write an equivalent expression for
$C \cdot C \cdot \overline{D} \cdot E \cdot \overline{D}$.

Solution:

$C \cdot C \cdot \overline{D} \cdot E \cdot \overline{D} = C \cdot C \cdot \overline{D} \cdot \overline{D} \cdot E$ ANDed terms are commutative
$C \cdot C \cdot \overline{D} \cdot \overline{D} \cdot E = C \cdot \overline{D} \cdot E$ by Theorem 7 (twice)

Example: Using Boolean algebra, write an equivalent expression for

$$X + Y + \overline{X} = X + \overline{X} + Y \quad = 1 + Y \quad \text{Theorem 8}$$
$$= 1 \quad \text{Theorem 5}$$

Example: Using Boolean algebra, write an equivalent expression for $\overline{AB} + AB$.

Solution:

$\overline{AB} + AB = 1$ by Theorem 8

Example: Using Boolean algebra, write an equivalent expression for $A\overline{B} + \overline{A}B$.

Solution:

$A\overline{B}$ and $\overline{A}B$ are not the complements of each other, as shown by this truth table. Avoid the trap of reducing this expression to 1. It does not reduce.

A	B	$\overline{A}$	$\overline{B}$	$A\overline{B}$	$\overline{A}B$
0	0	1	1	0	0
0	1	1	0	0	1
1	0	0	1	1	0
1	1	0	0	0	0

Not complements of
each other

Example: Using Boolean algebra, write an equivalent expression for $A \cdot B \cdot \overline{D} \cdot \overline{A}$.

Solution:

$A \cdot B \cdot \overline{D} \cdot \overline{A} = A \cdot \overline{A} \cdot B \cdot \overline{D} = 0 \cdot B \cdot \overline{D}$ Theorem 9
$= 0$ Theorem 2

10. $A \cdot B + A \cdot C = A(B + C)$ (Refer to Figure 3-27)

One way to prove this theorem is to examine its validity for all possible combinations of A, B, and C. An organized approach is to develop a truth table for each side of the equation and see if they are identical. To develop the truth table for a complex expression start with each term. Write a column for $A \cdot B$ then $A \cdot C$ and finally "OR" those two columns together to produce the left side of the equation. To develop the right side, first write a column for $B + C$ then "AND" with A. The truth tables are identical so the equation holds true. Notice the way that we "built up" the final expressions for each side.

FIGURE 3-25   **FIGURE 3-26**

A	B	C	A • B	A • C	A • B + A • C	B + C	A(B + C) AND Function
0	0	0	0	0	0	0	0
0	0	1	0	0	0	1	0
0	1	0	0	0	0	1	0
0	1	1	0	0	0	1	0
1	0	0	0	0	0	0	0
1	0	1	0	1	1	1	1
1	1	0	1	0	1	1	1
1	1	1	1	1	1	1	1

$$A \bullet B + A \bullet C = A(B + C)$$

FIGURE 3-27

A	B	$\overline{A}$	$\overline{A} \bullet B$	$A + \overline{A} \bullet B$	A + B
0	0	1	0	0	0
0	1	1	1	1	1
1	0	0	0	1	1
1	1	0	0	1	1

$$A + \overline{A} \bullet B = A + B$$

FIGURE 3-28

A	B	A • B	$\overline{A \bullet B}$	$\overline{A}$	$\overline{B}$	$\overline{A} + \overline{B}$
0	0	0	1	1	1	1
0	1	0	1	1	0	1
1	0	0	1	0	1	1
1	1	1	0	0	0	0

$$\overline{A \bullet B} = \overline{A} + \overline{B}$$

FIGURE 3-29

Example: Using Boolean algebra, write an equivalent expression for $\overline{A}B + \overline{A}C$.

Solution:

$\overline{A}B + \overline{A}C = \overline{A}(B + C)$ by Theorem 10

Example: Using Boolean algebra, write an equivalent expression for $XYZ + X\overline{Y}Z$.

Solution:

$$\begin{aligned} XYZ + X\overline{Y}Z &= XZ(Y + \overline{Y}) \quad \text{by Theorem 10} \\ &= XZ(1) \quad \text{by Theorem 8} \\ &= XZ \quad \text{by Theorem 4} \end{aligned}$$

Example: Using Boolean algebra, write an equivalent expression for

$AB + AC + AD$.

Solution:

$AB + AC + AD = A(B + C + D)$

11. $A + \overline{A} \cdot B = A + B$ (Refer to Figure 3-28)

Once again using truth tables, the left side of the equation is shown to be identical to the right side.

Example: Using Boolean algebra, write an equivalent expression for $\overline{X} + XZ$.

Solution:

$\overline{X} + XZ = \overline{X} + Z$ by Theorem 11

Example: Using Boolean algebra, write an equivalent expression for $A + \overline{A}B$.

Solution:

$A + \overline{A}B = A + B$ by Theorem 11

3.4 DE MORGAN'S THEOREMS

De Morgan's theorems are as follows:

1. $\overline{A \cdot B} = \overline{A} + \overline{B}$ (Refer to Figure 3-29)

The truth table shows that the left side of the equation is equal to the right side of the equation.

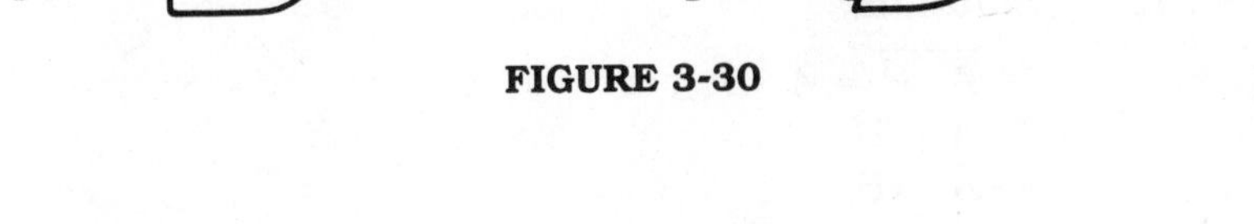

FIGURE 3-30

This theorem reinforces the fact that a NAND is the same as inverting the inputs into an OR. See Figure 3-30.

2. $\overline{A + B} = \overline{A} \cdot \overline{B}$ (Refer to Figure 3-31)

The truth table shows that the left side of the equation is equal to the right side of the equation.

This theorem supports the fact that a NOR is the same as inverting the inputs into an AND. See Figure 3-32.

A	B	A + B	$\overline{A + B}$	$\overline{A}$	$\overline{B}$	$\overline{A} \cdot \overline{B}$
0	0	0	1	1	1	1
0	1	1	0	1	0	0
1	0	1	0	0	1	0
1	1	1	0	0	0	0

$$\overline{A + B} \qquad = \qquad \overline{A} \cdot \overline{B}$$

FIGURE 3-31

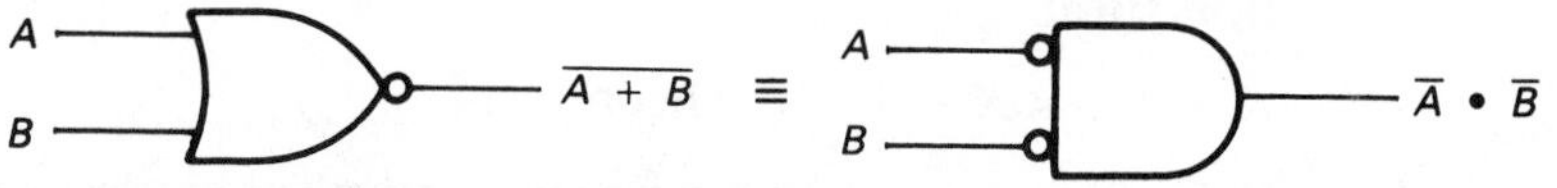

$$\overline{A + B} \quad \equiv \quad \overline{A} \cdot \overline{B}$$

FIGURE 3-32

To apply De Morgan's theorems successfully, you must be able to identify the terms of an expression. In the expression $AB + CDE + FG$, the three terms AB, CDE, and FG are ORed together. In the expression $(A + B)(C + D)$, the two terms $A + B$ and $C + D$ are ANDed together. In the expression $(A + B)(C + D) + EFG$, the two terms $(A + B)(C + D)$ and EFG are ORed together. In the expression ABC the three terms A, B, and C are ANDed together.

To implement De Morgan's theorems follow these three rules:

1. Complement the entire expression.
2. Change the function between each term.
3. Complement each term.

Example: Apply the three rules to $\overline{A \cdot B}$

Solution:

1. $\overline{\overline{A \cdot B}} = A \cdot B$
2. $A + B$
3. $\overline{A} + \overline{B}$

$$\overline{A \cdot B} = \overline{A} + \overline{B}$$

Example: Apply the three rules to $\overline{A} \cdot \overline{B}$

Solution:

1. $\overline{\overline{A} \cdot \overline{B}}$
2. $\overline{\overline{A}} + \overline{\overline{B}}$
3. $\overline{\overline{A}} + \overline{\overline{B}} = \overline{A + B}$

$\overline{A} \cdot \overline{B} = \overline{A + B}$

Example: Change the form of $A \cdot B \cdot C$ by using De Morgan's theorems.

Solution:

1. $\overline{A \cdot B \cdot C}$
2. $\overline{\overline{A + B + C}}$
3. $\overline{\overline{A} + \overline{B} + \overline{C}}$

$A \cdot B \cdot C = \overline{\overline{A} + \overline{B} + \overline{C}}$

Example: Change the form of $ABC + \overline{A}\,\overline{B}\,\overline{C}$ by using De Morgan's theorems.

Solution:

1. $\overline{\overline{ABC + \overline{A}\,\overline{B}\,\overline{C}}}$
2. $\overline{\overline{ABC} \cdot \overline{\overline{A}\,\overline{B}\,\overline{C}}}$
3. $\overline{\overline{ABC} \cdot \overline{\overline{A}\,\overline{B}\,\overline{C}}}$

Now apply De Morgan's theorems to each term.

4. $\overline{\overline{ABC} \cdot \overline{\overline{A}\,\overline{B}\,\overline{C}}} = \overline{\overline{ABC}} \cdot \overline{\overline{\overline{A}\,\overline{B}\,\overline{C}}}$
5. $(\overline{A} + \overline{B} + \overline{C}) \cdot (\overline{\overline{A}} + \overline{\overline{B}} + \overline{\overline{C}})$
6. $(\overline{A} + \overline{B} + \overline{C}) \cdot (A + B + C)$
7. $ABC + \overline{\overline{A}\overline{B}\overline{C}} = (\overline{A} + \overline{B} + \overline{C}) \cdot (A + B + C)$

3.5   DESIGNING LOGIC CIRCUITS

The following examples demonstrate a method used to develop the Boolean expression for a logic diagram that will behave according to a given truth table. An equation is written from the truth table and then reduced using the Boolean algebra theorems.

Example: Design a logic circuit that will behave according to the truth table shown in Figure 3-33.

Solution:

We want the output X to be 1 when $A = 0$ and $B = 0$ (1st line) OR when $A = 1$ and $B = 0$ (3rd line). Working with the first line, if $A = 0$ then $\overline{A} = 1$ (if we don't have A then we have $\overline{A}$) and if $B = 0$ then $\overline{B} = 1$. We want X to be 1 when we have $\overline{A}$ AND $\overline{B}$. The first line is represented by

$\overline{A} \cdot \overline{B}$. Working with the third line, if $A = 0$ then $\overline{A} = 1$. We want X to be 1 when we have $\overline{A}$ AND B. The third line is represented by $\overline{A} \cdot B$. The output X is 1 when the first line is 1 or when the third line is 1. This results in the equation $X = \overline{A} \cdot \overline{B} + \overline{A} \cdot B$. This equation can be reduced using the theorems.

$X = \overline{A} \cdot \overline{B} + \overline{A} \cdot B$
$X = \overline{A}(\overline{B} + B)$ by Theorem 10
$X = \overline{A}(1)$ by Theorem 8
$X = \overline{A}$ by Theorem 4

The desired output is the complement of A and can be produced by inverting the A input. See Figure 3-34.

Example: Given the truth table shown in Figure 3-35, design the logic circuit.

Inputs		Output
B	A	X
0	0	1
0	1	0
1	0	1
1	1	0

FIGURE 3-33

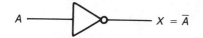

$X = \overline{A}$

FIGURE 3-34

Inputs			Output
C	B	A	X
0	0	0	0
0	0	1	0
0	1	0	0
0	1	1	1
1	0	0	0
1	0	1	0
1	1	0	1
1	1	1	1

FIGURE 3-35

Solution:

We want the output X to be a 1 when $C = 0$, $B = 1$, and $A = 1$ (4th line) OR when $C = 1$, $B = 1$, and $A = 0$ (7th line) OR when $C = 1$, $B = 1$, and $A = 1$ (8th line). Working with line 4, if $C = 0$ then $\overline{C} = 1$; also $B = 1$ and $A = 1$. We want X to be 1 when we have $\overline{C}$ AND B AND A. Line 7 is represented by $C \cdot B \cdot \overline{A}$, and line 8 is represented by $C \cdot B \cdot A$. This results in the equation $X = \overline{C}BA + CB\overline{A} + CBA$. This equation can be reduced using the theorems.

$X = \overline{C}BA + CB\overline{A} + CBA$
$X = \overline{C}BA + CB (\overline{A} + A)$ by Theorem 10
$X = \overline{C}BA + CB(1)$ by Theorem 8

$X = \overline{C}BA + CB$ by Theorem 4
$X = B(\overline{C}A + C)$ by Theorem 10
$X = B(C + \overline{C}A)$ ORed terms are commutative
$X = B(C + A)$ by Theorem 11
$X = B(A + C)$ ORed terms are commutative

Build the logic circuit so that $X = B(A + C)$. See Figure 3-36.

The truth table can be implemented using one two-input OR gate and one two-input AND gate.

This solution can be checked by hooking up the circuit and verifying that the output follows the truth table (a good exercise) or by developing a truth table for the expression $B(A + C)$. See Figure 3-37.

This is identical to the original truth table.

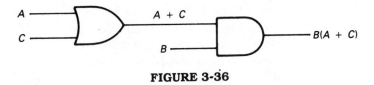

FIGURE 3-36

C	B	A	A + C	B(A + C)
0	0	0	0	0
0	0	1	1	0
0	1	0	0	0
0	1	1	1	1
1	0	0	1	0
1	0	1	1	0
1	1	0	1	1
1	1	1	1	1

FIGURE 3-37

Input			Output
C	B	A	X
0	0	0	1
0	0	1	1
0	1	0	0
0	1	1	1
1	0	0	1
1	0	1	0
1	1	0	1
1	1	1	1

FIGURE 3-38A

Try creating your own three-input truth table and designing the logic circuit to implement it.

Example: Design a circuit that will implement the truth table shown in Figure 3-38A.

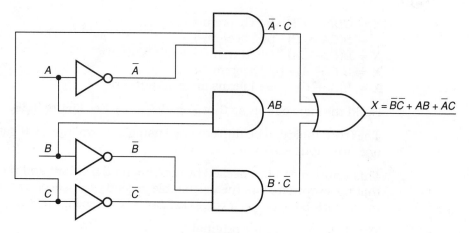

FIGURE 3-38B

Solution:

We want X to be 1 when we have

$\overline{A}\overline{B}\overline{C}$ (first line) or $A\overline{B}\overline{C}$ (second line) or
$AB\overline{C}$ (fourth line) or $\overline{A}\overline{B}C$ (fifth line) or
$\overline{A}BC$ (seventh line) or ABC (last line)

$X = \overline{A}\overline{B}\overline{C} + A\overline{B}\overline{C} + AB\overline{C} + \overline{A}\overline{B}C + \overline{A}BC + ABC$

$X = \overline{B}\overline{C}(\overline{A} + A) + AB\overline{C} + \overline{A}\overline{B}C + \overline{A}BC + ABC$ by Theorem 10

$X = \overline{B}\overline{C} + AB\overline{C} + \overline{A}\overline{B}C + \overline{A}BC + ABC$ by Theorem 8

$X = \overline{B}\overline{C} + AB(\overline{C} + C) + \overline{A}\overline{B}C + \overline{A}BC$ by Theorem 10

$X = \overline{B}\overline{C} + AB + \overline{A}\overline{B}C + \overline{A}BC$ by Theorem 8

$X = \overline{B}\overline{C} + AB + \overline{A}C(\overline{B} + B)$ by Theorem 10

$X = \overline{B}\overline{C} + AB + \overline{A}C$ by Theorem 8

The desired output can be produced as shown in Figure 3-38B.

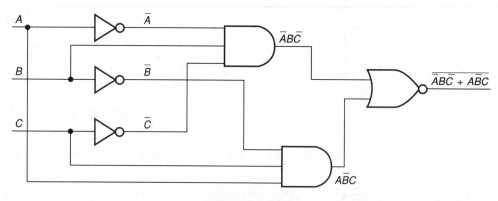

FIGURE 3-38C

Alternative Solution:

The first solution was a lot of work because there were six 1s in the output column of the truth table. An alternative approach is to work with the two 0s in the output column by writing an output expression for $\overline{X}$, simplifying it, and then taking the complement to obtain X.

When $\overline{X}$ is 1, X is 0. We want X to be 0 when we have $\overline{A}B\overline{C}$ (3rd line) OR $A\overline{B}C$ (6th line).

$X = \overline{A}B\overline{C} + A\overline{B}C$. Both sides of this expression must be complemented to obtain an expression for X.

$$\overline{X} = \overline{A}B\overline{C} + A\overline{B}C$$
$$X = \overline{\overline{A}B\overline{C} + A\overline{B}C}$$

See Figure 3-38C.

	Inputs		Outputs
C	B	A	X
0	0	0	0
0	0	1	1
0	1	0	0
0	1	1	0
1	0	0	1
1	0	1	1
1	1	0	1
1	1	1	0

FIGURE 3-39A

Example: Design a circuit that will implement the truth table shown in Figure 3-39A.

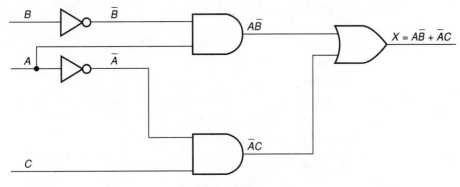

FIGURE 3-39B

Solution:

$$X = A\overline{B}\overline{C} + \overline{A}\overline{B}C + A\overline{B}C + \overline{A}BC$$
$$X = A\overline{B}(\overline{C} + C) + \overline{A}\overline{B}C + \overline{A}BC$$
$$X = A\overline{B} + \overline{A}\overline{B}C + \overline{A}BC$$
$$X = A\overline{B} + \overline{A}C(\overline{B} + B)$$
$$X = A\overline{B} + \overline{A}C$$

The required circuitry is shown in Figure 3-39B.

Example: Design a circuit that will behave according to the truth table shown in Figure 3-40A.

Inputs			Outputs
C	B	A	X
0	0	0	0
0	0	1	0
0	1	0	1
0	1	1	1
1	0	0	1
1	0	1	1
1	1	0	0
1	1	1	0

FIGURE 3-40A

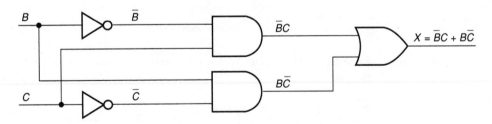

FIGURE 3-40B

Solution:

$$X = \overline{A}B\overline{C} + AB\overline{C} + \overline{A}\overline{B}C + A\overline{B}C$$
$$X = \overline{A}B\overline{C} + AB\overline{C} + \overline{B}C(\overline{A} + A)$$
$$X = \overline{A}B\overline{C} + AB\overline{C} + \overline{B}C$$
$$X = B\overline{C}(\overline{A} + A) + \overline{B}C$$
$$X = B\overline{C} + \overline{B}C$$

The circuit can be constructed as shown in Figure 3-40B.

Notice that the A input did not appear in the final circuit. It was not needed to produce the required output.

Example: Design a circuit that will implement the truth table shown in Figure 3-41A.

Inputs			Outputs
C	B	A	X
0	0	0	1
0	0	1	1
0	1	0	0
0	1	1	1
1	0	0	1
1	0	1	0
1	1	0	1
1	1	1	1

FIGURE 3-41A

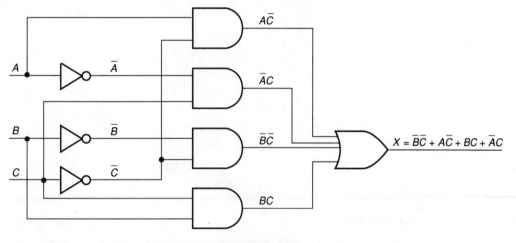

FIGURE 3-41B

Solution:

$$X = \overline{A}\overline{B}\overline{C} + A\overline{B}\overline{C} + AB\overline{C} + \overline{A}BC + \overline{A}BC + ABC$$
$$X = \overline{B}\overline{C}(\overline{A} + A) + AB\overline{C} + \overline{A}BC + \overline{A}BC + ABC$$
$$X = \overline{B}\overline{C} + AB\overline{C} + \overline{A}BC + BC(\overline{A} + A)$$
$$X = \overline{B}\overline{C} + AB\overline{C} + \overline{A}BC + BC$$
$$X = \overline{C}(\overline{B} + BA) + \overline{A}BC + BC$$
$$X = \overline{C}(\overline{B} + A) + \overline{A}BC + BC$$
$$X = \overline{B}\overline{C} + A\overline{C} + \overline{A}BC + BC$$
$$X = \overline{B}\overline{C} + A\overline{C} + C(\overline{A}B + B)$$

$$X = \overline{BC} + A\overline{C} + C(B + \overline{BA})$$
$$X = \overline{BC} + A\overline{C} + C(B + \overline{A})$$
$$X = \overline{BC} + A\overline{C} + BC + \overline{A}C$$

The required circuitry is shown in Figure 3-41B.

In the preceding examples, Boolean algebra was used to reduce an expression to its lowest terms so that circuitry could be created to implement a truth table. The following examples present an alternative method, called Karnaugh (Car'no) maps, for reducing Boolean expressions to lowest terms.

The Karnaugh method uses a table or "map" to reduce its expressions. Each position in the tables is called a cell. Cells are filled with ones and zeros according to the expression to be reduced. Adjacent ones are grouped together in clusters, called subcubes, following definite rules. A subcube must be of size 1, 2, 4, 8, 16, etc. All ones must be included in a subcube of maximum size. These rules will be expounded upon and explained by working several examples.

Example: Design a circuit that will behave according to the truth table shown in Figure 3-42.

Inputs			Output
C	B	A	X
0	0	0	0
0	0	1	0
0	1	0	1
0	1	1	1
1	0	0	0
1	0	1	1
1	1	0	1
1	1	1	1

FIGURE 3-42

Solution:

Step 1. Draw the table. Choose two of the variables to use as column headings across the top. We will use C and B. Form all combinations of C and $\overline{C}$ with B and $\overline{B}$. Each column heading should differ from the adjacent column by one variable only.

Part 1

$\overline{C}\overline{B}$	$\overline{C}B$	CB	$C\overline{B}$

Start with $\overline{C}\overline{B}$ and change $\overline{B}$ to B to form the heading for column 2, $\overline{C}B$. Then change $\overline{C}$ to C for the third column CB, and finally $C\overline{B}$. The fourth column wraps around to the first column and should differ by one variable only, which it does.

Part 2	$\overline{C}\overline{B}$	$\overline{C}B$	CB	$C\overline{B}$
$\overline{A}$				
A				

Use the third variable, A, for row headings $\overline{A}$ and A.

Step 2. Fill the table with ones and zeros from the truth table. The output X is 1 in line 3 when we have $\overline{C}$ and B and $\overline{A}$. Place a 1 in the table in cell $\overline{C}B\overline{A}$. The output X is also 1 on line 4, which is represented by $\overline{C}BA$, on line 6, which is $CB\overline{A}$, on line 7, which is $CB\overline{A}$, and on line 8, which is CBA. Fill those cells with ones and the remaining cells with zeros.

	$\overline{C}\overline{B}$	$\overline{C}B$	CB	$C\overline{B}$
$\overline{A}$	0	1	1	0
A	0	1	1	1

Step 3. Combine adjacent cells that contain ones in subcubes of maximum size. The four ones in the center of the table compose a subcube of size 4.

	$\overline{C}\overline{B}$	$\overline{C}B$	CB	$C\overline{B}$
$\overline{A}$	0	1	1	0
A	0	1	1	1

The 1 in cell $CB\overline{A}$ has not been included in a subcube so it is used with its adjacent 1 in a subcube of size 2.

Step 4. Write the expression that each subcube represents. In the subcube of size 4 find the variables that occur in all four cells. In this case B is the only variable that appears in all four cells. The subcube of size 4 represents B. In the subcube of size two, A and C appear in each cell, so the subcube represents AC.

Step 5. Form the output expression. The output X is the expression from each subcube ORed together. In this case $X = B + AC$. The truth table can be implemented by the logic diagram:

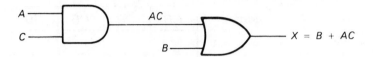

Example: Use a Karnaugh map to design a logic diagram to implement the following truth table.

Inputs			Output
C	B	A	X
0	0	0	0
0	0	1	1
0	1	0	0
0	1	1	0
1	0	0	0
1	0	1	1
1	1	0	1
1	1	1	1

Solution:

Step 1. Draw the table.

	$\bar{C}\bar{B}$	$\bar{C}B$	CB	$C\bar{B}$
$\bar{A}$	0	0	1	0
A	1	0	1	1

Step 2. Fill the table with 1s and 0s from the truth table.

Step 3. Combine adjacent cells that contain 1s into subcubes (size 1, 2, 4, or 8).

$\bar{C}\bar{B}$	$\bar{C}B$	CB	$C\bar{B}$
0	0	1	0
1	0	1	1

The side of the table "wraps around" to the other side so that the table is continuous. The 1s in the lower corners form a subcube of size 2. The two subcubes "cover" the map in that all 1s are contained in subcubes. Any additional subcubes drawn would add unneeded terms to the final expression.

Step 4. Write the expression that each subcube represents. In the vertical subcube C and B remain constant. In the horizontal subcube $\bar{B}$ and A are constant.

Step 5. Form the output expression.

$$X = BC + A\bar{B}$$

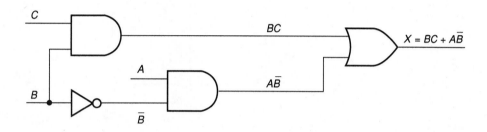

Example: Use a Karnaugh map to design a logic diagram to implement the following truth table.

Inputs				Output
D	C	B	A	X
0	0	0	0	1
0	0	0	1	0
0	0	1	0	1
0	0	1	1	0
0	1	0	0	0
0	1	0	1	1
0	1	1	0	0
0	1	1	1	0
1	0	0	0	1
1	0	0	1	0
1	0	1	0	1
1	0	1	1	0
1	1	0	0	1
1	1	0	1	1
1	1	1	0	1
1	1	1	1	1

Solution:

Step 1. Draw the table. Since four variables are needed, use two across the top and two down the side.

	$\overline{D}\overline{C}$	$\overline{D}C$	DC	$D\overline{C}$
$\overline{B}\overline{A}$	1	0	1	1
$\overline{B}A$	0	1	1	0
BA	0	0	1	0
$B\overline{A}$	1	0	1	1

Step 2. Fill the table with 1s and 0s from the truth table.

Step 3. Combine adjacent cells that contain 1s into subcubes of size 1, 2, 4, 8, or 16.

	$\overline{D}\overline{C}$	$\overline{D}C$	DC	$D\overline{C}$
$\overline{B}\overline{A}$	1	0	1	1
$\overline{B}A$	0	1	1	0
BA	0	0	1	0
$B\overline{A}$	1	0	1	1

Since the map is continuous top to bottom and side to side, the four corners are adjacent and form a subcube of size 4. The DC column forms another subcube of size 4. One cell remains uncovered. $\overline{D}C\overline{B}A$ forms a subcube of size 2 with the cell on its right.

Step 4. Write the expression that each subcube represents. The subcube formed by the four corners represents the term $\overline{A}\overline{C}$. The vertical subcube represents the expression CD, and the subcube of size 2 represents the expression $A\overline{B}C$.

Step 5. Form the output expression.

$$X = CD + \overline{A}\overline{C} + A\overline{B}C$$

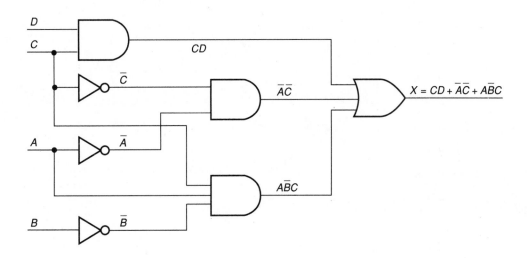

Exercise 3

1. In Figure 3-43, use the waveforms A, B, and C to determine the waveforms for each expression shown.
2. In Figure 3-44, use the waveforms A, B, and C to determine the waveforms for each expression shown.

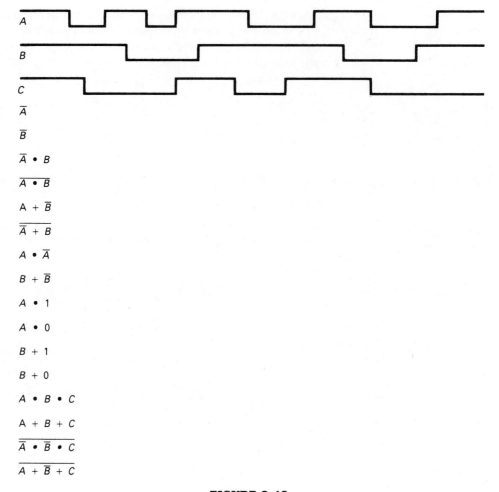

A

B

C

$\overline{A}$

$\overline{B}$

$\overline{A} \cdot B$

$\overline{A \cdot B}$

$A + \overline{B}$

$\overline{A + B}$

$A \cdot \overline{A}$

$B + \overline{B}$

$A \cdot 1$

$A \cdot 0$

$B + 1$

$B + 0$

$A \cdot B \cdot C$

$A + B + C$

$\overline{\overline{A} \cdot \overline{B} \cdot C}$

$\overline{A + \overline{B} + C}$

FIGURE 3-43

3. Use the delayed clock, shift counter waveforms in Figure 3-5 (page 80) to determine the missing inputs and outputs for the figures listed.

    a. Figure 3-45   e. Figure 3-49

    b. Figure 3-46   f. Figure 3-50

    c. Figure 3-47   g. Figure 3-51

    d. Figure 3-48

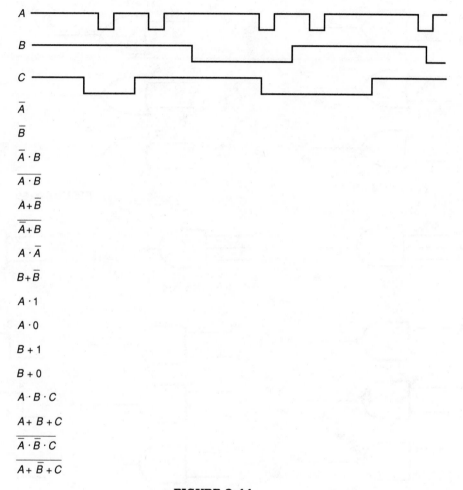

FIGURE 3-44

4. Use the waveforms in Figure 3-5 (page 80) to determine the missing inputs and outputs for the figures listed.

    a. Figure 3-52   e. Figure 3-56
    b. Figure 3-53   f. Figure 3-57
    c. Figure 3-54   g. Figure 3-58
    d. Figure 3-55

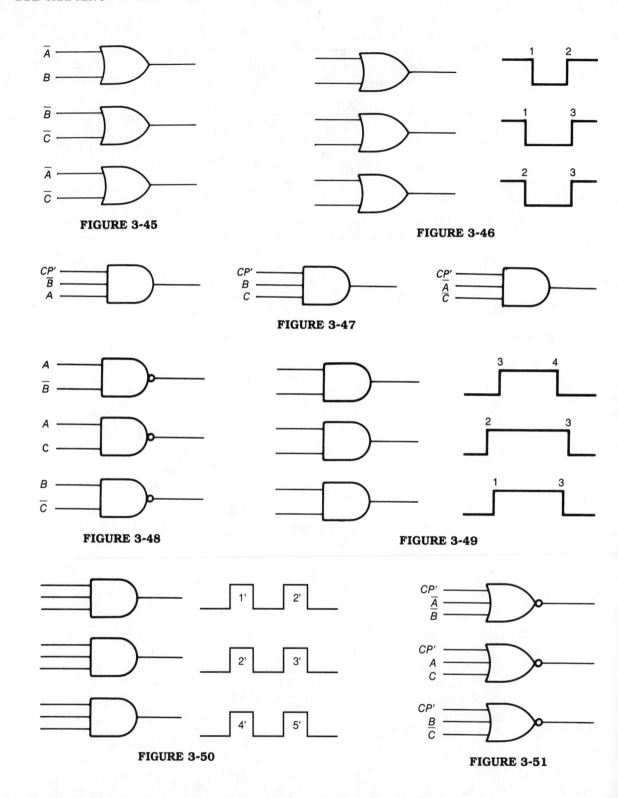

FIGURE 3-45

FIGURE 3-46

FIGURE 3-47

FIGURE 3-48

FIGURE 3-49

FIGURE 3-50

FIGURE 3-51

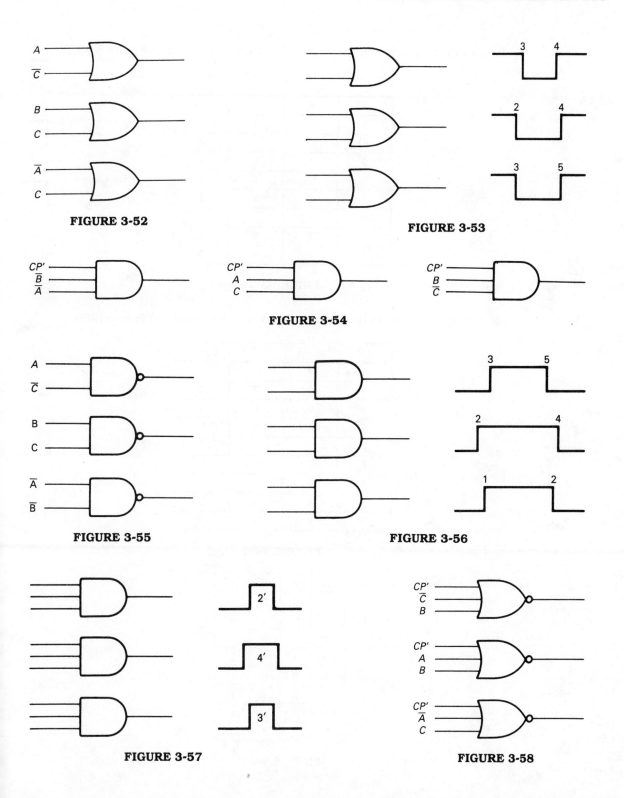

FIGURE 3-52

FIGURE 3-53

FIGURE 3-54

FIGURE 3-55

FIGURE 3-56

FIGURE 3-57

FIGURE 3-58

5. Design a circuit to implement the truth table shown in Figure 3-59.

Inputs			Output
A	B	C	X
0	0	0	0
0	0	1	0
0	1	0	1
0	1	1	1
1	0	0	0
1	0	1	0
1	1	0	1
1	1	1	0

FIGURE 3-59

6. Design a circuit to implement the truth table shown in Figure 3-60.

Inputs			Output
A	B	C	X
0	0	0	1
0	0	1	1
0	1	0	0
0	1	1	1
1	0	0	0
1	0	1	0
1	1	0	0
1	1	1	1

FIGURE 3-60

C	B	A	X
0	0	0	1
0	0	1	0
0	1	0	1
0	1	1	0
1	0	0	1
1	0	1	1
1	1	0	1
1	1	1	0

FIGURE 3-61

7. Design a circuit to implement the truth table shown in Figure 3-61.
   a. Use Boolean algebra.
   b. Use the Karnaugh map method.
8. Design a circuit to implement the truth table shown in Figure 3-62.
   a. Use Boolean algebra.
   b. Use the Karnaugh map method.

C	B	A	X
0	0	0	1
0	0	1	0
0	1	0	0
0	1	1	0
1	0	0	1
1	0	1	1
1	1	0	1
1	1	1	0

FIGURE 3-62

9. Reduce these expressions.
   a. $\overline{A}B + \overline{A}C$
   b. $AB + A\overline{B}$
   c. $A + \overline{A}D$
   d. $A \cdot \overline{A} \cdot B \cdot C$
   e. $\overline{A} + AB$
10. Reduce these expressions.
   a. $XYZ + X\overline{Y}$
   b. $BCD + B\overline{C}$
   c. $B + \overline{B}E$
   d. $A \cdot C \cdot \overline{A}$
   e. $AB + C + 1$
11. Reduce these expressions.
   a. $ABC + AB\overline{C} + \overline{B}$
   b. $A\overline{B}C + A\overline{C} + C$
   c. $AB\overline{C} + A\overline{B}\,\overline{C} + ABC$
   d. $A\overline{B}C + AB\overline{C} + A\overline{B}\,\overline{C}$
12. Reduce these expressions.
   a. $\overline{A}\overline{B}C + A\overline{B}C + \overline{C}$
   b. $ABC + \overline{A}\,\overline{B}C + \overline{A}BC$
   c. $A + A\overline{B}C + AB$
   d. $A\overline{C} + A\overline{B} + AB$

13. Group the 1s into subcubes and write the output expression.

	$\bar{C}\bar{B}$	$\bar{C}B$	CB	$C\bar{B}$
$\bar{A}$	1	0	1	1
A	1	0	0	1

	$\bar{C}\bar{B}$	$\bar{C}B$	CB	$C\bar{B}$
$\bar{A}$	1	1	0	0
A	0	1	1	0

14. Group the 1s into subcubes and write the output expressions.

	$\bar{C}\bar{B}$	$\bar{C}B$	CB	$C\bar{B}$
$\bar{A}$	0	1	1	1
A	0	0	1	1

	$\bar{C}\bar{B}$	$\bar{C}B$	CB	$C\bar{B}$
$\bar{A}$	0	1	1	1
A	1	0	0	0

15. Group the 1s into subcubes and write the output expression.

	$\bar{D}\bar{C}$	$\bar{D}C$	DC	$D\bar{C}$
$\bar{B}\bar{A}$	1	0	0	1
$\bar{B}A$	0	1	1	0
BA	0	1	1	0
$B\bar{A}$	1	0	0	1

16. Group the 1s into subcubes and write the output expression.

	$\bar{D}\bar{C}$	$\bar{D}C$	DC	$D\bar{C}$
$\bar{B}\bar{A}$	0	1	1	1
$\bar{B}A$	0	0	0	0
BA	0	1	0	0
$B\bar{A}$	0	1	1	1

3 Boolean Algebra

COMPONENTS NEEDED

1	7400 IC
1	7402 IC
1	7404 IC
1	7408 IC
1	7411 IC
3	7432 IC
4	LEDs
4	330-Ω resistors

PREPARATION

In Part I and Part II of Lab 3 you will be asked to write the Boolean expression and truth table for two circuits. To develop the Boolean expression, write the output expression for the first gates encountered. Use those expressions as inputs to the following gates. For example,

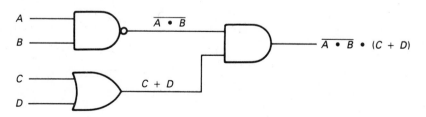

Assign pin numbers to your logic diagram before you begin wiring. To develop the truth table, list all the inputs and all the possible combinations, (count in binary). Write a column for each term in the expression. Combine the terms into a final expression in the circuit shown previously.

A	B	C	D	A • B	$\overline{A • B}$	C + D	$\overline{A • B}$ • (C + D)
0	0	0	0	0	1	0	0
0	0	0	1	0	1	1	1
	etc.						

To verify the truth table, connect the inputs to switches to supply the required 1s and 0s.

Part IV asks you to design and build an encoder. An encoder changes a decimal number into another number system or code. A decoder converts a number or code back into decimal. On your encoder, your 10 inputs represent the ten decimal digits. Only one should go HIGH at a time. The outputs are active HIGH also and should produce the corresponding BCD number. For example, if 5 goes HIGH on the input, then outputs 1 and 4 should go HIGH in response.

Draw a logic diagram with pin numbers and IC numbers before beginning construction. Insert your hookup wires away from the IC pins so that you have room to check voltages on the pins of the IC.

PROCEDURE

Part I

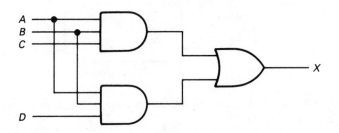

A. Write the Boolean expression and the truth table for the output.
B. Connect the circuit and verify the output.

Part II

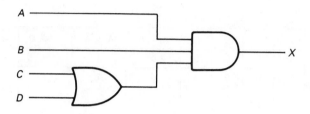

A. Write the Boolean expression and truth table for the output.
B. Connect the circuit and verify the output.

Part III

A. Write the truth table for the Boolean expression, $X = \overline{(A + B)C}$.
B. Connect the logic circuit and verify the output.

Part IV

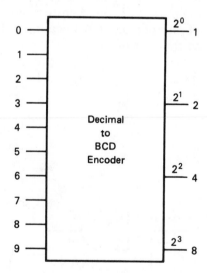

A. Construct a decimal-to-BCD encoder using three two-input OR gate integrated circuits.

Hint: The 1 output should go HIGH for each odd decimal number input, 1 OR 3 OR
        7 OR 9, so OR those inputs together to produce the 1 output (expand an OR
        with another OR). The 2 output should go HIGH for inputs 2, 3, 6, 7, so OR
        these inputs together to produce the 2 output, etc.

Part V

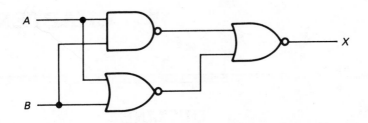

A. Write a Boolean expression and write a truth table for this circuit.
B. Reduce the circuit. Construct and verify.

Part VI

Inputs			Output
A	B	C	X
0	0	0	0
0	0	1	1
0	1	0	0
0	1	1	0
1	0	0	1
1	0	1	0
1	1	0	0
1	1	1	0

A. Design a circuit to implement this truth table.
B. Construct and verify.

CHAPTER 4

OUTLINE

Exclusive-OR Gates

OBJECTIVES

After completing this chapter, you should be able to:

■ draw the symbol and write the truth table for an exclusive-OR gate and an exclusive-NOR gate.

■ construct an exclusive-OR gate from the basic gates.

■ predict the output waveforms from given inputs.

■ use an exclusive-OR to invert data.

■ use an exclusive-OR to pass data unaltered.

■ construct a comparator using exclusive-OR gates.

■ define even and odd parities.

■ construct a parity generator using exclusive-OR gates.

■ use a 74180 as a parity generator and a parity checker.

4.1   EXCLUSIVE-OR

The *exclusive-OR* is a two-input gate that produces a 1 on its output when its inputs are different and a 0 if they are the same. The symbol and the truth table for an exclusive-OR are shown in Figure 4-1. Notice that the output X is 1 if A is 1 or if B is 1, but not if both A and B are 1. If A and B are both 1s, the 1s are excluded from the output; hence the name of the gate is the exclusive-OR. The output X is sometimes written $A \oplus B$ which is read "A exclusive-OR B."

An exclusive-OR gate is not one of the basic gates, but is constructed from a combination of the basic gates. To design an exclusive-OR, first write a Boolean expression for the truth table in Figure 4-1. The output is 1 for the conditions on line 2 or line 3. On line 2 we have $\overline{A} \cdot B$ and on line 3 we have $A \cdot \overline{B}$. The output X is 1 for $\overline{A} \cdot B + A \cdot \overline{B}$. The logic diagram for this expression is shown in Figure 4-2. This solution requires two AND gates, one OR gate, and, if the complemented inputs are not available, two inverters.

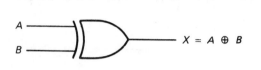

Inputs		Output
A	B	X
0	0	0
0	1	1
1	0	1
1	1	0

FIGURE 4-1   Exclusive-OR symbol and truth table

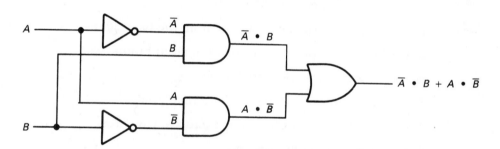

FIGURE 4-2   Exclusive-OR logic diagram

An equivalent logic diagram can be developed by using Boolean algebra and De Morgan's Theorems.

$$X = \overline{\overline{\overline{A} \cdot B} \cdot \overline{A \cdot \overline{B}}}$$
$$X = \overline{(A + \overline{B}) \cdot (\overline{A} + B)}$$
$$X = \overline{A\overline{A} + AB + \overline{A}\,\overline{B} \cdot \overline{B}B}$$
$$X = AB + \overline{A}\,\overline{B}$$

The logic diagram for this expression is shown in Figure 4-3. This solution requires two NOR gates and an AND gate.

The pinouts for a 7486 and a 4070 quad 2-input exclusive-OR are shown in Figure 4-4.

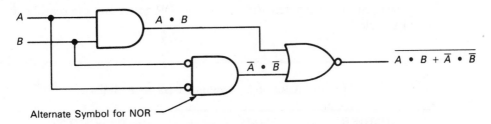

FIGURE 4-3   Equivalent exclusive-OR logic diagram

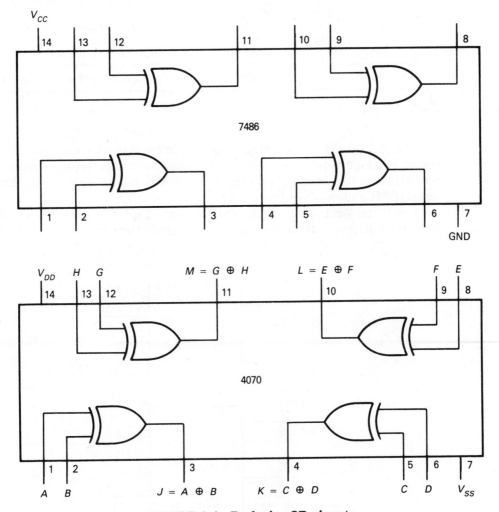

FIGURE 4-4   Exclusive-OR pinouts

Some exclusive-OR gates and exclusive-NOR gates that are readily available are listed in Table 4-1.

TABLE 4-1   Exclusive-OR/NOR Gates

NUMBER	FAMILY	DESCRIPTION
7486	TTL	Quad 2-input exclusive-OR
74135	TTL	Quad exclusive-OR/NOR
74C86	CMOS	Quad 2-input exclusive-OR
4030	CMOS	Quad 2-input exclusive-OR
4070	CMOS	Quad 2-input exclusive-OR

The IEC symbol for a 7486 quad two-input exclusive-OR gate integrated circuit is shown in Figure 4-5. The = 1 sign indicates that exactly one input must be active. Since there are no triangles on inputs or outputs, all are active HIGH. Exactly one input HIGH makes the output HIGH.

The 1 sign was used to signify the inverter (Chapter 2), the ≥ 1 to signify the OR gate and NOR gate (Chapter 2), and = 1 to signify the exclusive-OR gate.

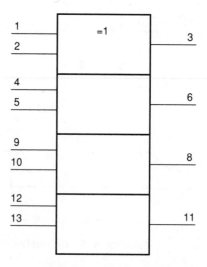

FIGURE 4-5   IEC symbol—7486 quad excluse-OR gate

4.2 ENABLE/INHIBIT

The truth table for the exclusive-OR is rewritten in Figure 4-6. When the control input is 0, data passes through the gate unaltered. When the control input is 1, the data passes through inverted. Although there is no real inhibit mode, as there was with the basic gates, it is quite useful to be able to invert or not invert a signal by changing the control input. We will see applications of the exclusive-OR used in this mode in this and later chapters.

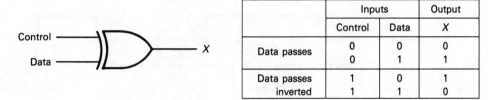

		Inputs		Output
		Control	Data	X
Data passes		0	0	0
		0	1	1
Data passes inverted		1	0	1
		1	1	0

FIGURE 4-6   Enable/inhibit

Example: Predict the output of each gate.

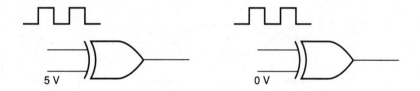

Solution:

Use the waveform as the data input and the static signal as the control. For the first gate the control input is 1, and the data passes through inverted. For the second gate the control signal is 0, and the data passes through unaltered.

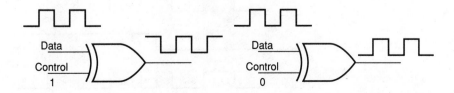

4.3   WAVEFORM ANALYSIS

The output of an exclusive-OR is 1 when the inputs are unlike and 0 when the inputs are the same. This makes it easy to predict the output waveform from given inputs. In Figure 4-7, the intervals in which A and B are unlike are shaded. The output, X, is HIGH during these times and LOW elsewhere.

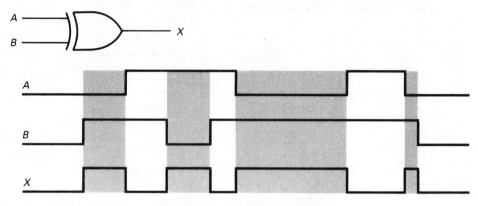

FIGURE 4-7   Waveform analysis

Example: Predict the output of the exclusive-OR.

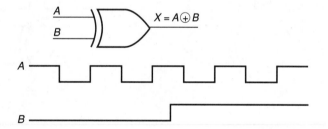

Solution:

Find the times when the inputs are different. The output is HIGH during those times. It is LOW at all other times.

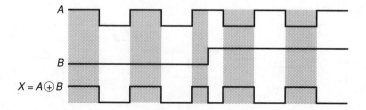

Another way to analyze this problem is to consider the B input as the control. In the first half of the waveform B is 0 and the data passes to the output unaltered. When B goes HIGH, the data on A passes through inverted.

4.4 EXCLUSIVE-NOR

An exclusive-NOR, sometimes called nonexclusive-OR, has a truth table and symbol as shown in Figure 4-8. The output, X, is HIGH when the outputs are alike and LOW when they are unlike.

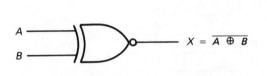

$$X = \overline{A \oplus B}$$

Inputs		Output
A	B	X
0	0	1
0	1	0
1	0	0
1	1	1

FIGURE 4-8   Exclusive-NOR

Example: Predict the output of each gate.

Solution:

The first two gates are exclusive-ORs. Since the inputs to the first gate are different, its output is 1. The last two gates are exclusive-NORs. Since the inputs to the third gate are different, its output is 0.

4.5 EXCLUSIVE-OR/NOR

The pinout and truth table of a 74135 quad exclusive-OR/NOR is shown in Figure 4-9.

The signal on the control input C determines whether the combination will function as an exclusive-OR or as an exclusive-NOR. If C is 0 (first four lines of the truth table) then the second gate, see Figure 4-10, allows the data to pass unaltered and the combination functions as an exclusive-OR. If C is 1 (last four lines of the truth table) then the second gate inverts the data and the combination will function as an exclusive-NOR.

Inputs			Output
A	B	C	Y
L	L	L	L
L	H	L	H
H	L	L	H
H	H	L	L
L	L	H	H
L	H	H	L
H	L	H	L
H	H	H	H

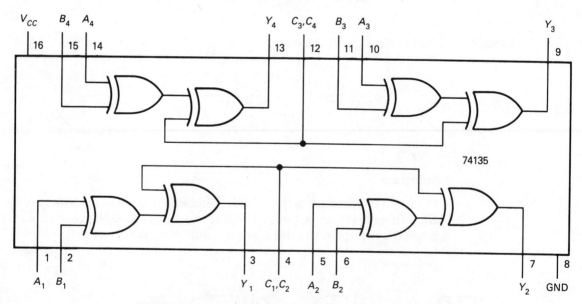

FIGURE 4-9   Exclusive-OR/NOR pinout and truth table

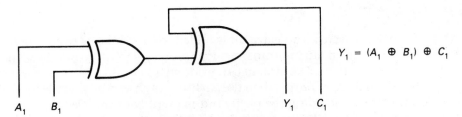

$$Y_1 = (A_1 \oplus B_1) \oplus C_1$$

FIGURE 4-10   Exclusive-OR/NOR

Example: Predict the output of the exclusive-OR/NOR.

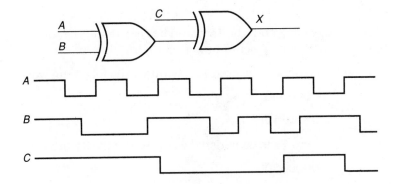

Solution:

When C is HIGH, the circuit functions as an exclusive-NOR. Find the times when A and B are different and draw the output LOW. For the rest of the time that C is HIGH, draw the output HIGH. When C is LOW, the circuit functions as an exclusive-OR. Find the times when A and B are different and draw the output HIGH. For the rest of the time that C is LOW, draw the output LOW.

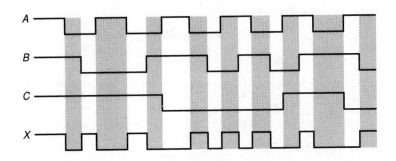

4.6 PARITY

In some systems an extra bit is added to the data bits to ensure that a bit does not get changed during transmission from one circuit to another. The extra bit is called a *parity bit*. The system can work on an even-parity or odd-parity system. If the system is *even parity*, then the parity bit is chosen so that the total number of 1s in the word, including the parity bit, is even. For example, suppose that we have seven data bits with the eighth bit (most significant bit) as an even parity bit.

1011101

With the data shown, the proper parity bit can be generated. There are five 1s in the data bits so a 1 must be generated by the even-parity generator to make the total number of 1s even. The word would be transmitted as:

11011101

At the receiving end, the 1s would be counted. If the total number of 1s was not even, an alarm or flag would be set, notifying that an error has occurred.

Example: Generate the even-parity bit in the following word:

0000000

Solution:

Zero 1s is considered even, so the parity bit would be 0.

00000000

Example: Generate the even-parity bit in the following word:

1111111

Solution:

Seven 1s is odd, so the even-parity bit must be a 1.

11111111

Example: Generate the even-parity bit in the following word:

1010101

Solution:

Four 1s is even, so the even-parity bit must be a 0.

01010101

In an *odd-parity system*, the parity generator supplies a parity bit that makes the total number of 1s odd.

Example: Set the odd-parity bit in the following word:

1100110

Solution:

Four 1s is even, so the odd-parity bit must be a 1.

11100110

Example: Set the odd-parity bit in the following word:

1100111

Solution:

Five 1s is odd, so the odd-parity bit must be a 0.

01100111

Example: Set the odd-parity bit in the following word:

0000000

Solution:

Zero 1s is considered even, so the odd-parity bit must be a 1.

10000000

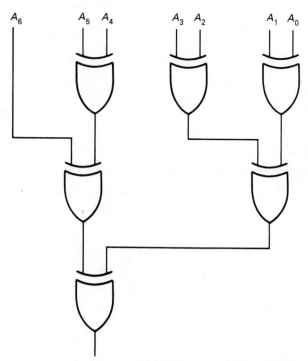

Parity Bit   **FIGURE 4-11   Even parity generator**

4.7 EVEN-PARITY GENERATOR

A circuit that can determine whether the parity bit is a 1 or 0 is called a *parity generator*. Exclusive-OR gates can be used to construct a parity generator. Figure 4-11 shows exclusive-OR gates being used as an even-parity generator. Each exclusive-OR gate checks for unlike inputs. Unlike inputs into the last exclusive-OR means that an odd number of 1s has been encountered. The 1 output would be used as a parity bit, making the total number of 1s even.

Example: Use the even-parity generator, in Figure 4-11, to generate the even-parity bit for 1011101.

Solution:

Five 1s is odd. The generator supplies a 1 as the even-parity bit.

11011101

See Figure 4-12.

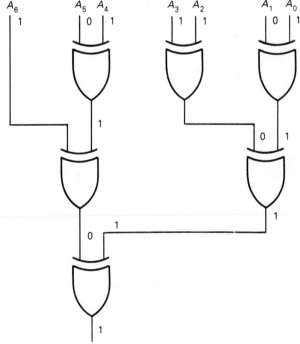

FIGURE 4-12   Parity Bit

Example: Use the even-parity generator, in Figure 4-11, to supply the even-parity bit for 1000001.

Solution:

Two 1s is even. The generator must supply an even-parity bit of 0.

01000001

See Figure 4-13.

Example: Use exclusive-OR gates to construct a 6-bit even-parity generator. The parity bit will become the seventh bit.

Solution:

See Figure 4-14.

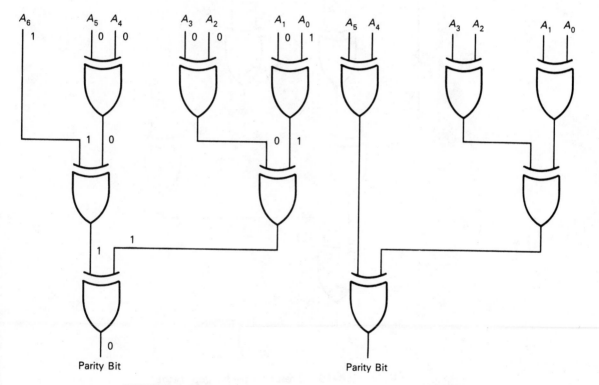

FIGURE 4-13 FIGURE 4-14

4.8 EVEN/ODD-PARITY GENERATOR

By adding an extra exclusive-OR gate, the circuit can be made more versatile. By using a 1 on the control input, you can invert the output and change the circuit into an odd-parity generator. See Figure 4-15.

Example: Use the even/odd parity generator, in Figure 4-15, to supply the odd-parity bit for 0110101.

Solution:

The control input must be 1 for an odd-parity generator. Four 1s is even, so the generator supplies an odd-parity bit of 1.

10110101

See Figure 4-16.

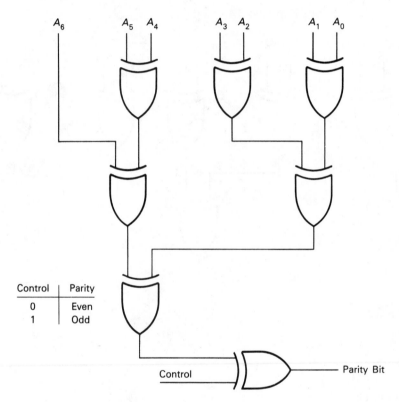

FIGURE 4-15   Even/odd-parity generator

Example: Use the parity generator, in Figure 4-15, to supply the even-parity bit for 1111110.

Solution:

The control input must be 0 for an even-parity generator. Six 1s is even, so the even-parity bit must be 0.

01111110

See Figure 4-17.

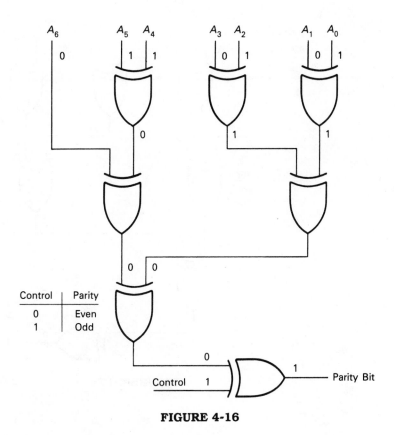

FIGURE 4-16

4.9 PARITY CHECKER

A circuit that can determine whether the total number of 1s is even or odd is called a *parity checker.* Figure 4-18 shows eight bits (seven data and one parity) fed into an exclusive-OR gate circuit.

Example: 01001101 has been received as seven data bits and an even-parity bit. Use the circuit, in Figure 4-18, to check for parity errors.

Solution:

The 0 out indicates that an even number of 1s has been received. No parity error has been detected.
See Figure 4-19.

Example: 11000110 has been received as seven data bits and an odd-parity bit. Use the circuit, in Figure 4-18, to check for parity error.

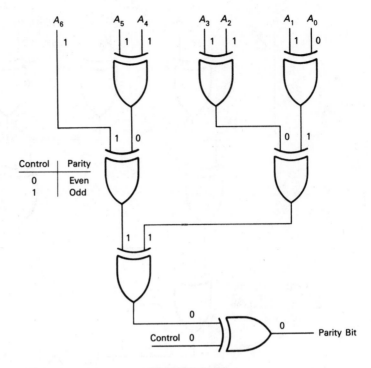

Control	Parity
0	Even
1	Odd

FIGURE 4-17

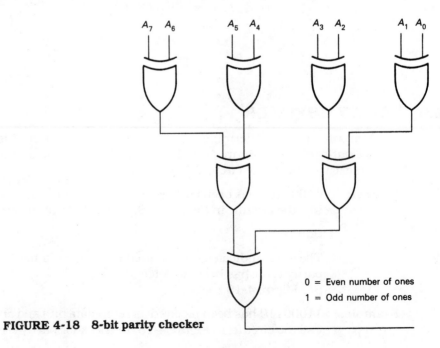

0 = Even number of ones
1 = Odd number of ones

FIGURE 4-18 8-bit parity checker

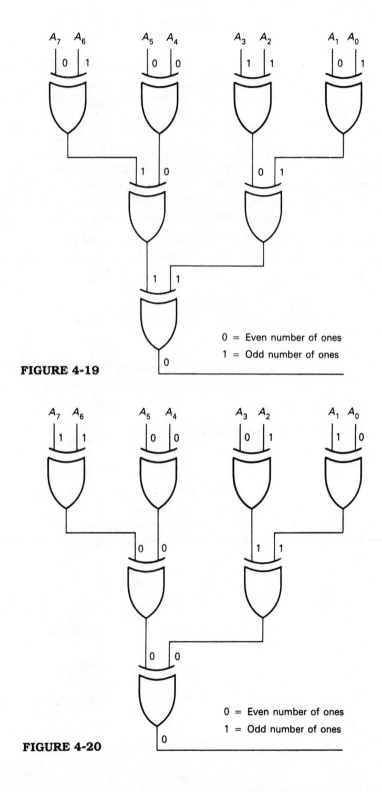

FIGURE 4-19

0 = Even number of ones
1 = Odd number of ones

FIGURE 4-20

0 = Even number of ones
1 = Odd number of ones

Solution:

The 0 out indicates that an even number of 1s has been received. An error has occurred.
One of the bits has probably been inverted during transmission or reception.
See Figure 4-20.

The 74S280 is a medium-scale integrated circuit which functions as a 9-bit parity generator/checker. The pinout and truth table are shown in Figure 4-21. If the number of inputs (A through I) that are HIGH is even, then the Σ even output goes HIGH and the Σ odd output goes LOW (line 1 of the truth table). To use the IC as an even-parity generator, the Σ odd output could generate the parity bit.

Example: Use the 74S280 as an 8-bit even-parity generator (seven data bits and one parity bit). Generate the parity bit for 0101010.

Solution:

A through G supply the seven data inputs. H and I are not used and are grounded. Pin 3 is not connected internally (NC).

0101010

Three inputs HIGH is the second line of the truth table.
Σ odd goes HIGH and sets the even-parity bit to 1.

10101010

The total number of 1s is even.

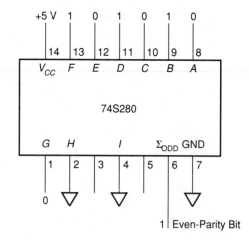

Number of Inputs	Outputs	
(A–I) that are HIGH	Σ Even	Σ Odd
0, 2, 4, 6, 8	H	L
1, 3, 5, 7, 9	L	H

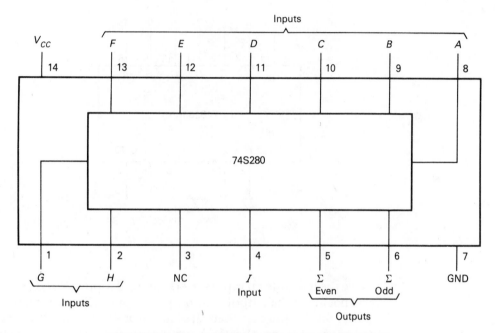

FIGURE 4-21   74S280 9-bit generator/checker

Example: Use the 74S280 as an 8-bit even-parity generator (seven data bits and one parity bit). Generate the parity bit for 1111110.

Solution:

Six inputs HIGH is the first line of the truth table. Σ odd goes LOW and resets the even-parity bit to 0.

01111110

The total number of 1s is even.

To use the 74S280 as an odd-parity generator, use the Σ even output to generate the parity bit.

Example: Use the 74S280 as an 8-bit odd-parity generator (seven data bits and one parity bit). Generate the parity bit for 1010101.

Solution:

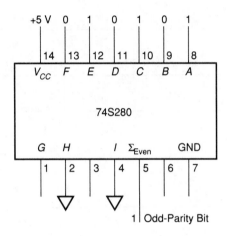

Place the data bits on inputs $A - G$. Ground inputs H and I so that they will not affect the output. The output on pin 5 (Σ even) will be the odd-parity bit. In this case there are four 1s on the inputs. The first line of the truth table says that the Σ even output will be HIGH, as it should be for our odd-parity bit 11010101.

To use the 74S280 as an odd-parity checker, feed up to nine data bits into inputs A through I. If the total number of inputs that are HIGH is odd, as it should be (bottom line of the truth table), then Σ odd goes HIGH and Σ even goes LOW. LOW-level outputs on this IC can sink (provide a path to ground) 20 mA so Σ odd could drive an LED as shown, Figure 4-22. If the total number of 1s received were even, then Σ odd would go LOW and the LED would light, signifying that a parity error had occurred.

Example: Use the 74S280 as an 8-bit odd-parity checker (Figure 4-22) to check for parity error on the data 10111010.

Solution:

The 8 bits are fed into the 74S280 on inputs A through H. I is grounded so that it will not influence the output. The second line of the truth table shows that five 1s in produces a HIGH on the Σ odd output, and the LED does not light. The LED off indicates lack of a parity error.

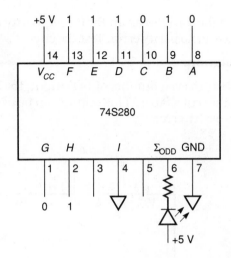

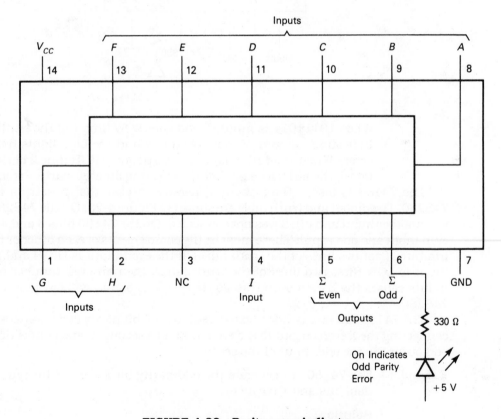

FIGURE 4-22 Parity error indicator

Example: Use the 74S280 as an 8-bit even-parity checker. Have an LED light if there is no parity error. Test the circuit with 10111011 and 10111010.

Solution:

When an even number of 1s is input, the Σ odd output goes LOW (line 1 of the truth table). This output can be used to light an LED, indicating no parity error.

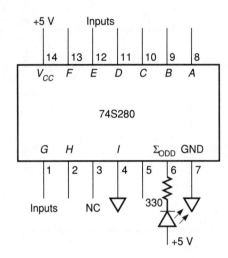

When 10111011 is input, Σ odd goes LOW (line 1 of the truth table), indicating an even number of inputs and the LED lights (no parity error). When 10111010 is input, Σ odd goes HIGH (line 2 of the truth table), the LED does not light, indicating an even-parity error.

The 74180 is also a 9-bit parity generator/checker that functions like the 74S280. The pinout and truth table are shown in Figure 4-23. On the 74S280 pin 4 was another input and pin 3 was not connected. On the 74180 pins 3 and 4 are the even input and odd input. These must be the complements of each other or the IC is inhibited (last two lines of the truth table). If the even input is HIGH and the odd input is LOW (first two lines of the truth table), then an even number of HIGH inputs causes the Σ even output to go HIGH and Σ odd to go LOW just as in the 74S280.

The 74180 can be cascaded (expanded) to a 17-bit parity generator/checker by connecting the Σ even output to the even input of a second IC and connecting the Σ odd output to the odd input of the second IC.

Example: Use 74180s to generate the odd-parity bit for an 18-bit number (17 data bits and 1 parity bit).

Solution:

Use the Σ odd output to generate the odd-parity bit.
See Figure 4-24.

Inputs			Outputs	
Σ of H's at A – H	Even	Odd	Σ Even	Σ Odd
Even	H	L	H	L
Odd	H	L	L	H
Even	L	H	L	H
Odd	L	H	H	L
X	H	H	L	L
X	L	L	H	H

H = HIGH Level   L = LOW Level   X = Don't Care

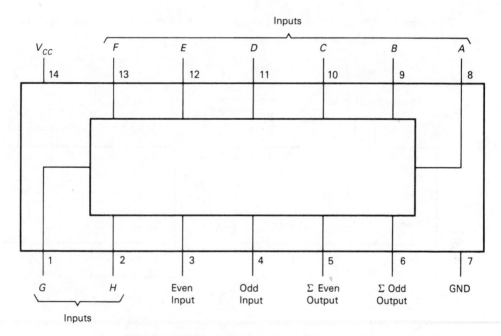

FIGURE 4-23   74180 9-bit parity generator

Once again an extra exclusive-OR gate on the output would allow the user to invert the signal if the user desired to create an even- or odd-parity checker.

Example: Use the parity generator of Figure 4-24 to generate the odd-parity bit for the data $1A239_{16}$.

Solution:

$1A239_{16} = 1\ 1010\ 0010\ 0011\ 1001_2$.

Use these bits as the 17 inputs. It does not matter which goes where, but assume the most significant bit is placed on input 17 and that the lower 8 bits, 0011 1001, are placed on inputs 1 to 8. Since the even input pin is HIGH, we are working on the top two lines of the truth table (even = *H*, odd = *L*). There are an even number of 1s into the first IC (on the right). The truth table (first line) says that Σ even goes HIGH and Σ odd goes LOW. These are wired to the even input and odd inputs of the next IC (the 74180 on the left). In this case the second IC is behaving according to the first two lines of the truth table. The second IC has three 1s input (line 2 of the truth table). Σ odd goes HIGH to set the odd-parity bit to 1. The eight 1s in the data plus the 1 on the parity bit make the total number of 1s odd.

The 74S280 can be expanded by tying the Σ even output to a data input of a following stage.

Example: Use two 74S280s to produce a 17-bit odd-parity checker. Have an LED light on parity error.

Solution:

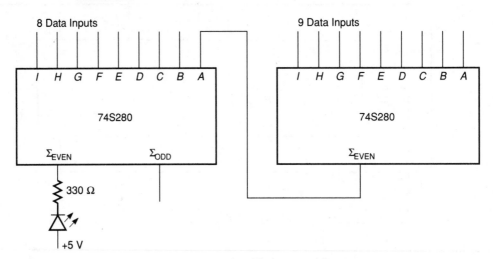

Check: Place all 1s into the first 74S280 (right) and all 0s into the second (left). The output on Σ even would be LOW. All 0s into the second 74S280 would cause its Σ even to go HIGH, and the LED would not light. This is correct since we had an odd number of 1s in.

In summary, in a parity generator, data bits are fed into the generator, and the output of the generator provides the parity bit. In a parity checker, the parity bit and data bits are fed into the checker, and the output indicates whether an error has been detected.

The IEC symbol for a 74S280 is shown in Figure 4-25. The 2k indicates that an even number of 1s must be present on the input pins to cause the output on pin 5 (Σ even) to go HIGH (no triangle) and pin 6 (Σ odd) to go LOW (triangle).

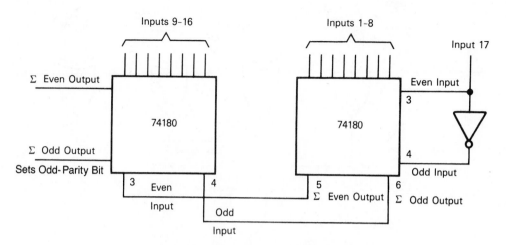

FIGURE 4-24   Odd-parity generator

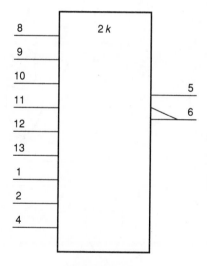

FIGURE 4-25

4.10   COMPARATOR

The exclusive-OR can also be used to compare two numbers and decide if they are equal. Figure 4-26 shows a circuit that compares each bit of two numbers. If any corresponding bits are unequal, a 1 is fed into the NOR gate for a 0 out. So, a 0 out indicates that the numbers are not equal and a 1 out means that the numbers are equal.

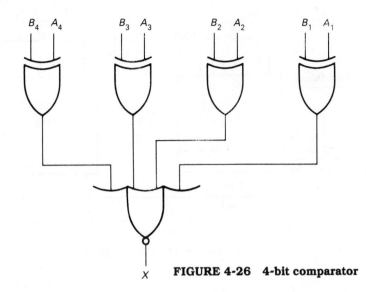

FIGURE 4-26   4-bit comparator

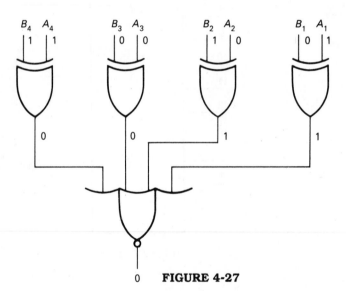

0   **FIGURE 4-27**

Example: Compare 1010 and 1001.

Solution:

See Figure 4-27. 0 out indicates that the numbers are not equal.

The 7485 and 74C85 are 4-bit magnitude comparators whose outputs indicate whether $A = B$, $A < B$, or $A > B$. The comparator in Figure 4-26 indicates only whether $A = B$ or $A \neq B$.

The pinout and functional diagram for the 74LS85 4-bit magnitude comparator are shown in Figure 4-28. In the functional diagram, the pins have been grouped together according to function. Pins 2, 3, and 4 are expansion inputs that are used to expand the IC for use in circuits with more than 4 bits. When used as a 4-bit comparator, pins 2 and 4 ($I_{A<B}$ and $I_{A>B}$) must be grounded and pin 3 ($I_{A=B}$) must be HIGH. Figure 4-29 shows the truth table of the '85. The IC compares the two 4-bit numbers on its inputs and sends output 5, 6, or 7 HIGH, depending on the relative magnitude of the 4-bit numbers input.

Example: Use the 74LS85 to compare the numbers $A = 1011$ and $B = 1100$.

Solution:

Since $A < B$, pin 7 will go HIGH, pins 6 and 5 will go LOW.

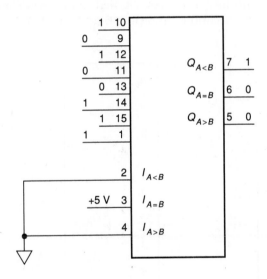

Example: Use two 74LS85s to compare the 8-bit numbers $A = 9D_{16}$ and $B = B6_{16}$.

Solution:

$9D_{16} = 10011101_2$ and $B6_{16} = 10110110_2$

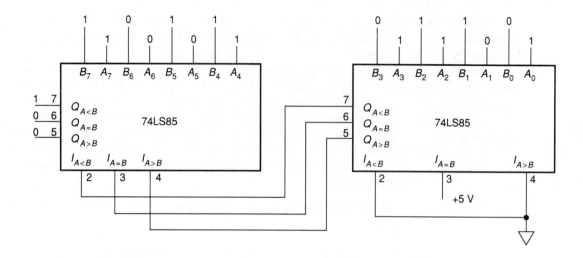

The outputs of the first IC (right) are wired into the expansion inputs of the second IC (left). Since A is less than B, pin 7 goes HIGH on the second IC and pins 6 and 5 go LOW.

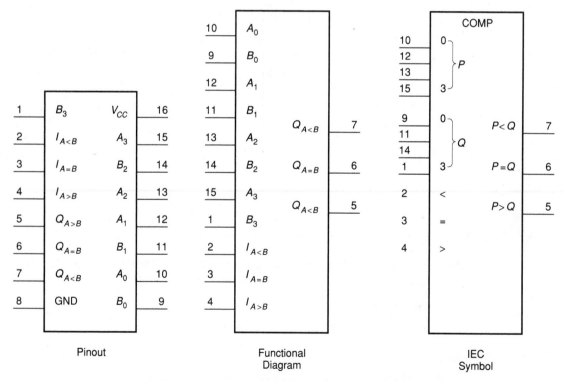

| Pinout | Functional Diagram | IEC Symbol |

FIGURE 4-28   74LS85 4-Bit magnitude comparator

	Comparing Inputs			Cascading Inputs			Outputs		
A_3, B_3	A_2, B_2	A_1, B_1	A_0, B_0	$A > B$	$A < B$	$A = B$	$A > B$	$A < B$	$A = B$
$A_3 > B_3$	X	X	X	X	X	X	H	L	L
$A_3 < B_3$	X	X	X	X	X	X	L	H	L
$A_3 = B_3$	$A_2 > B_2$	X	X	X	X	X	H	L	L
$A_3 = B_3$	$A_2 < B_2$	X	X	X	X	X	L	H	L
$A_3 = B_3$	$A_2 = B_2$	$A_1 > B_1$	X	X	X	X	H	L	L
$A_3 = B_3$	$A_2 = B_2$	$A_1 < B_1$	X	X	X	X	L	H	L
$A_3 = B_3$	$A_2 = B_2$	$A_1 = B_1$	$A_0 > B_0$	X	X	X	H	L	L
$A_3 = B_3$	$A_2 = B_2$	$A_1 = B_1$	$A_0 < B_0$	X	X	X	L	H	L
$A_3 = B_3$	$A_2 = B_2$	$A_1 = B_1$	$A_0 = B_0$	H	L	L	H	L	L
$A_3 = B_3$	$A_2 = B_2$	$A_1 = B_1$	$A_0 = B_0$	L	H	L	L	H	L
$A_3 = B_3$	$A_2 = B_2$	$A_1 = B_1$	$A_0 = B_0$	L	L	H	L	L	H

FIGURE 4-29   Truth table of the 74LS85

Exercise 4

1. Draw the symbol and write the truth table for an exclusive-OR.
2. Draw the symbol and write the truth table for an exclusive-NOR.
3. With a 1 on the control input of an exclusive-OR, will the data pass through unaltered or inverted?
4. Using Figure 4-30, sketch the output waveforms for X and Y.

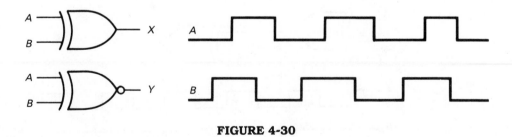

FIGURE 4-30

5. Predict the output of each gate.

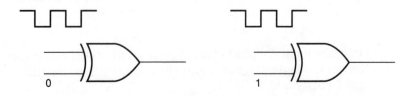

6. Predict the output of each gate.

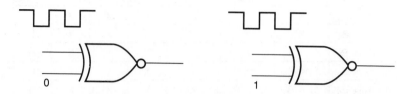

7. Sketch the output of this exclusive-OR/NOR.

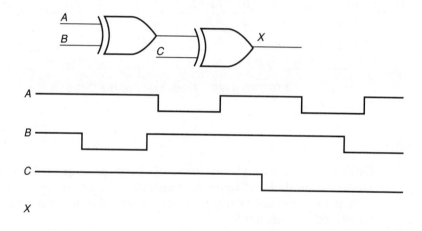

8. Sketch the output of this exclusive-OR/NOR.

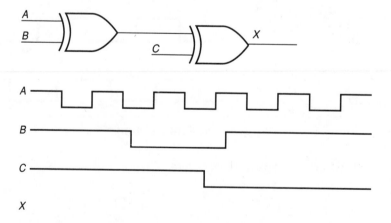

9. Draw the logic diagram of an exclusive-OR, using basic gates.
10. Sketch the logic diagram of an even-parity generator that uses 5 data bits and 1 parity bit.

11. Supply the parity bit
    a. Even: 101101
    b. Odd: 110000
    c. Even: 000011
    d. Odd: 110010
12. Sketch the logic diagram of a 5-bit comparator and indicate the meaning of the output.
13. Sketch the pinout of a 7485 4-bit magnitude comparator.
14. Describe, in your own words, the operation of a 4-bit magnitude comparator.
15. Draw the logic diagram of an 8-bit magnitude comparator, using two 7485s.
16. Draw the logic diagram of a 74S280 used as an 8-bit (7 data bits plus 1 parity bit) even-parity generator.
17. Draw the logic diagram of a 74180 used as an 8-bit (7 data bits plus 1 parity bit) odd-parity generator.
18. Draw the logic diagram of a 74S280 used as an 8-bit (7 data bits plus 1 parity bit) even-parity checker. Use an LED to indicate that a parity error has occurred. (On = Error)
19. Draw the logic diagram of a 74180 used as an 8-bit (7 data bits plus 1 parity bit) odd-parity checker. Use an LED to indicate that an error has occurred. (On = Error)
20. Draw the logic diagram of two 74S280s being used as a 16-bit (15 data bits plus 1 parity bit) odd-parity generator.
21. Draw the logic diagram of two 74180s being used as a 16-bit (15 data bits plus 1 parity bit) even-parity generator.
22. Draw the logic diagram of two 74S280s being used as a 16-bit (15 data bits and 1 parity bit) odd-parity checker. Use an LED to indicate that a parity error has occurred. (On = Error)
23. Draw the logic diagram of two 74180s being used as a 16-bit (15 data bits plus 1 parity bit) even-parity checker. Use an LED to indicate that a parity error has occurred. (On = Error)
24. Draw the pinout of a CMOS comparator.
25. How do the IEC symbols differ for an inverter, OR gate, and an exclusive-OR gate?
26. Draw the IEC logic symbol for a 74S280 parity generator/checker.
27. Draw the IEC logic symbol for a 7485 4-bit magnitude comparator.

LAB **4** **Exclusive-OR**

OBJECTIVES

After completing this lab, you should be able to:

- use 7486s to construct a parity generator.
- use a 74180 to generate parity bits.
- use a 74180 to check for parity errors.
- cascade two 74180s to make a 16-bit parity generator/checker.
- use exclusive-OR gates to construct a 4-bit comparator.

COMPONENTS NEEDED

2	7485 IC
2	74180 IC
1	4009
1	4012
1	4070
1	LED
1	330-Ω resistor

PREPARATION

In a parity generator, the magnitude bits are input and the circuit generates the parity bit. The same circuit can be used as a parity checker by inputting the parity bit along with the magnitude bits. The output becomes a signal or flag declaring whether there is a parity error.

PROCEDURE

1. Check each gate on a 7486 IC by determining its truth table.
2. Use 7486 ICs to construct an 8-bit parity generator (8th bit is parity bit).

Input 136_8

What is the output? Is this an even- or an odd-parity generator? How can it be converted to the other?

Input 063_8

What is the output? Predict the output of your parity generator for the following numbers. Input these numbers and verify your conclusions.

135_8

056_8

060_8

177_8

3. Complete the truth table of a 74180 9-bit parity generator/checker.

Inputs			Outputs	
Σ of H's at 0-7	Even	Odd	Σ Even	Σ Odd
Even	H	L		
Odd	H	L		
Even	L	H		
Odd	L	H		
X	H	H		
X	L	L		

H = HIGH Level   L = LOW Level   X = Don't Care

4. Use the 74180 as an odd-parity generator to determine the parity bits of the numbers in step 2 of this lab.
5. Use the 74180 as an odd-parity checker. Have an LED light if there is no parity error. Input your results from step 4.
6. Use two 74180s to make a 16-bit odd-parity generator. The 16th bit will be the parity bit. Use your circuit to determine the parity bit of the following words:

$2D6B_{16}$, $6F50_{16}$, $3BD4_{16}$

7. Use your circuit as a 16-bit even-parity checker by feeding all 16 bits into the inputs. Have an LED light if there is a parity error. Check these numbers for parity error:

$F809_{16}$, $400A_{16}$, $CD13_{16}$

8. Use a 4070, 4012, and 4009 to construct a 4-bit comparator. Input 0110 and 1010. What is the output? What does it indicate? Input 0110 and 0110. What is the output? What does it indicate? Try several other combinations of inputs.

CHAPTER **5**

Adders

OBJECTIVES

After completing this chapter, you should be able to:

- define half adder and draw the block diagram and truth table.
- develop the logic circuitry and construct a half adder.
- define full adder and draw the block diagram and truth table.
- develop the logic circuitry and construct a full adder.
- design the circuitry required to use a full adder as a 1's complement adder/subtractor.
- design the circuitry required to use a full adder as a 2's complement adder/subtractor.
- add in binary-coded decimal.
- design the circuitry required to use a full adder as a BCD adder.

5.1 HALF ADDER

A *half adder* is a circuit that has two inputs, A and B, and two outputs, sum and carry. It adds A and B according to the rules for binary addition and outputs the sum and carry. The block diagram and truth table for a half adder are shown in Figure 5-1. The truth table follows the rules for binary addition. The last line shows 1 plus 1 is 10, as it must be.

The sum output has a truth table identical to the exclusive-OR and the carry output has a truth table identical to an AND gate. One way to construct a half adder is shown in Figure 5-2. Another way to construct a half adder is shown in Figure

157

5-3. In Figure 5-3, the exclusive-OR is constructed from one AND gate and two NOR gates. *A* is ANDed with *B* as part of the exclusive-OR and can be used as the carry signal.

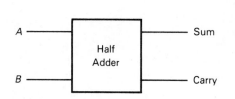

Inputs		Outputs	
A	*B*	Sum	Carry
0	0	0	0
0	1	1	0
1	0	1	0
1	1	0	1

FIGURE 5-1   Half adder

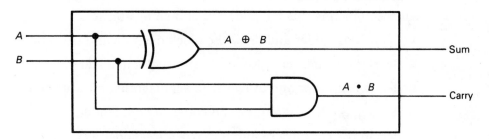

FIGURE 5-2   Logic diagram of a half adder

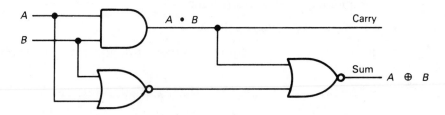

FIGURE 5-3   Using an AND gate and two NOR gates to construct a half adder

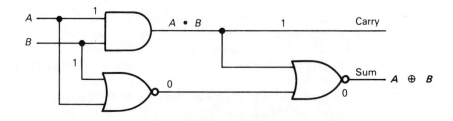

FIGURE 5-4

Example: Add $A = 1$, $B = 1$.

Solution:

See Figure 5-4.

1 plus 1 has a sum of 0 and a carry of 1.

5.2   FULL ADDER

Whereas the half adder added two inputs, A and B, the *full adder* adds three inputs together, A, B, and a carry from a previous addition, and outputs a sum and carry. The truth table follows the rules for binary addition. The block diagram and truth table for a full adder are shown in Figure 5-5.

The sum is 1 each time the total number of 1s on inputs A, B, and carry in is odd. This is analogous to an even-parity generator as shown in Figure 5-6. The output is 1 for an odd number of 1s in.

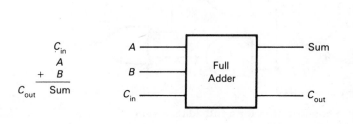

Inputs			Outputs	
A	B	Carry In	Sum	Carry Out
0	0	0	0	0
0	0	1	1	0
0	1	0	1	0
0	1	1	0	1
1	0	0	1	0
1	0	1	0	1
1	1	0	0	1
1	1	1	1	1

FIGURE 5-5   Full adder

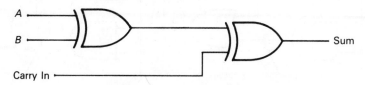

FIGURE 5-6   Full adder—sum

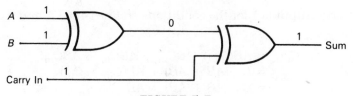

FIGURE 5-7

Example: Add $A = 1$, $B = 1$, Carry in $= 1$.

Solution:

See Figure 5-7.

1 plus 1 plus 1 has a sum of 1.

Each of the exclusive-OR gates in Figure 5-6 can be replaced with two NOR gates and an AND gate, Figure 5-8.

Example: Add $A = 1$, $B = 0$, Carry in $= 1$.

Solution:

See Figure 5-9.

1 plus 0 plus 1 has a sum of 0.

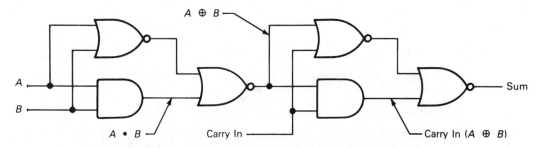

FIGURE 5-8   A second construction for a full adder—sum

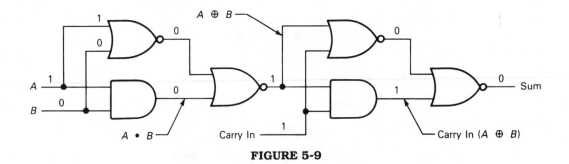

FIGURE 5-9

The carry output is 1 for the conditions on lines 4, 6, 7, and 8 of the truth table in Figure 5-5.

$$
\begin{aligned}
C_{out} &= \overline{A}BC_{in} + A\overline{B}C_{in} + AB\overline{C}_{in} + ABC_{in} \\
&= C_{in}(\overline{A}B + A\overline{B}) + AB(\overline{C}_{in} + C_{in}) \\
&= C_{in}(\overline{A}B + A\overline{B}) + AB \\
&= C_{in}(A \oplus B) + AB
\end{aligned}
$$

In Figure 5-8, *A* has already been exclusive-ORed with *B* and the result has been ANDed with C_{in}. Also, *A* has been ANDed with *B*. A two-input OR gate will combine these two signals to produce the carry-out signal as shown in Figure 5-10. The full adder in Figure 5-10 is constructed from two half adders and an OR gate. Each half adder is enclosed in broken lines.

Example: Add *A* = 1, *B* = 1, Carry in = 0.

Solution:

See Figure 5-11.

Sum = 0; Carry = 1
1 plus 1 plus 0 = 10

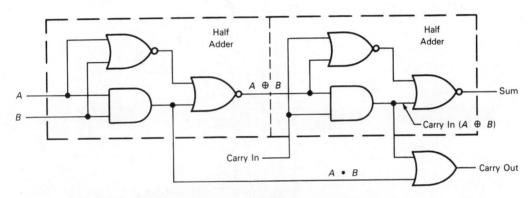

FIGURE 5-10   Full adder—sum and carry

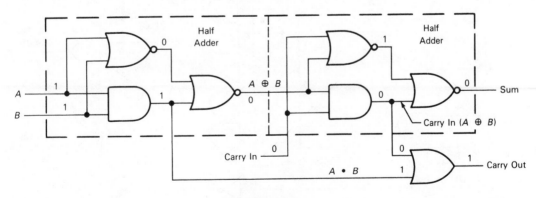

FIGURE 5-11

$$
\begin{array}{r}
C_3 \; C_2 \; C_1 \; C_0 \\
A_4 \; A_3 \; A_2 \; A_1 \\
+ \; B_4 \; B_3 \; B_2 \; B_1 \\
\hline
C_4 \; \Sigma_4 \; \Sigma_3 \; \Sigma_2 \; \Sigma_1
\end{array}
$$

FIGURE 5-12   4-bit full adder inputs and outputs

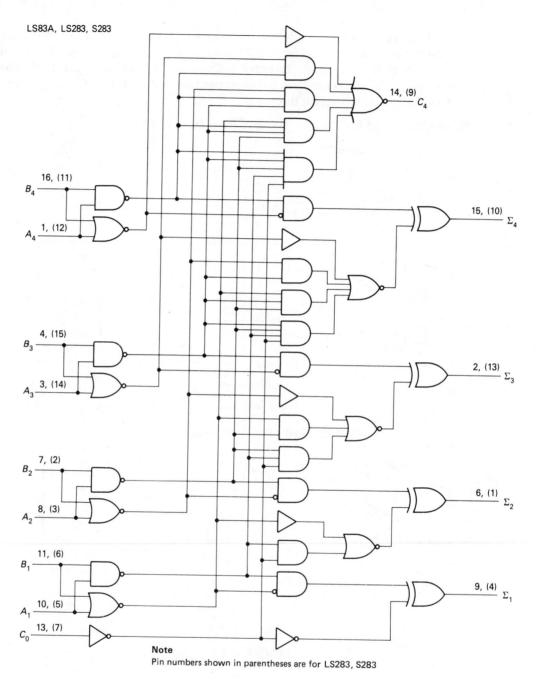

LS83A, LS283, S283

16, (11)
B_4

1, (12)
A_4

4, (15)
B_3

3, (14)
A_3

7, (2)
B_2

8, (3)
A_2

11, (6)
B_1

10, (5)
A_1

13, (7)
C_0

14, (9) C_4

15, (10) Σ_4

2, (13) Σ_3

6, (1) Σ_2

9, (4) Σ_1

Note
Pin numbers shown in parentheses are for LS283, S283

FIGURE 5-13   7483 logic diagram

TABLE 5-1   Medium Scale Integration Adder Circuits

DEVICE NO.	FAMILY	DESCRIPTION
7483	TTL	4-bit binary adder with fast carry
74C83	CMOS	4-bit binary adder with fast carry
4008	CMOS	4-bit full adder with fast carry

Table 5-1 lists some of the available 4-bit full adder ICs. The internal circuitry contains enough components for the IC to be classified as medium-scale integration. The 7483 was used in Lab 1 to add two 4-bit numbers, A and B, and a carry in, called C_0. The outputs are a 4-bit sum and carry out, C_4, as shown in Figure 5-12. Carries C_1, C_2, and C_3 are handled internally and do not appear on the pins of the IC.

When working the addition problem longhand, C_4 is not determined until each of the columns has been added. The carry has to "ripple-through" four stages of addition. The logic diagram in Figure 5-13 shows how a 7483 produces C_4 from the inputs without waiting for the "ripple-effect" to take place. This results in a "fast carry" or "look ahead carry." The result is a faster operation; in fact, C_4 appears before the Σ outputs are established.

Two of the gates in the logic diagram shown in Figure 5-13 are not basic gates, but are combinations of basic gates. The two gates function as shown in Figure 5-14 and Figure 5-15. The gate in Figure 5-14 functions as 0 AND 1 in, 1 out. The gate in Figure 5-15 functions as 0 AND 1 in, 0 out.

Example: Add these numbers using a 7483. Follow the logic levels through the logic diagram.

$A = 1001$, $B = 1010$, and $C_0 = 1$

Solution:

See Figure 5-16.

$\overline{A} \cdot B$

FIGURE 5-14

$\overline{\overline{A} \cdot B}$

FIGURE 5-15

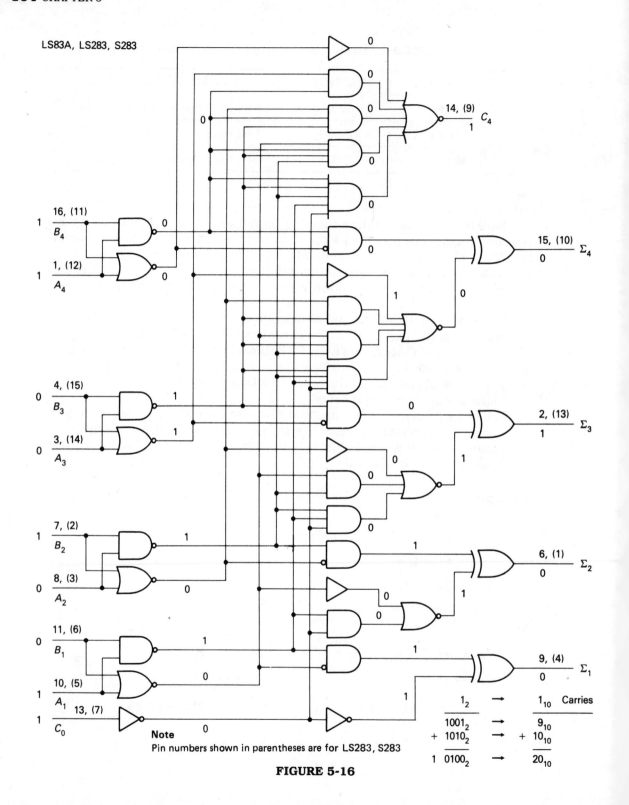

FIGURE 5-16

The truth table and connection diagram (pinout) for the 7483 and 74S283 are shown in Figure 5-17. Note this is not a standard pinout. V_{CC} is pin 5 and ground is on pin 12.

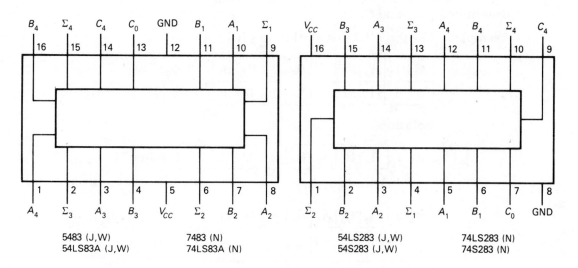

5483 (J,W)
54LS83A (J,W)

7483 (N)
74LS83A (N)

54LS283 (J,W)
54S283 (J,W)

74LS283 (N)
74S283 (N)

Input				Output					
				When $C_0 = L$		When $C_2 = L$	When $C_0 = H$		When $C_2 = H$
A_1 / A_3	B_1 / B_3	A_2 / A_4	B_2 / B_4	Σ_1 / Σ_3	Σ_2 / Σ_4	C_2 / C_4	Σ_1 / Σ_3	Σ_2 / Σ_4	C_2 / C_4
L	L	L	L	L	L	L	H	L	L
H	L	L	L	H	L	L	L	H	L
L	H	L	L	H	L	L	L	H	L
H	H	L	L	L	H	L	H	H	L
L	L	H	L	L	H	L	H	H	L
H	L	H	L	H	H	L	L	L	H
L	H	H	L	H	H	L	L	L	H
H	H	H	L	L	L	H	H	L	H
L	L	L	H	L	H	L	H	H	L
H	L	L	H	H	H	L	L	L	H
L	H	L	H	H	H	L	L	L	H
H	H	L	H	L	L	H	H	L	H
L	L	H	H	L	L	H	H	L	H
H	L	H	H	H	L	H	L	H	H
L	H	H	H	H	L	H	L	H	H
H	H	H	H	L	H	H	H	H	H

H = HIGH Level, L = LOW Level

Note

Input conditions at A_1, B_1, A_2, B_2 and C_0 are used to determine outputs Σ_1 and Σ_2 and the value of the internal carry C_2. The values at C_2, A_3, B_3, A_4, and B_4 are then used to determine outputs Σ_3, Σ_4, and C_4.

FIGURE 5-17   Truth table and connection diagrams

Writing a truth table for nine inputs creates a table of 512 lines (2^9). The truth table shown has been reduced to 16 lines. The note below the truth table explains that the table is used in two steps. A_1, B_1, A_2, B_2, and C_0 determine the outputs Σ_1, Σ_2, and C_2 that are internal to the IC. C_2 is then used with A_3, B_3, A_4, and B_4 to determine Σ_3, Σ_4, and C_4.

Example: Use the 7483 to add 0110 to 1101 with $C_0 = 1$.

A_4	A_3	A_2	A_1
0	1	1	0

B_4	B_3	B_2	B_1
1	1	0	1

C_0
1

Solution:

Step 1. | A_1 | B_1 | A_2 | B_2 | = | L | H | H | L | (line 7)

with $C_0 = $ H, $\Sigma_1 = $ L, $\Sigma_2 = $ L, $C_2 = $ H

Step 2. | A_3 | B_3 | A_4 | B_4 | = | H | H | L | H | (line 12)

with $C_2 = $ H, $\Sigma_3 = $ H, $\Sigma_4 = $ L, $C_4 = $ H

| Σ_4 | Σ_3 | Σ_2 | Σ_1 | = | L | H | L | L | = 0100

with $C_4 = $ H $= 1$.

$0110 + 1101 + 1 = 10100$
$6 + 13 + 1 = 20$

By implementing the complement system presented in Chapter 1, an adder circuit can double as a subtractor.

The IEC logic symbol for a 7483 is shown in Figure 5-18. A capital sigma, Σ, is used to denote addition. This symbol uses P_3, P_2, P_1, P_0 and Q_3, Q_2, Q_1, Q_0 to represent the two 4-bit numbers to be added and Σ_3, Σ_2, Σ_1, Σ_0 to represent the result. Note that C_I is used for carry in and C_O for carry out.

5.3 1'S COMPLEMENT ADDER/SUBTRACTOR

Design a circuit that will use a 7483 to add the 4-bit number B_4,B_3,B_2,B_1 to the 4-bit number A_4,A_3,A_2,A_1 or subtract B_4,B_3,B_2,B_1 from A_4,A_3,A_2,A_1. Use the 1's complement method for subtraction.

To use a 7483 4-bit full adder as a 1's complement adder/subtractor, the following details must be considered.

1. Leave the number B_4,B_3,B_2,B_1 unaltered for an addition problem, but take the 1's complement of the subtrahend for a subtraction problem. An exclusive-OR inverts data (1's complement) when the control input is HIGH.

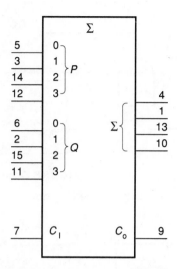

FIGURE 5-18   IEC logic symbol 7483

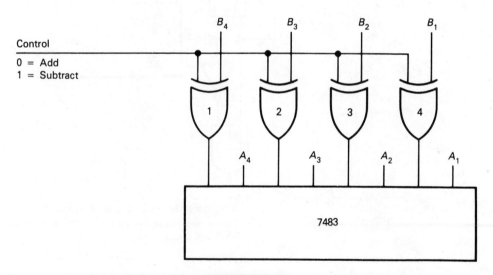

FIGURE 5-19

Exclusive-OR gates will be used to invert B_4, B_3, B_2, B_1 for subtraction. A control signal is needed that will be 1 for subtraction and 0 for addition. A_4, A_3, A_2, A_1 will be fed directly into the 7483. Refer to Figure 5-19.

2. If the problem is subtraction and if there is overflow ($C_4 = 1$), perform an EAC (end-around-carry). To detect when subtraction and overflow occur, AND the control line with C_4. The output of the AND gate number 1 is 1 when an EAC results. But in this case, the output of AND gate number 1 can be fed directly into C_0. Refer to Figure 5-20.

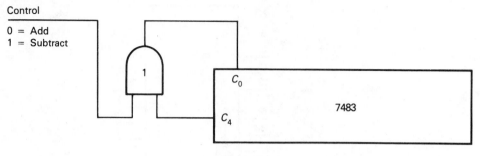

FIGURE 5-20

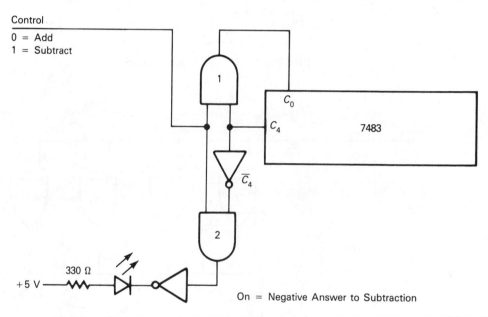

On = Negative Answer to Subtraction

FIGURE 5-21

3. If the problem is subtraction and if there is no overflow ($C_4 = 0$), indicate the answer is negative and take the 1's complement of the result to obtain the true magnitude of the answer. If C_4 is inverted, then an AND gate can be used to detect when a subtraction process is to be performed and when C_4 is 0. The control input and C_4 are fed into AND gate number 2. A HIGH output indicates a subtraction problem is being performed and the answer is negative. This signal could be used to light an LED to indicate the answer is negative. The LED requires approximately 12 mA to burn brightly. As will be seen in Chapter 6, TTL can handle more current in the 0 mode than in the 1 mode. This signal will be inverted to drive the LED in the active LOW mode. A red LED drops about 1.6 V or 1.7 V when lit (LED voltage drops very greatly

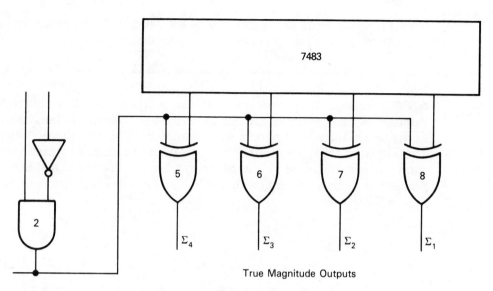

True Magnitude Outputs

FIGURE 5-22

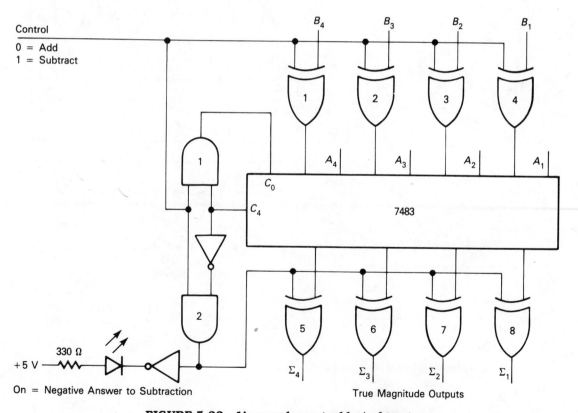

FIGURE 5-23   1's complement adder/subtractor

with different colors). This leaves 5 V − 1.7 V or 3.3 V to be dropped across the resistor. Ohm's Law dictates the resistor should be about

$$\frac{3.3 \text{ V}}{12 \times 10^{-3} \text{ A}} = 275 \text{ }\Omega$$

A 330-Ω resistor, the nearest standard size, will be used to limit the current through the LED. Refer to Figure 5-21. The 7404 inverter can handle 16 mA in the zero mode which is more than enough to light the LED as designed. The output of AND gate number 2 can be used to control four exclusive-OR gates to invert (take the 1's complement) when the answer to a subtraction problem is negative. Refer to Figure 5-22. The full schematic is drawn in Figure 5-23.

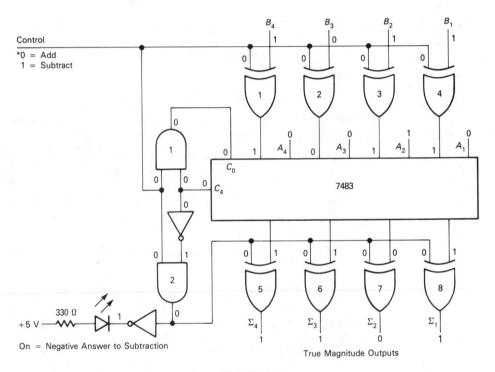

FIGURE 5-24

Example: Add 1011 plus 0010.

Solution:

See Figure 5-24.

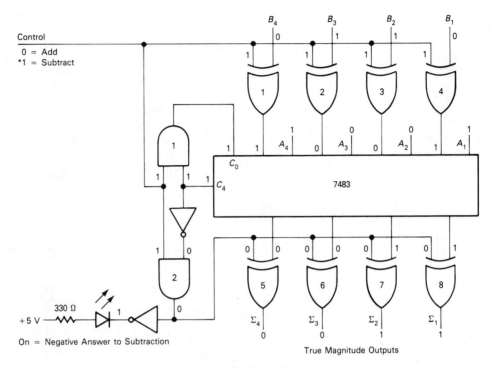

FIGURE 5-25

Example: Subtract 0110 from 1001.

Solution:

See Figure 5-25.

Example: Subtract 1010 from 0011.

Solution:

See Figure 5-26.

5.4 2'S COMPLEMENT ADDER/SUBTRACTOR

Design a circuit that will use a 7483 to add the 4-bit number B_4,B_3,B_2,B_1 to the 4-bit number A_4,A_3,A_2,A_1 and to subtract B_4,B_3,B_2,B_1 from A_4,A_3,A_2,A_1. Use the 2's complement method for subtraction.

To use the 7483 4-bit full adder as a 2's complement adder/subtractor, the following details must be considered.

1. Leave the number B_4,B_3,B_2,B_1 unaltered for an addition problem, but take the 2's complement of the subtrahend for a subtraction problem. The 2's complement can be formed by taking the 1's complement and adding 1. The 1's complement can be formed by using exclusive-OR gates as we did in the

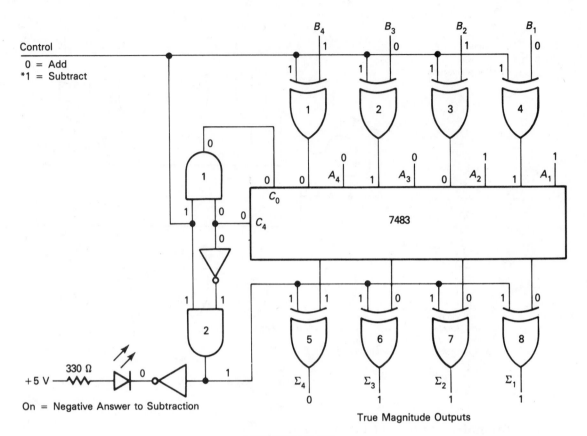

FIGURE 5-26

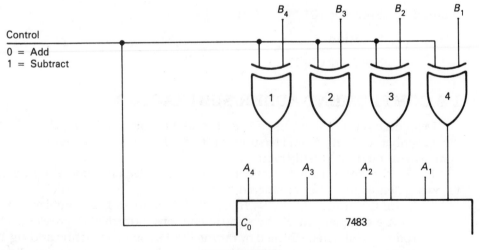

FIGURE 5-27

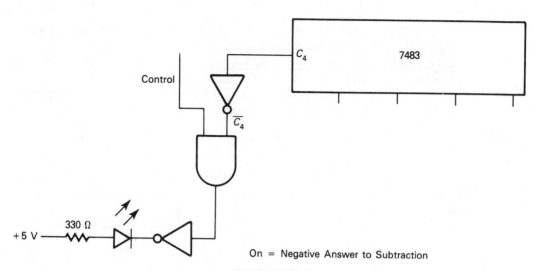

On = Negative Answer to Subtraction

FIGURE 5-28

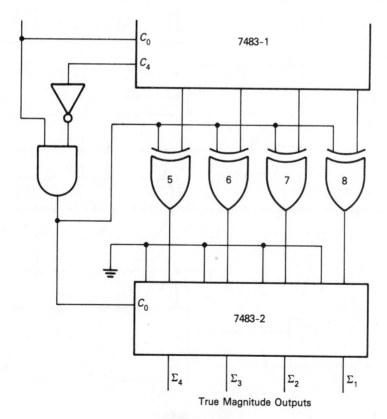

True Magnitude Outputs

FIGURE 5-29

1's complement subtractor. 1 can be added to form the 2's complement by wiring the control signal directly to C_0. Refer to Figure 5-27.

2. If the problem is subtraction and if there is no overflow ($C_4 = 0$), indicate the answer is negative and take the 2's complement of the result to obtain the true magnitude of the answer. As in the 1's complement subtraction circuit, C_4 can be inverted to form $\overline{C}_4$. $\overline{C}_4$ can be ANDed with the control signal. A

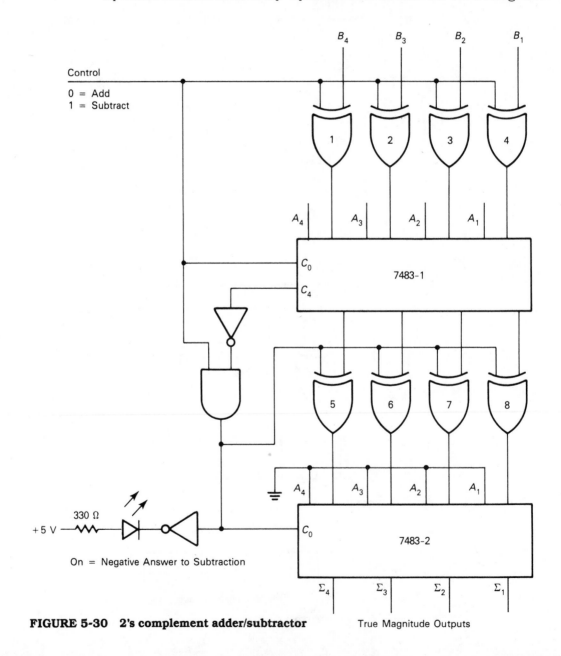

FIGURE 5-30   2's complement adder/subtractor   True Magnitude Outputs

HIGH out of the AND gate indicates that a subtraction problem is being performed and the answer is negative. This signal will be inverted to drive an LED in the active LOW mode. Refer to Figure 5-28. A HIGH output of the AND gate also indicates the result of the addition should be 2's complemented to obtain the true magnitude of the answer. The 1's complement can be formed by exclusive-ORing the results with the output of the AND gate. To add 1 to form the 2's complement, another 7483 must be used. The output of the AND

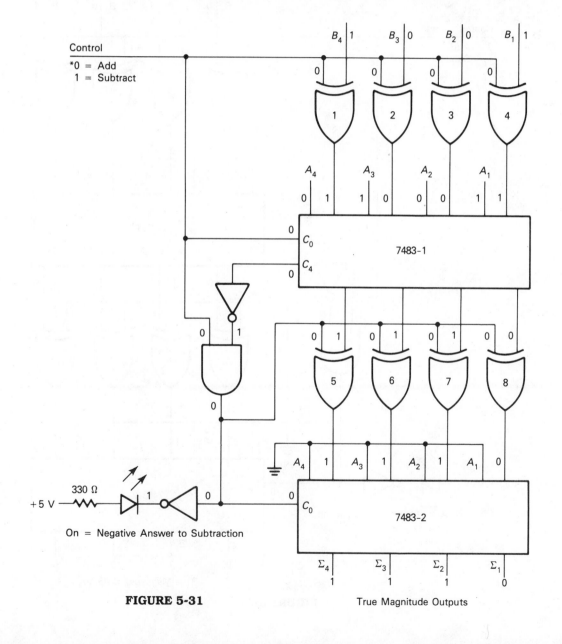

FIGURE 5-31   True Magnitude Outputs

can be fed directly into C_0 of 7483-2 to complete the 2's complement process. The true magnitude outputs appear at $\Sigma_4,\Sigma_3,\Sigma_2,\Sigma_1$ of 7483-2. Refer to Figure 5-29.

3. If the problem is subtraction and if there is overflow ($C_4 = 1$), do not take the 2's complement of the result from 7483-1. The answer is already in true

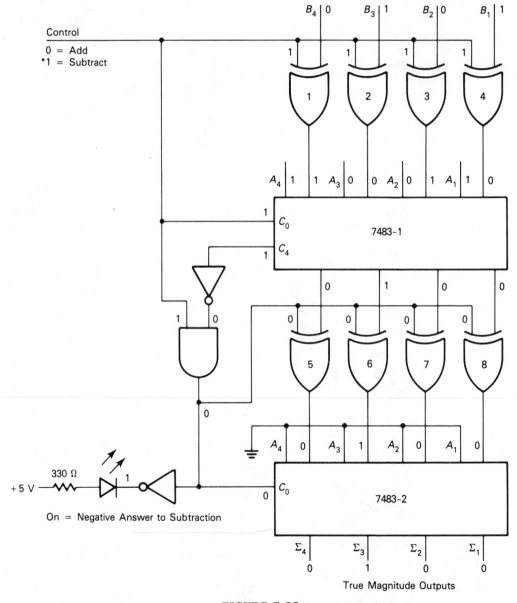

FIGURE 5-32

magnitude form and should not be altered by the following circuitry. In this situation the output of the AND gate will be 0. With a zero on the control inputs of exclusive-OR gates 5, 6, 7, and 8, the result from 7483-1 passes through 7483-2 unaltered and the true magnitude outputs appear at $\Sigma_4, \Sigma_3, \Sigma_2, \Sigma_1$ of 7483-2. This detail has been taken care of by the circuitry developed in step 2. The full schematic is drawn in Figure 5-30.

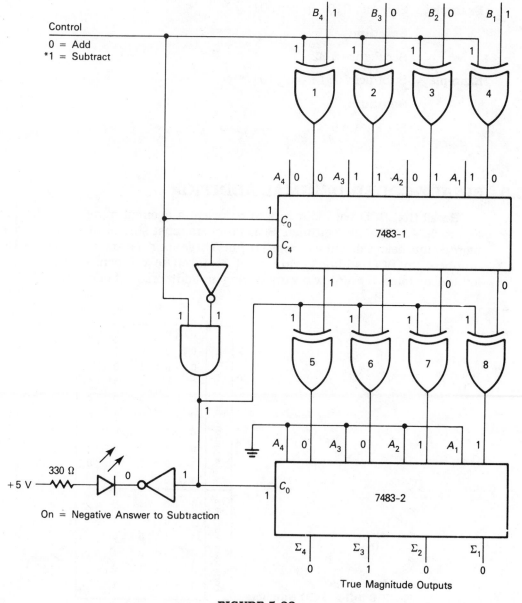

FIGURE 5-33

Try several addition and subtraction problems to fully understand the operation of the circuit.

Example: Add 1001 to 0101.

      Solution:

      See Figure 5-31.

Example: Subtract 0101 from 1001.

      Solution:

      See Figure 5-32.

Example: Subtract 1001 from 0101.

      Solution:

      See Figure 5-33.

5.5 BINARY-CODED-DECIMAL ADDITION

Recall that BCD uses four bits to represent a decimal number as shown in Figure 5-34. Although legitimate BCD numbers must stop at nine, there are six more counts before all four columns are full. These six steps are not legitimate BCD numbers. In BCD addition, care must be taken to compensate for these six forbidden states. If overflow occurs during an addition, or if one of the forbidden

0	0	0	0	0
0	0	0	1	1
0	0	1	0	2
0	0	1	1	3
0	1	0	0	4
0	1	0	1	5
0	1	1	0	6
0	1	1	1	7
1	0	0	0	8
1	0	0	1	9
1	0	1	0	
1	0	1	1	
1	1	0	0	
1	1	0	1	
1	1	1	0	
1	1	1	1	

Legitimate BCD Numbers / Forbidden Numbers

FIGURE 5-34 BCD Numbers

states occurs as a result of an addition, then 6 must be added to the result to "flip through" the unwanted states. In Figure 5-34, start at 7 and add 5. The result is 1100. To flip out of the forbidden states, count 6 more. The answer is 0010 or 2 with a carry to the next column. When you reach 1111 the next count is 0000 and a carry has occurred.

$$7 + 5 = 12$$

Example: Add 3 plus 5.

   Solution:

```
    0011
 +  0101
    1000
```

There is no overflow and the result is a legitimate BCD number, so it is correct. The answer is 8.

Example: Add 8 plus 5.

   Solution:

```
    1000
 +  0101
    1101
```

There is no overflow but the result is not a legitimate number.
Six must be added to compensate for the six forbidden numbers.

```
    1101
 +  0110
   10011
```

The answer is 13.

Example: Add 8 plus 9.

   Solution:

```
    1000
 +  1001
   10001
```

The result is a legitimate BCD number, but there was overflow.
Six must be added to compensate for the forbidden states.

```
   10001
 +  0110
   10111
```

The answer is 17.

Example: Add 167 plus 396.

Solution:

1	1		Carries
0001	0110	0111	
+0011	1001	0110	
0101	0000	1101	

Six is added to the least significant digit because the result is not a legitimate BCD number. Six must also be added to the middle digit because of the overflow. The most significant digit result produced no overflow, and it is a legitimate BCD number, so it is not necessary to add 6.

0101	0000	1101
+	0110	0110
0101	0110	0011

The answer is 563.

To convert a binary adder into a BCD adder, logic must be provided that will produce a signal when 6 should be added to the result of an addition. The carry out

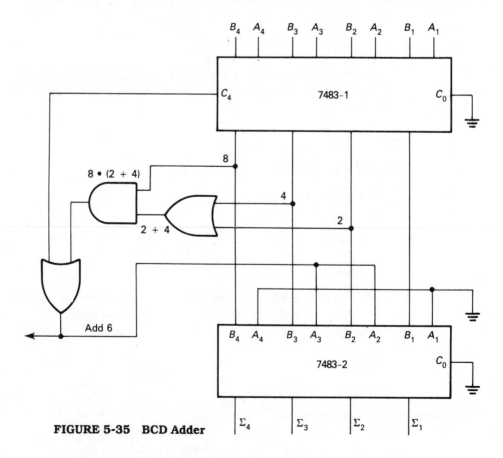

FIGURE 5-35 BCD Adder

of the binary adder can be monitored to see if overflow resulted. But how do you distinguish a legitimate BCD number from a forbidden one? (See Figure 5-34.)

All 4-bit numbers above nine have 1's in the eight's column and a 1 in the four's column or two's column. Written in Boolean, this is $8(4 + 2)$. This signal can be produced with a 2-input OR gate and a 2-input AND gate. If this signal is 1 or if overflow occurs $(C_4 = 1)$, 6 must be added to compensate for the 6 forbidden states. $8(2 + 4)$ is ORed with C_4 to produce the ADD 6 signal. Another 7483 will be used to add 6 to the result from 7483-1 when the ADD 6 signal is 1. Since 6 is 0110 in binary, B_4 and B_1 will be grounded, while the ADD 6 signal will be wired to B_3 and

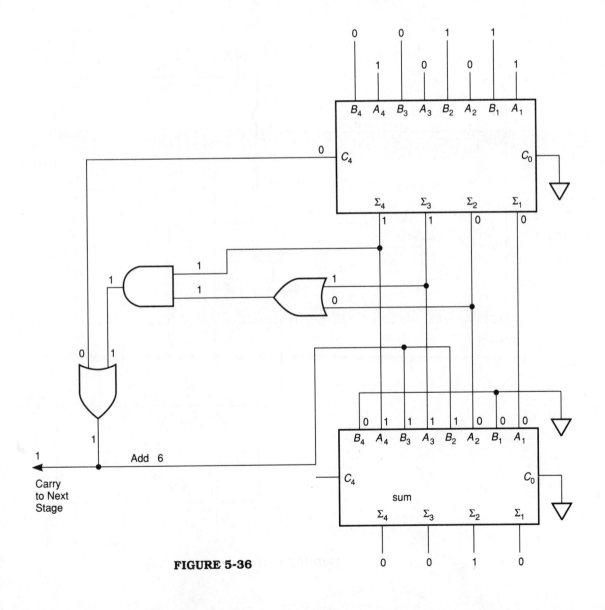

FIGURE 5-36

B_2. When ADD 6 is 0, B_4,B_3,B_2,B_1 will be 0000. When ADD 6 is 1, B_4,B_3,B_2,B_1 will be 0110 or 6. C_0 of 7483-2 must be grounded, or 7 will be added instead of 6. The complete schematic is shown in Figure 5-35.

Example: Use the BCD adder to add 9 and 3.

Solution:

See Figure 5-36. The sum from the first adder is 1100, which generates an ADD 6 signal and a carry to the next stage. The second adder adds 1100 + 0110 for a sum of 0010 or 2. The answer is 12.

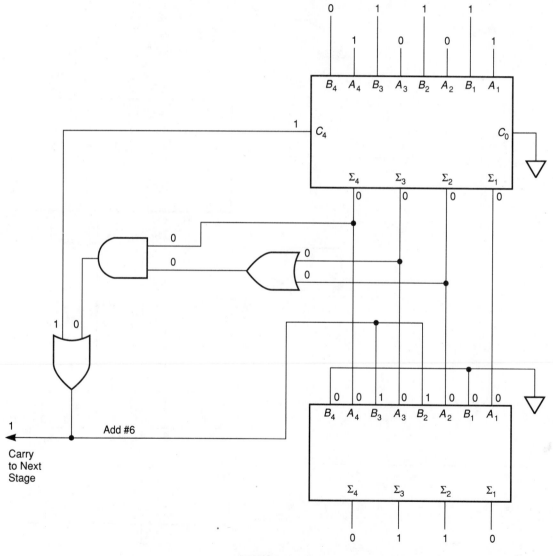

FIGURE 5-37

Example: Use the BCD adder to add 9 and 7.

Solution:

See Figure 5-37. The sum from the first adder is 0000 with a 1 out on C_4. This time C_4 generates the ADD 6 signal and the carry to the next stage. The answer is 16.

Follow several examples through until you understand the functioning of the ADD 6 circuit.

Exercise 5

1. Draw the truth table and logic diagram of a half adder. Show IC numbers and pin numbers.
2. Draw the truth table and logic diagram of a full adder. Show IC numbers and pin numbers.
3. Work the following problems. Use the 1's complement method on the subtraction problems. Confirm Figure 5-23 by following these problems through the circuit.

   a.   0111   b.   1010   c.   0011
        $+1000$ -0111 -1000

4. Work the following problems. Use the 1's complement method on the subtraction problems. Confirm Figure 5-23 by following these problems through the circuit.

   a.   0101   b.   1000   c.   0100
        $+1011$ $+0110$ -1100

5. Work the following problems. Use the 2's complement method on the subtraction problems. Confirm Figure 5-30 by following these problems through the circuit.

   a.   0101   b.   1001   c.   0011
        $+1000$ -0111 -1000

6. Work the following problems. Use the 2's complement method on the subtraction problems. Confirm Figure 5-30 by following these problems through the circuit.

   a.   0110   b.   1011   c.   101
        $+1000$ $-\ 101$ -1010

7. Work the following BCD problems. Confirm Figure 5-34 by following these problems through the circuit.

   a.   0100   b.   1001   c.   1001
        $+0101$ 0110   0111

8. Work the following BCD problems. Confirm Figure 5-35 by following these problems through the circuit.

   a. $\quad$ 101 $\qquad$ b. $\quad$ 1001 $\qquad$ c. $\quad$ 0101
      $\underline{+1001}$ $\qquad\quad$ $\underline{+1000}$ $\qquad\quad$ $\underline{+0010}$

9. Draw the logic diagram of an 8-bit 1's complement adder/subtractor.
10. Draw the logic diagram of an 8-bit 2's complement adder/subtractor.
11. Draw the logic diagram of an 8-bit BCD adder.
12. Draw the logic diagram of a CMOS adder IC. Describe in your own words its function.
13. Work the following problems. Use the 1's complement method. Confirm your design by following the problem through the circuit that you designed in number 9.

   a. $\quad$ 01101100 $\qquad$ b. $\quad$ 10101011 $\qquad$ c. $\quad$ 00011100
      $\underline{+00111010}$ $\qquad\quad$ $\underline{-01001100}$ $\qquad\quad$ $\underline{-10110101}$

14. Work the following problems. Use the 1's complement method. Confirm your design by following the problem through the circuit that you designed in number 9.

   a. $\quad$ 10000001 $\qquad$ b. $\quad$ 10001111 $\qquad$ c. $\quad$ 00111111
      $\underline{+00111010}$ $\qquad\quad$ $\underline{-10000111}$ $\qquad\quad$ $\underline{-01000110}$

15. Work the following problems. Use the 2's complement method on the subtraction problems. Confirm your design by following the problem through the circuit that you designed in number 10.

   a. $\quad$ 01010001 $\qquad$ b. $\quad$ 11010011 $\qquad$ c. $\quad$ 11000000
      $\underline{+01111010}$ $\qquad\quad$ $\underline{-00101101}$ $\qquad\quad$ $\underline{-11000001}$

16. Work the following problems. Use the 2's complement method on the subtraction problems. Confirm your design by following the problem through the circuit that you designed in number 10.

   a. $\quad$ 10011000 $\qquad$ b. $\quad$ 01111101 $\qquad$ c. $\quad$ 01101100
      $\underline{+01100110}$ $\qquad\quad$ $\underline{-00110010}$ $\qquad\quad$ $\underline{-10010010}$

17. Work the following problems. Confirm your design by following the problem through the circuit that you designed in number 11. All the numbers are BCD.

   a. $\quad$ 10000010 $\qquad$ b. $\quad$ 00101000 $\qquad$ c. $\quad$ 10010101
      $\underline{+00000111}$ $\qquad\quad$ $\underline{-01001001}$ $\qquad\quad$ $\underline{+01010001}$

18. Work the following problems. Confirm your design by following the problem through the circuit you designed in problem 11. All the numbers are BCD.

   a. $\quad$ 01110100 $\qquad$ b. $\quad$ 10000110 $\qquad$ c. $\quad$ 01010001
      $\underline{+00111001}$ $\qquad\quad$ $\underline{+01111000}$ $\qquad\quad$ $\underline{+00110100}$

These questions refer to the 1's complement subtractor circuit in Figure 5-23.

19. How is the end-around-carry accomplished?
20. Explain the function of exclusive-OR gates 1, 2, 3, and 4.
21. Explain the function of exclusive-OR gates 5, 6, 7, and 8.
22. What is the function of AND gate 2?

These questions refer to the 2's complement adder/subtractor in Figure 5-30.

23. What is the function of exclusive-OR gates 5, 6, 7, and 8?
24. When is the C_0 input on 7483-2 a 1 level?
25. What are the four pins that are grounded on 7483-2?
26. Why are they grounded?
27. When is the output of the AND gate a 1?

These questions refer to the BCD adder in Figure 5-35.

28. What is the function of the 7483-2?
29. Why does the C_4 output of 7483-1 need to be included in the development of the ADD 6 signal?
30. What would happen if C_0 on the 7483-2 were not grounded?
31. Draw the IEC symbol for a 4-bit full adder.

5 Adders

OBJECTIVES

After completing this lab, you should be able to:

- draw the logic diagram for a BCD adder.

- construct and use a BCD adder.

- draw the logic diagram for a 1's complement adder/subtractor.

- construct and use a 1's complement adder/subtractor.

COMPONENTS NEEDED

1	7408 IC
1	7432 IC
2	7483 ICs
2	7486 ICs
5	LEDs
5	330-Ω resistors

PREPARATION

If your circuit does not work properly, consider these points:

1. Check all power supply voltages and connections.
2. Check all input and output voltages for proper voltages.
3. Troubleshoot the circuit using the following basic steps. These steps can be used to troubleshoot any electronic circuit.
   A. First understand the electronic circuit's operation theory.
   B. Determine or set the input value to the circuits.
   C. Measure the circuit's output value.
   D. Based on the input values, output values, and your understanding of how the circuits work, determine which part of the circuit is faulty. Then test the part you suspect.

Use a BCD adder as follows.

- Put 0000 and 0011 into the BCD adder input. Check to see if the output of the first 7438 IC is 0011. If it is, check the second 7483 IC for the sum of 0011.
- If the adder works with sums of 9 or less but does not work with sums of 10 or more, then the ADD 6 part of the adder is not functioning.
- If the sum of the two inputs 0000 and 0011 is 1001, then the ADD 6 part of the BCD adder is turned on when it should be off.

Troubleshoot the rest of Lab 5 if it is needed. It is a good idea to write the steps you use to troubleshoot a circuit in a notebook for reference during the troubleshooting procedure and for use at a later time.

PROCEDURE

1. Draw the logic diagram of a BCD adder. Show pin numbers. Use two 7483s and additional gates as needed. Use LEDs to monitor the outputs. Have your instructor approve your drawing.
2. Construct the circuit. Let $A = 0101$ and $B = 0011$. What is the sum? Is there a carry out of the first 7483?
3. Let $A = 0101$ and $B = 1001$. What is the sum? Is there a carry out of the first 7483?
4. What is the purpose of the ADD 6 signal?
5. Add these combinations:

   a. 0110   b. 1001   c. 1001
      +0110   +1001   +0001

6. Draw the logic diagram of a 1's complement adder/subtractor. Use one 7483 and additional gates as needed. Use LEDs to monitor the outputs. Have your instructor approve your drawing.
7. Construct the circuit. Add $A = 0101$ and $B = 0011$. What is the sum? Is there a carry out? Does the first set of exclusive-OR gates complement B? Does the last set invert the answer? Is an EAC performed?
8. Let $A = 0101$ and $B = 1001$. What is the difference $(A - B)$? Is the result of the addition complemented by the 2nd set of exclusive-OR gates? Is B complemented by the first set of exclusive-OR gates? Is an EAC performed?
9. Let $A = 1001$ and $B = 0101$. What is the difference $(A - B)$? Is B complemented by the first set of exclusive-OR gates? Is the result of the addition complemented by the second set of exclusive-OR gates? Is an EAC performed?
10. Try these combinations:

    a. 0110   b. 1001   c. 1001
       -0110   -1010   -0001

CHAPTER 6

OUTLINE

Specifications and Open-Collector Gates

OBJECTIVES

After completing this chapter, you should be able to:

- define fan-out.

- use a data book to determine fan-out, maximum zero-level input voltage, maximum zero-level output voltage, minimum one-level input voltage, minimum one-level output voltage, and noise margin for TTL and CMOS ICs.

- describe the operation of a totem-pole output in TTL and CMOS circuits.

- interface CMOS to TTL and TTL to CMOS.

- describe the operation of open-collector TTL and CMOS circuits.

- use open-collector gates in a wired NOR circuit.

6.1 TTL SUBFAMILIES

One popular, readily available family of digital ICs is Transistor Transistor Logic (TTL). The TTL family is identified by the first two digits of the device number. 74XX denotes that the IC meets commercial specifications and has an operating range of 0°C to 70°C. 54XX denotes that the IC meets full military-range specifications and can operate between $-55°C$ and $+125°C$. The 54XX series can withstand harsher environments than the 74XX series. The pinouts for corresponding ICs such as 5404 and 7404 are the same.

Any letters following 54 or 74 denote the subfamily of the IC. Common TTL subfamilies are listed as follows:

No letters	Standard TTL
LS	Low-power Schottky
S	Schottky
L	Low power
ALS	Advanced low-power Schottky
AS	Advanced Schottky
F	Fairchild advanced Schottky TTL (FAST)

The numbers following the subfamily designation indicate the function of the IC. The 54L10 is a low-power triple 3-input NAND gate that meets military specifications. The 74LS32 is a low-power Schottky quad 2-input OR gate that meets commercial specifications.

Manufacturers guarantee that a 74 series can be operated with a supply voltage that ranges between 4.75 V and 5.25 V, and that a 54 series IC can be operated between 4.50 V and 5.50 V.

6.2 TTL ELECTRICAL CHARACTERISTICS

In TTL, the high level is usually represented by a 1 and the low level by a 0. When used in this manner, the system is called *positive logic*. If the high level is represented by a 0 and the low level by a 1, the system is called *negative logic*. Positive logic will be used in this text for TTL.

Figure 6-1 is reproduced from the National Semiconductor Data Book. It shows the recommended operating conditions, electrical characteristics, and switching characteristics for a 5400 and 7400 quad two-input NAND gate. These values are typical of all the standard TTL gates.

After these specifications have been studied they will be compared to those of the L, S, AS, LS, and ALS subfamilies. In this table, conventional current flowing out of the gate is considered negative and conventional current flowing into the gate is positive. The units for each parameter are listed in the last column.

V_{IH}, high level input voltage, is listed as a minimum of 2 V. An input must be at least 2 V to be recognized as a 1. A 1-level input can range from 2 V to V_{CC}. V_{OH}, high level output voltage, is listed as a minimum of 2.4 V. A 1 output can range from 2.4 V to V_{CC}.

If a gate supplies at least 2.4 V for a 1-level and a following gate can recognize down to 2.0 V as a 1, then there is a 0.4 V difference in levels. This safety margin is called the noise margin of the IC. As shown in Figure 6-2, 0.4 V of noise can be riding on a 1-level output and the signal will still be recognized as a 1 by the following IC.

Recommended Operating Conditons

Symbol	Parameter	DM5400			DM7400			Units
		Min	Nom	Max	Min	Nom	Max	
V_{CC}	Supply Volage	4.5	5	5.5	4.75	5	5.25	V
V_{IH}	High Level Input Voltage	2			2			V
V_{IL}	Low Level Input Voltage			0.8			0.8	V
I_{OH}	High Level Output Current			-0.4			-0.4	mA
I_{OL}	Low Level Output Current			16			16	mA
T_A	Free Air Operating Temperature	-55		125	0		70	°C

Electrical Characteristics over recommended operating free air temperature (unless otherwise noted)

Symbol	Parameter	Conditions		Min	Typ (Note 1)	Max	Units
V_I	Input Clamp Voltage	V_{CC} = Min, $I_I = -12mA$				-1.5	V
V_{OH}	High Level Output Voltage	V_{CC} = Min, I_{OH} = Max V_{IL} = Max		2.4	3.4		V
V_{OL}	Low Level Output Voltage	V_{CC} = Min, I_{OL} = Max V_{IH} = Min			0.2	0.4	V
I_I	Input Current @ Max Input Voltage	V_{CC} = Max, V_I = 5.5V				1	mA
I_{IH}	High Level Input Current	V_{CC} = Max, V_I = 2.4V				40	μA
I_{IL}	Low Level Input Current	V_{CC} = Max, V_I = 0.4V				-1.6	mA
I_{OS}	Short Circuit Output Current	V_{CC} = Max (Note 2)	DM54	-20		-55	mA
			DM74	-18		-55	
I_{CCH}	Supply Current With Outputs High	V_{CC} = Max			4	8	mA
I_{CCL}	Supply Current With Outputs Low	V_{CC} = Max			12	22	mA

Switching Characteristics at V_{CC} = 5V and T_A = 25°C (See Section 1 for Text Waveforms and Output Load)

Parameter	Conditions	C_L = 15 pF R_L = 400Ω			Units
		Min	Typ	Max	
I_{PLH} Propagation Delay Time Low to High Level Output			12	22	ns
I_{PHL} Propagation Delay Time High to Low Level Output			7	15	ns

Note: All typicals are at V_{CC} = 5V, T_A = 25°C.

Note: Not more than one ouput should be shorted at a time.

FIGURE 6-1 TTL NAND gate specifications

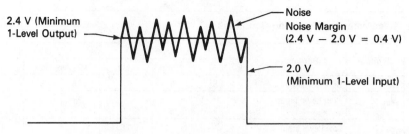

FIGURE 6-2   TTL 1-level noise margin

V_{IL}, low level input voltage, is listed as 0.8 V maximum. The highest voltage that an IC will accept as a 0 is 0.8 V. A 0-level input can range from 0 to 0.8 V. V_{OL}, low level output voltage, is listed as 0.4 V maximum. A 0-level output can range from 0 to 0.4 V. If the highest 0 level that an IC will supply is 0.4 V, but a following IC can recognize up to 0.8 V as a 0, then once again there is a noise margin of 0.4 V, as shown in Figure 6-3.

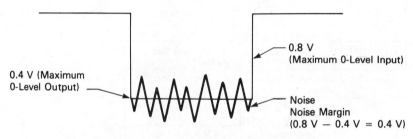

FIGURE 6-3   TTL 0-level noise margin

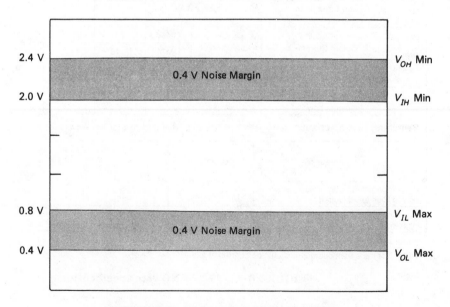

FIGURE 6-4   Voltage levels for TTL logic

DM5400/DM7400 2-INPUT NAND		TTL	L-TTL	LS	ALS	S	AS	Units
MII	V_{OH}	2.4	2.4	2.5	$V_{CC}-2$	2.5	$V_{CC}-2$	V
Com	V_{OH}	2.4	2.4	2.7	$V_{CC}-2$	2.7	$V_{CC}-2$	V
MII	V_{OL}	0.4	0.3	0.4	0.4	0.5	0.5	V
Com	V_{OL}	0.4	0.4	0.5	0.5	0.5	0.5	V
	V_{IH}	2	2	2	2	2	2	V
MII	V_{IL}	0.8	0.7	0.7	0.8	0.8	0.8	V
Com	V_{IL}	0.8	0.7	0.8	0.8	0.8	0.8	V

FIGURE 6-5   Input and output voltages: TTL subfamilies

Standard TTL noise margin is summarized in Figure 6-4.

Input and output voltages for each of the subfamilies are summarized in Figure 6-5.

Example: Calculate the noise margin of a 54S00.

Solution:

Figure 6-1 shows that V_{OH} = 2.5 V minimum, V_{IL} = 0.8 V maximum, V_{OL} = 0.5 V maximum, V_{IH} = 2.0 V minimum.

High level noise margin = $V_{OH} - V_{IH}$ = 2.5 V − 2.0 V = 0.5 V

Low level noise margin = $V_{IL} - V_{OL}$ = 0.8 V − 0.5 V = 0.3 V

Example: Calculate the noise margin of a 74ALS00 operating at 5 V.

Solution:

Figure 6-5 shows that V_{OH} = V_{CC} = − 2 V minimum.

V_{IL} = 0.8 V maximum, V_{OL} = 0.5 V maximum,

V_{IH} = 2 V minimum.

High level noise margin = $V_{OH} - V_{IH}$ = 5 − 2 − 2 = 1 V.

Low level noise margin = $V_{IL} - V_{OL}$ = 0.8 − 0.5 = 0.3 V.

I_{OL}, low level output current, is listed as a maximum of 16 mA in Figure 6-1 (page 191). This conventional current is flowing into the gate, and the gate is said to be "sinking" current. The manufacturer guarantees that the 7400 can "sink" 16 mA of current without the zero level output voltage rising above 0.4 V. I_{IL}, low level input current, is listed as a maximum of − 1.6 mA. This current is flowing out of the gate. Figure 6-6 shows a 7400 NAND gate sinking current from ten other gates, each with a low level input current of − 1.6 mA. The 1.6 mA is said to be "one standard TTL load." Fan-out is a measure of the number of loads that a gate can drive.

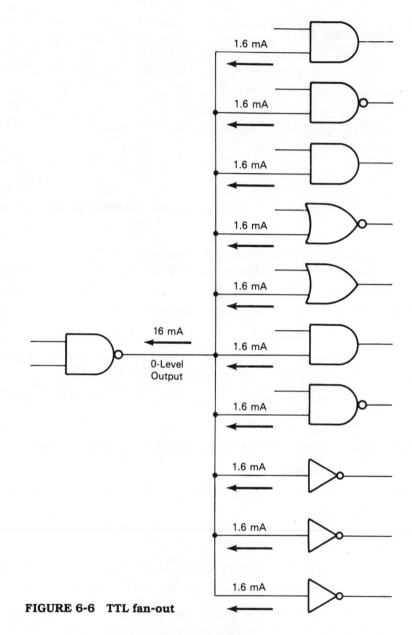

FIGURE 6-6   TTL fan-out

$$\text{Fan-out} \; = \; \frac{I_{OL} \text{ of the driving gate}}{I_{IL} \text{ of the driven gate}}$$

For a NAND gate driving other NAND gates or inverters,

$$\text{Fan-out} \; = \; \frac{I_{OL}}{I_{IL}} \; = \; \frac{16 \text{ mA}}{1.6 \text{ mA}} \; = \; 10 \text{ standard loads}$$

Each of the standard TTL gates can drive ten other standard gates.

Figure 6-7 shows the input and output currents for a NAND gate from each subfamily. I_{OL} is a measure of the drive capability of each subfamily. Schottky and advanced Schottky are highest at 20 mA, followed by standard TTL at 16 mA, low-power Schottky and advanced low-power Schottky at 8 mA (4 mA for military), and finally 3.6 mA for low-power TTL (2 mA for military).

Example: How many 54ALS00 gates can a 54L00 drive?

Solution:

Figure 6-7 lists I_{OL} for a 54L00 as 2 mA and I_{IL} for a 54ALS00 as -0.2 mA.

$$\text{Fan-out} = \frac{I_{OL}}{I_{IL}} = \frac{2}{0.2} = 10$$

		TTL	L-TTL	LS	ALS	S	AS	Units
	I_{OH}	-400	-200	-400	-400	-1000	-2000	μA
	I_{OL}	16	2	4	4	20	20	mA
	I_{IH}	40	10	20	20	50	20	μA
Military	I_{IL}	-1.6	-0.18	-0.36	-0.20	-2	-0.50	mA
Commercial	I_{OL}		3.6	8	8			mA

FIGURE 6-7   Input and output currents: TTL subfamilies

Example: How many standard TTL loads can a 54LS00 drive?

Solution:

Figure 6-7 lists I_{OL} for a 54LS00 as a 4 mA maximum.

$$\text{Fan-out} = \frac{I_{OL}}{I_{IL}} = \frac{4 \text{ mA}}{1.6 \text{ mA}} = 2.5$$

Since the 0.5 load is not a complete load, drop the 0.5. The 54LS00 has a fan-out of two standard TTL loads.

Example: How many 74LS04 inverters can a 74LS00 drive?

Solution:

I_{OL} for a 74LS00 is 8 mA.
I_{IL} for a 74LS04 is 0.36 mA.

$$\text{Fan-out} = \frac{I_{OL}}{I_{IL}} = \frac{8 \text{ mA}}{0.36 \text{ mA}} = 22.22$$

A 74LS00 can drive 22 other 74LS04 gates.

A gate from any of the TTL subfamilies can drive at least ten other gates from the same subfamily.

I_{OH}, high level output current, is listed as -0.4 mA for the 7400. The negative sign implies that current is flowing out of the gate. The gate supplies or "sources" currents when outputting a one. This group of gates can sink 16 mA on a low level output, but can only source 0.4 mA on a high level output. We can take advantage of the greater low level current by using the gate in an active LOW mode as shown in Figure 6-8. When A goes HIGH, B goes LOW. Since the 7404 can sink 16 mA when outputting a zero, the LED can burn brightly.

In the arrangement in Figure 6-9, a high level output should light the LED. However, the data book lists I_{OH} for a 7408 as -800 μA maximum. The LED draws more than 800 μA and the 1 level could fall below 2.4 V. The output would no longer be a legitimate 1 level.

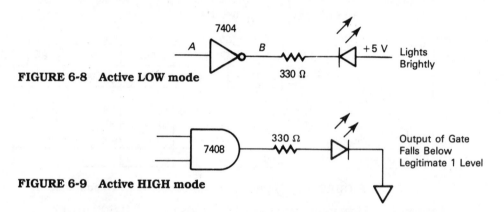

FIGURE 6-8   **Active LOW mode**

FIGURE 6-9   **Active HIGH mode**

I_{IH}, high level input current, is listed as 40 μA for the standard gates in Figure 6-1. Since a standard NAND gate can source 400 μA on a high level output, the gate can drive ten other gates. This result is compatible with what we learned from the low level signals.

6.3   TTL SUPPLY CURRENTS

Figure 6-10, reproduced from the National Semiconductor Data Book, lists the current drawn from the power supply per IC for the 7400 NAND gates. I_{CCH} represents the total collector current drawn with all outputs HIGH, and I_{CCL} represents those with outputs LOW. The standard TTL 7400 draws a maximum of 8 mA from the supply with outputs HIGH and 22 mA maximum with outputs LOW.

		S	TTL	AS	LS	ALS	L	Units
I_{CCH}	Supply current with outputs HIGH (maximum)	16	8	3.2	1.6	0.85	0.8	mA
I_{CCL}	Supply current with outputs LOW (maximum)	36	22	17.4	4.4	3.0	2.04	mA

FIGURE 6-10   **Supply currents: TTL quad 2-input NAND**

Figure 6-10 compares the supply currents drawn by the TTL subfamilies. A Schottky 74S00 draws the most current at 36 mA (outputs LOW), followed by TTL, AS, LS, ALS, and finally the low-power 74L00 at 2.04 mA.

To calculate the current that a power supply will need to provide for a circuit, follow these rules:

1. Total the worse-case currents for all the ICs in the circuit.
2. Add currents drawn by all other devices such as LEDs and displays.
3. As a rule of thumb, double the total and design your power supply accordingly.

6.4 TTL SWITCHING CHARACTERISTICS

Figure 6-11 lists the switching characteristics for TTL gates. The t_{PLH} (time propagation LOW-to-HIGH) is a measure of the time it takes for a change in the input to cause a LOW-to-HIGH change in the output. As shown in Figure 6-12, the t_{PLH} is measured from when the input reaches 1.5 V to where the output reaches 1.5 V on a LOW-to-HIGH transition. These parameters are measured with the gates driving a load equivalent to 10 gates of the same subfamily. Standard TTL has a maximum t_{PLH} of 22 nanoseconds, while Schottky is the fastest with only 4.5 nanoseconds delay.

	L	TTL	LS	ALS	S	AS	Volts
t_{PLH}	60	22	15	11	7	4.5	ns
t_{PHL}	60	15	15	8	8	4	ns

FIGURE 6-11   Maximum switching characteristics: TTL quad 2-input NAND

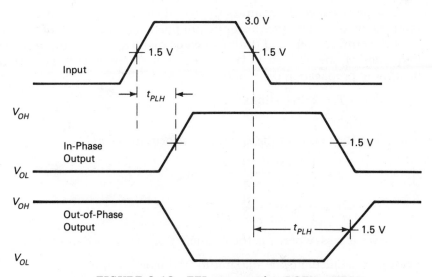

FIGURE 6-12   TTL propagation LOW-to-HIGH

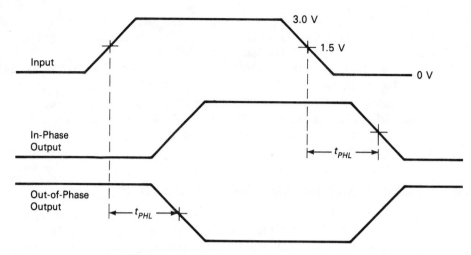

FIGURE 6-13   TTL propagation HIGH-to-LOW

The t_{PHL} (time propagation HIGH-to-LOW) is a measure of the time it takes for a change in the input to cause a HIGH-to-LOW transition in the output. As shown in Figure 6-13, the t_{PHL} is measured from when the input reaches 1.5 V to when the output reaches 1.5 V on a HIGH-to-LOW transition.

Standard TTL has a maximum t_{PHL} of 15 nanoseconds. Note that standard TTL turns off more quickly than it turns on. Schottky is once again the fastest with a maximum t_{PHL} of 4 nanoseconds, and low-power TTL is slowest at 60-nanosecond delay.

The propagation delay limits the speed at which the IC can operate. When the propagation delay becomes a significant portion of the period of the applied signal, then the output levels and timing become distorted. As a rule of thumb, limit the frequency of the applied waveform so that its period is more than twice the maximum propagation delay of the IC.

$$T \geq 2 \cdot t_{PLH} \text{ (max) or}$$
$$T \geq 2 \cdot t_{PHL} \text{ (max)}$$

where T is the period of the applied waveform in seconds.

$$f = \frac{1}{T} \leq \frac{1}{2 \cdot t_{PLH} \text{ (max)}} \text{ or}$$
$$f = \frac{1}{T} \leq \frac{1}{2 \cdot t_{PHL} \text{ (max)}}$$

whichever is less, where f is the frequency of the applied waveform in hertz.

Example: Calculate the maximum frequency that can be applied to standard TTL NAND gates.

Solution:

t_{PLH} (max) will be used in the calculation since it is greater than t_{PHL} (max).

$$f = \frac{1}{T} \le \frac{1}{2 \cdot t_{PLH} \text{ (max)}} \le \frac{1}{2 \cdot 22 \cdot 10^{-9} \text{ s}} \le 22.7 \text{ MHz}$$

Standard TTL NAND gates should be operated at less than 22.7 MHz.

Example: Calculate the maximum frequency that can be applied to a low-power Schottky NAND gate.

Solution:

$$f \le \frac{1}{2 \cdot 10 \cdot 10^{-9} \text{ s}} \le 50 \text{ MHz}$$

Example: Calculate the maximum frequency that should be applied to an advanced Schottky NAND gate.

Solution:

t_{PLH} will be used in the calculation since it is greater than t_{PHL}.

$$f = \frac{1}{T} \le \frac{1}{2t_{PLH}(\text{max})} \le \frac{1}{2 \cdot 4.5 \cdot 10^{-9} \text{ s}} = 111 \text{ MHz}$$

When a logic family is selected for use in a circuit, both speed and power consumption should be considered. Figure 6-14 summarizes these properties. Low-power TTL dissipates the least power but is the slowest. This is the classic trade-off, power consumption versus speed. Schottky is one of the fastest subfamilies, but it dissipates the most power. ALS is among the fastest, while dissipating less power than all subfamilies except L. These qualities make it a good choice for many applications.

Speed		Power Consumption	
Fastest	AS	Low	L
	S		ALS
	ALS		LS
	LS		AS
	TTL		TTL
Slowest	L	High	S

FIGURE 6-14   TTL relative speed and power consumption

Because low-power Schottky is faster than standard TTL and draws less current from the power supply, it is the most used subfamily.

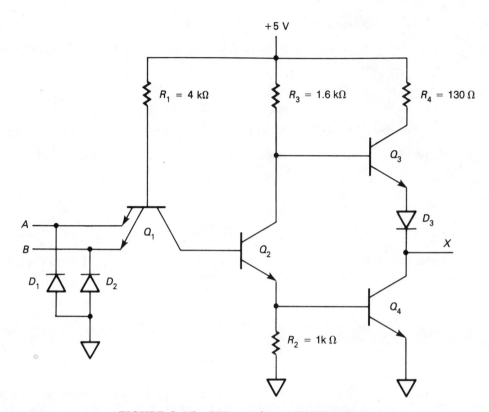

FIGURE 6-15   TTL two-input NAND gate

Figure 6-15 shows the internal circuit of a TTL NAND gate. Although the gate can be used without knowledge of the internal circuitry, the TTL characteristics can be better understood by investigating the circuitry. All of the transistors in Figure 6-15 are N-P-N silicon. Remember that in an N-P-N silicon transistor, the voltage on the base with respect to the emitter must be about $+0.7$ V to forward bias the transistor and turn it on. When the transistor is turned on hard and saturated, the voltage on the collector with respect to the emitter is less than $+0.4$ V. Also, when a transistor is turned on, collector current flows and a drop occurs across the collector resistor. For example, when Q_2 in Figure 6-15 is on, current flows through R_3, and most of the supply voltage is dropped across R_3. The voltage at the collector of Q_2 goes LOW. When Q_2 turns off, no collector current flows and the collector voltage rises.

Q_1 is a multiemitter transistor, one emitter for each input, connected in a collector-follower circuit. Any 0 into a NAND should yield a 1 out. Suppose A is a 0 as shown in Figure 6-16. The base-emitter junction of Q_1 is forward-biased and conventional current flows through R_1. This current is I_{IL} and is -1.6 mA maximum. The negative indicates flow out of the gate as shown in Figure 6-16.

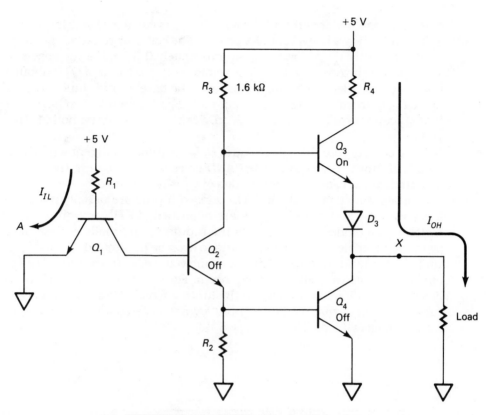

FIGURE 6-16   NAND gate: Any 0 in, 1 out

With the emitter grounded and the emitter-base junction forward-biased, the voltage on the base drops to about 0.7 V. Since 0.7 V on the base is not sufficient to forward-bias the base-collector junction of Q_1 and the base-emitter junction of Q_2, Q_2 turns off. With Q_2 off there is no emitter current through R_2. With no voltage developed across R_2, the base-emitter junction of Q_4 will be turned off. With Q_4 off, the output X will not be connected to ground.

Since Q_2 is off, there is no collector current flowing through R_3 and the collector voltage at Q_2 is HIGH. The emitter-base junction of Q_3 and diode D_3 are forward-biased. Q_3 is turned on and X is connected to +5 V through a saturated transistor and a forward-biased diode. The output X is HIGH, or 1. A 0 in at A causes X to go HIGH. The current flowing out of the gate at X is I_{OH} and is 400 µA maximum.

If both A and B are HIGH as shown in Figure 6-17, then the base-emitter junction of Q_1 is not forward-biased. Q_2 can now be forward-biased by +5 V dropped across R_1, the collector-base junction of Q_1, the base-emitter junction of Q_2 and R_2 to ground. The arrow in Figure 6-17 shows this path. Q_2 turns on and saturates. With Q_2 on, emitter current through R_2 causes a voltage drop across R_2 that forward-biases the base-emitter junction of Q_4 and saturates it. With Q_4 on, output X has a path to ground through Q_4. Q_4 is capable of sinking 16 mA, I_{OL}, and still maintaining a 0 level of 0.4 V or less. The collector voltage at Q_2 is equal to the collector-emitter drop across Q_2, approximately 0.3 V, plus the base-emitter drop across Q_4, approximately 0.7 V. The voltage at the collector of Q_2 is about 1.0 V. One volt is not sufficient to forward-bias both the base-emitter junction of Q_3 and the diode D_3, Q_3 turns off and X does not have a path to $+5$ V through Q_3. D_3 insures that Q_3 cannot turn on while Q_4 is on. With A and B inputs both 1, the output X is 0.

The current into the gate at A and B, I_{IH}, is leakage current with a maximum of 40 μA. If inputs A and B to the NAND gates are left floating (not connected to a signal) then the base-emitter junction of Q_1 is not forward-biased. Q_1 behaves as if the inputs were tied HIGH. In TTL, unused inputs are usually interpreted as one levels by the ICs. Unused inputs are often tied HIGH through a 1-kΩ to 10-kΩ resistor. The diodes D_1 and D_2 on the inputs are normally reverse-biased. When switching transients cause the inputs to drop below ground, D_1 and/or D_2 turn on to clamp the input voltage. V_1, input clamp voltage, is a parameter that specifies the amount of negative swing that can occur. For the NAND gates and inverters, V_I is listed as -1.5 V maximum when the input current, I_I, is -12 mA. The manufacturer guarantees that when the input swings LOW enough to draw -12 mA, the input voltage will not fall below -1.5 V.

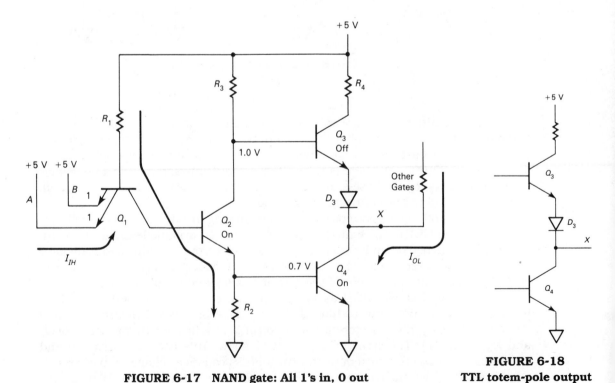

FIGURE 6-17   NAND gate: All 1's in, 0 out

**FIGURE 6-18
TTL totem-pole output**

R_4, Q_3, D_3, and Q_4 constitute the output circuit as shown in Figure 6-18. Their configuration is called the "totem-pole" output. Normally either Q_3 or Q_4 is on, but not both. With Q_3 on, X is pulled up to a one level, and with Q_4 on, X is pulled down to a zero level. However, during switching Q_3 and Q_4 are on simultaneously for a short time and a heavy load is placed on the power supply. TTL circuits are designed to switch Q_3 and Q_4 quickly to minimize the effects on the power supply.

To filter the noise induced on the power supply by the switching action, wire a 0.01-µF capacitor across the power supply near the pins on the IC. One capacitor for every two ICs is a good rule of thumb to follow.

Two totem-pole outputs should not be connected directly together as shown in Figure 6-19. If the upper transistor of one totem-pole output turns on and the lower transistor of the other turns on, then the power supply is shorted out and heavy currents will flow. Damage to the gates and power supply can result. Totem-pole outputs should be connected together through other gates.

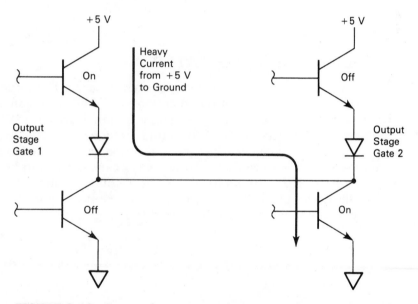

FIGURE 6-19   Danger from connecting totem-pole outputs together

From the standard TTL circuits studied so far, several subfamilies have evolved. The low-power subfamily L was developed from TTL by increasing the resistances of the resistors in the internal circuitry by a factor of 10. The power dissipation of the L device is reduced by a factor of 10 but at a sacrifice of speed. L devices have a propagation delay that is three times longer than standard TTL.

The Schottky subfamily S uses Schottky diodes as clamps to prevent the transistors from becoming saturated, and uses resistors whose resistances are about half those of standard TTL. This unsaturated logic switches three times more quickly than standard TTL but also consumes more power. Gates in this family dissipate about 20 milliwatts with propagation delays of 3 nanoseconds.

The low-power Schottky subfamily LS uses increased resistor valves and diode inputs instead of the multiemitter inputs seen in standard TTL circuitry. These diode inputs switch more quickly to yield a propagation delay of 10 nanoseconds and a power dissipation of 2 milliwatts per gate.

The advanced low-power Schottky subfamily, ALS, uses refined fabricating techniques to increase switching speeds and lower power consumption over LS devices. ALS device gates have a propagation delay of about 4 nanoseconds and a power dissipation of 1 mW per gate.

The advanced Schottky subfamily AS is designed for speed. It uses networks in the output circuits that reduce rise time. AS gates dissipate about 7 nW and have a propagation delay of 1.5 s. Fairchild's advanced Schottky gates use F to indicate the subfamily. 74F04 is a Fairchild advanced Schottky TTL quad 2-input NAND gate.

6.5 TTL OPEN-COLLECTOR GATES

A special type of gate, called the open-collector gate, has a modified output circuit. The upper transistor of the totem-pole pair has been omitted so that the output has no internal path to $+5$ V. When the output is driven LOW, the transistor turns on and connects the output X to ground through a saturated transistor. When the output is driven into the HIGH state, X is no longer connected to ground, nor is it connected to $+5$ V since that path no longer exists. The output enters a high impedance state, "HiZ," in which the gate has no influence on the output at all. The output is floating. Open-collector outputs can be connected together since there is no danger of the power supply being shorted. Figure 6-20 shows the output circuits

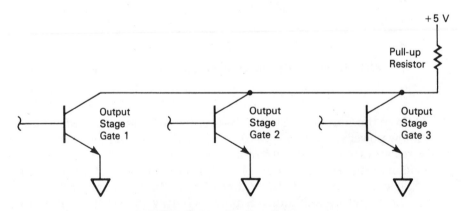

FIGURE 6-20 Open-collector gates

of three open-collector gates with outputs connected together. The outputs are tied to a common resistor that is connected to $+5$ V.

The resistor is called a "pull-up" resistor since it supplies the output with a path to $+5$ V and "pulls" the output up to a one level. If the output of any of the gates goes LOW, then the output X goes LOW. Conventional current flows from $+5$ V through the pull-up resistor, and through the saturated transistor to ground. Most of the 5 V is dropped across the pull-up resistor.

Figure 6-21 shows three open-collector inverters with outputs tied together. When A, B, and C are all LOW, the outputs are not connected to ground internally. The 1-kΩ resistor pulls the output X up to a one level. If any of the inputs goes HIGH, the corresponding output transistor turns on and pulls X down to a LOW level. The truth table for the circuit is also shown in Figure 6-21. The truth table is identical to that of a three-input NOR gate. This configuration is called a *wired-NOR* or a *dot-NOR* circuit.

Open-collector outputs are denoted in IEC and IEEE symbols by the symbol on each output.

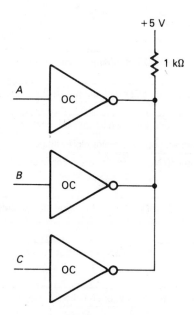

Inputs			Output
A	*B*	*C*	*X*
0	0	0	1
0	0	1	0
0	1	0	0
0	1	1	0
1	0	0	0
1	0	1	0
1	1	0	0
1	1	1	0

FIGURE 6-21   Open-collector inverter and truth table

Open-collector gates that are available for use are listed in Table 6-1.

TABLE 6-1   Open-collector gates

Device Number	Description
5401/7401 54ALS01/74ALSO1 54L01/74L01 54LS01/74LS01 5403/7403 54L03/74L03 54LS03/74LS03 54ALS03/74ALS03 54S03/74S03	Quad 2-input NAND gates with open-collector outputs
5405/7405 54L05/74L05 54ALS05/74ALS05 54S05/74S05	Hex inverters with open-collector outputs
5406/7406	Hex inverter buffers with high-voltage open collectors
5407/7407	Hex buffers with high-voltage open-collector outputs
5409/7409 54L09/74L09 54LS09/74LS09 54ALS09/74ALS09 54S09/74S09	Quad 2-input AND with open-collector outputs
54LS12/74LS12 54ALS12/74ALS12	Triple 3-input NAND with open-collector outputs
54LS15/74LS15 54ALS15/74ALS15	Triple 3-input AND gates with open-collector outputs
5416/7416	Hex inverter buffers with open-collector high-voltage outputs
5417/7417	Hex buffers with open-collector high-voltage outputs
54ALS22/74ALS22 54LS22/74LS22 54S22/74S22	Dual 4-input NAND gates with open-collector outputs
5438/7438 54ALS38/74ALS38 54LS38/74LS38	Quad 2-input NAND buffers with open-collector outputs
54LS136/74LS136 54AS136/74AS136 54S136/74S136	Quad exclusive-OR with open-collector outputs
54LS266/75LS266	Quad exclusive-NOR with open-collector outputs

Example: Draw the IEC logic symbol for 74LS136 quad exclusive-OR with open-collector outputs.

Solution:

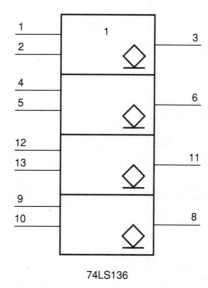

74LS136

The 1 indicates that exactly 1 input must be HIGH for the output to go into the high-impedance state.

6.6 OPEN-COLLECTOR APPLICATIONS

The 7406, 7407, 7416, and 7417 are open-collector gates with high-voltage outputs. Although the ICs themselves operate at 5 V, the open-collector outputs can be pulled up to a higher voltage; 30 V for the 06 and 07, and 15 V for the 16 and 17. The 7406 and 7416 invert the input signal and the 7407 and 7417 do not. Along with the higher voltages on the outputs, these gates can sink more low level output current than the totem-pole output gates can. The 5406, 07, 16, and 17 gates can sink 30 mA and the 7406, 07, 16, and 17 can sink 40 mA. These open-collector gates are used in high-voltage applications, and for tying outputs together.

Figure 6-22 shows a 7406 hex inverter with outputs pulled-up to 30 V through a 1-kΩ resistor. When the output goes LOW about 30 mA will be sinked by the gate. Since this is less than the 40 mA maximum listed by the specs, the output voltage will not rise above 0.4 V. This IC will be used in later chapters to step-up to higher supply voltages, and in this chapter, to interface TTL with CMOS.

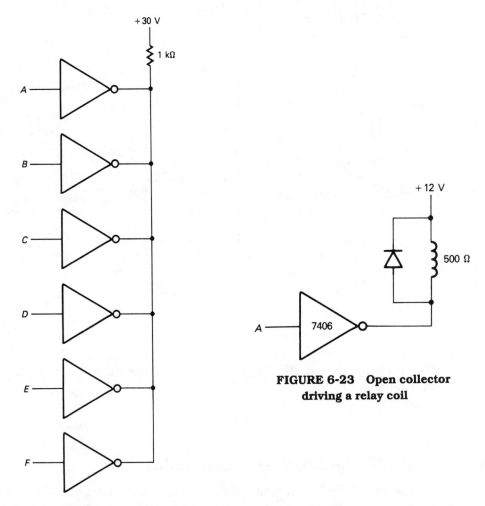

FIGURE 6-23   Open collector driving a relay coil

FIGURE 6-22   7406 open-collector inverter

Figure 6-23 shows a 7406 open-collector high-voltage inverter driving a 12-V, 500-Ω relay coil. When the output of the inverter goes LOW, approximately 24 mA flows through the coil and the contacts are activated. The 7406 can sink up to 40 mA. The diode clamps the reverse voltage that is induced across the coil when the output goes HIGH and current through the coil is interrupted.

6.7   CMOS TRANSISTOR REVIEW

To understand CMOS specifications a review of enhancement mode N-channel and P-channel MOS transistors is helpful.

The supply voltages for CMOS ICs are often called V_{DD} for drain voltage and V_{SS} for source voltage. Sometimes supply voltages are called V_{CC} for collector voltage and ground as they were in TTL. V_{DD} can range from 3 V to 15 V above V_{SS}. V_{SS} is usually 0 V.

The symbols for the N-channel and P-channel enhancement mode transistors are shown in Figure 6-24. Channel refers to the path through the transistor from drain to source. The symbol shows the channel broken into three parts.

The channel has to be completed or "enhanced" for conduction through the transistor to take place. For an N-channel device the drain and source are N-type material. The substrate is P-type. Note that the arrow points from the P-type substrate toward the N-type channel. The gate is isolated from the channel by a thin layer of silicon dioxide insulator. The gate, channel, and insulator form a small capacitor. This capacitive input determines many of the characteristics of CMOS ICs. If the substrate and source are connected to ground and the drain to a positive voltage as shown in Figure 6-25, then the gate can control the amount of current that flows through the channel.

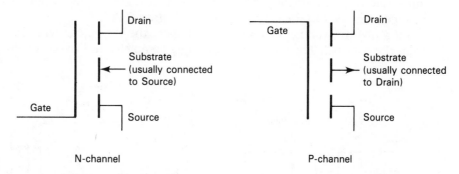

FIGURE 6-24 Enhancement mode MOS schematic symbols

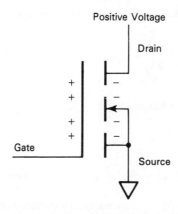

FIGURE 6-25 N-channel MOS

If the gate is held at a voltage near ground, the channel remains incomplete and only leakage current flows through the channel. If a positive voltage is applied to the gate, Figure 6-25, free electrons in the P-type substrate are attracted to the channel. The N-channel is completed or enhanced and conventional current can flow from V_{DD} through the channel to ground. As the voltage on the gate becomes more positive, the number of free electrons that are attracted into the channel region increases and the amount of drain current that can flow through the transistor increases.

For digital applications the gate inputs are either near V_{DD} for a 1 or near V_{SS} for a 0. The transistor is either completely enhanced (saturated) or off.

The symbol for a P-channel enhancement mode transistor (Figure 6-24) differs from the N-channel in two respects. The arrow on the substrate lead points away from the P-type channel toward the N-type substrate, and the gate is drawn from top to bottom. The drain and source are P-type material.

If the substrate and drain are connected to a positive voltage as shown in Figure 6-26, and the source is grounded, then the gate can control the amount of current that flows through the transistor. To enhance the channel, P-type carriers in the substrate must be attracted into the channel region. This happens when a low voltage is applied to the gate as shown in Figure 6-26.

If a positive voltage is applied to the gate, the P-type carriers are driven away from the channel. The channel is no longer complete, and the transistor turns off.

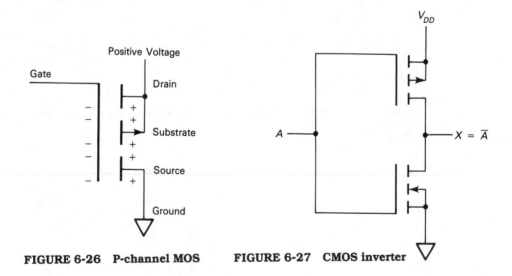

FIGURE 6-26   P-channel MOS   FIGURE 6-27   CMOS inverter

6.8   CMOS

CMOS stands for complementary metal-oxide semiconductor. Complementary means that a P-channel transistor and an N-channel transistor work together in a totem-pole arrangement as shown in Figure 6-27. Metal-oxide refers to the silicon dioxide layer between the gate and channel.

In Figure 6-27, when A is HIGH, the N-channel on the bottom is enhanced and the output X is connected to ground through a completed channel. The P-channel MOS on the top is turned off. When A is LOW the P-channel is enhanced and the N-channel is turned off. X is connected to V_{DD} through the P-channel. These two transistors produce an inverter.

Figure 6-28 shows the simplicity of a four-input CMOS NAND gate. Each input controls a P-channel transistor and an N-channel transistor. The four N-channel transistors are connected in series. All four have to be enhanced by a 1 for X to be connected to ground. All 1s in, 0 out. The four P-channel transistors are connected in parallel. If any one of the four is enhanced by a low level input then X is pulled up to V_{DD} through the enhanced channel. Any 0 in, 1 out.

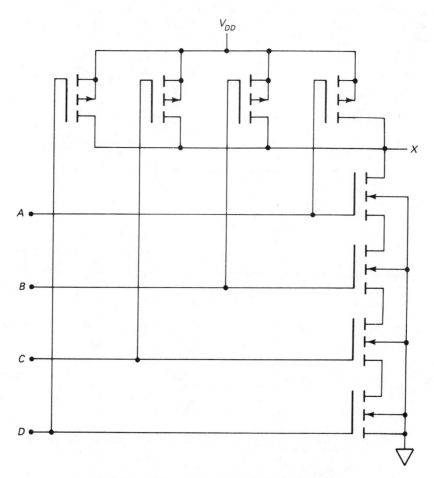

FIGURE 6-28   Four-input CMOS NAND gate

6.9 CMOS SPECIFICATIONS

Six series of CMOS ICs are 4000M, 4000C, 4000BM, 4000BC, 54C00 and 74C00. The M suffix indicates that the IC meets military specifications and can operate between $-55°C$ and $125°C$. The C suffix indicates that the IC meets commercial specifications and can operate between $-40°C$ and $+85°C$. The commercial series ICs draw more current from the power supply and dissipate more power than their military series counterparts. The input and output signals, though, have the same voltage, current, and noise specifications. The B suffix indicates that the IC has an additional buffering stage.

In CMOS, the high level is usually represented by a 1 and the low level by a zero. Therefore, CMOS is usually used as positive logic. Positive logic will be used in this text for CMOS.

Figure 6-29 is reproduced from National Semiconductor's CMOS Data Book. It lists the absolute maximum ratings and dc electrical characteristics for a 4001M quad two-input NOR gate and a 4011M quad two-input NAND gate. Limits for the parameters are listed for supply voltages of 5 V and 10 V for three temperature ranges, $-55°C$, $25°C$, and $125°C$. The limits listed for $25°C$ will be used in the following discussion.

DC Electrical Characteristics — CD4001M, CD4011M

Parameter	Conditions	−55°C Min.	−55°C Max.	25°C Min.	25°C Typ.	25°C Max.	125°C Min.	125°C Max.	Units
I_L Quiescent Device Current	$V_{DD} = 5.0V$		0.05		0.001	0.05		3.0	μA
	$V_{DD} = 10V$		0.1		0.001	0.1		6.0	μA
P_D Quiescent Device Dissipation/Package	$V_{DD} = 5.0V$		0.25		0.005	0.25		15	μW
	$V_{DD} = 10V$		1.0		0.01	1.0		60	μW
V_{OL} Output Voltage Low Level	$V_{DD} = 5.0V, V_I = V_{DD}, I_O = 0A$		0.05		0	0.05		0.05	V
	$V_{DD} = 10V, V_I = V_{DD}, I_O = 0A$		0.05		0	0.05		0.05	V
V_{OH} Output Voltage High Level	$V_{DD} = 5.0V, V_I = V_{SS}, I_O = 0A$	4.95		4.95	5.0		4.95		V
	$V_{DD} = 10V, V_I = V_{SS}, I_O = 0A$	9.95		9.95	10		9.95		V
V_{NL} Noise Immunity (All Inputs)	$V_{DD} = 5.0V, V_O = 3.6V, I_O = 0A$	1.5		1.5	2.25		1.4		V
	$V_{DD} = 10V, V_O = 7.2V, I_O = 0A$	3.0		3.0	4.5		2.9		V
V_{NH} Noise Immunity (All Inputs)	$V_{DD} = 5.0V, V_O = 0.95V, I_O = 0A$	1.4		1.5	2.25		1.5		V
	$V_{DD} = 10V, V_O = 2.9V, I_O = 0A$	2.9		3.0	4.5		3.0		V
I_DN Output Drive Current N-Channel (4001)	$V_{DD} = 5.0V, V_O = 0.4V, V_I = V_{DD}$	0.5		0.40	1.0		0.28		mA
	$V_{DD} = 10V, V_O = 0.5V, V_I = V_{DD}$	1.1		0.9	2.5		0.65		mA
I_DP Output Drive Current P-Channel (4001)	$V_{DD} = 5.0V, V_O = 2.5V, V_I = V_{SS}$	−0.62		−0.5	−2.0		−0.35		mA
	$V_{DD} = 10V, V_O = 9.5V, V_I = V_{SS}$	−0.62		−0.5	−1.0		−0.35		mA
I_DN Output Drive Current N-Channel (4011)	$V_{DD} = 5.0V, V_O = 0.4V, V_I = V_{DD}$	0.31		0.25	0.5		0.175		mA
	$V_{DD} = 10V, V_O = 0.5V, V_I = V_{DD}$	0.63		0.5	0.6		0.35		mA
I_DP Output Drive Current P-Channel (4011)	$V_{DD} = 5.0V, V_O = 2.5V, V_I = V_{SS}$	−0.31		−0.25	−0.5		−0.175		mA
	$V_{DD} = 10V, V_O = 9.5V, V_I = V_{SS}$	−0.75		−0.6	−1.2		−0.4		mA
I_I Input Current					10				pA

Note 1: "Absolute Maximum Ratings" are those values beyond which the safety of the device cannot be guaranteed. Except for "Operating Temperature Range" they are not meant to imply that the devices should be operated at these limits. The table of "Electrical Characteristics" provides conditions for actual device operation.

FIGURE 6-29 CMOS NOR gates and NAND gates

V_{OL}, output voltage low level, is listed as a maximum of 0.05 V. Note that the conditions list I_O, output current, as 0 A. V_{OH}, output voltage high level, is listed as a minimum of 4.95 V when $V_{DD} = 5.0$ V and 9.95 V when $V_{DD} = 10$ V. Both high and low level output voltages are maintained within 0.05 V of the supply voltages.

V_{NL}, noise immunity (low level), with V_{DD} equal to 5.0 V, is given as 1.5 V minimum. The manufacturer guarantees that 0.0 V to 1.5 V on the inputs will be taken as a zero.

V_{NH}, noise immunity (high level), with V_{DD} equal to 5.0 V is also listed as 1.5 V minimum. This means that a 1-level input can range from V_{DD} down to $V_{DD} - 1.5$ V or 3.5 V. The manufacturer guarantees that 3.5 V to 5.0 V on the input will be taken as a 1 level.

Another way to look at noise specifications is to consider the noise margin as we did in the TTL section. CMOS manufacturers guarantee that for an output to remain within one-tenth of V_{DD} from V_{DD} or ground, the input can be as much as one-tenth of V_{DD} plus one volt away from V_{DD} or ground. In other words, CMOS has a noise margin of one volt.

For example, suppose that $V_{DD} = 5.0$ V. One-tenth of V_{DD} is 0.5 V. For a high level output to remain within 0.5 V of V_{DD}, 4.5 V or greater, the input can be as much as 0.5 V plus 1.0 V away from ground or V_{DD}. The high level input can drop as low as 3.5 V and a low level input can rise as high as 1.5 V. These results are consistent with those obtained from the noise immunity specs.

Figure 6-30 summarizes noise margin for the range of possible V_{DD} voltages. The noise margin is one volt for all supply voltages.

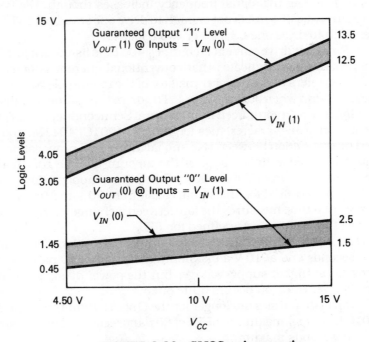

FIGURE 6-30 CMOS noise margin

I_I, input current, is the high level or low level input current drawn by a gate. It is listed as 10 pA, typically. CMOS gates have an extremely high input impedance and draw virtually no current in a static situation. The input impedance is equivalent to 10^{12} Ω in parallel with a 5-pF capacitor. When the gate changes states it conducts current momentarily to charge or discharge the input capacitors formed by the gate, channel and silicon dioxide insulator. As the frequency of the signal increases, these input capacitors charge and discharge at a faster rate and the input impedance decreases.

P_D, quiescent device dissipation per package, lists the maximum power dissipated with 10 V applied as 1.0 μW and 0.25 μW with 5 V applied. CMOS ICs dissipate more power as the power supply voltage increases. This low power consumption makes CMOS attractive for use in battery powered circuits. As the gate changes state both upper and lower transistors conduct at the same time and current is drawn from the power supply. As the frequency of the signal increases, the power consumed by the IC increases. Above 1 MHz, CMOS consumes more power than low-power Schottky.

I_L, quiescent device current, lists the maximum current drawn from the power supply by a 4001M, or 4011M IC as 0.05 μA with 5 V applied. A TTL NAND gate IC draws a maximum of 22 mA, about 440,000 times more current.

I_DN, output drive current N-channel (4001), is listed as a minimum 0.4 mA for V_{DD} = 5.0 V. N-channel refers to the lower transistor of the totem-pole pair. The 4001 can sink at least 0.4 mA while outputting a 0. Since the input current I is very low at low frequencies, a CMOS NOR gate can drive an unlimited number of other CMOS gates. As the signal frequency increases though, the fan-out decreases. A drive current of 0.4 mA is not enough to drive one standard TTL load (1.6 mA) but it is enough to drive one LS load or two L loads.

I_DP, output drive current P-channel (4001), is listed as a minimum of -0.5 mA. The negative sign indicates that conventional current is flowing out of the gate. P-channel refers to the upper transistor of the totem-pole pair. The 4001 can source at least 0.5 mA when outputting a 1. The drive capabilities are discussed under I_DN.

Figure 6-31, reproduced from National Semiconductor's CMOS Data Book, lists the ac electrical characteristics for 4001M, 4001C, 4011M, and 4011C.

The distinction between rise and fall times and propagation delays is shown in Figure 6-32. Rise time, t_{RISE}, is the amount of time that it takes for an input waveform to rise from 10% to 90% of the power supply voltage V_{DD}. Fall time, t_{FALL}, is the time required for the signal to fall from 90% to 10% of the applied voltage. Propagation time measures the lag between a change in input and a corresponding change in output as was defined for TTL.

The 4001M at 5 V has a maximum propagation delay HIGH-to-LOW of 50 nanoseconds and at 10 V a maximum delay t_{PHL} of 40 nanoseconds. CMOS speed increases at higher supply voltages but the power dissipation also increases. This speed/power trade-off occurs quite often.

Propagation times are longer for the C series than they are for the M series. The 4001C has a t_{PHL} maximum of 5 V of 80 nanoseconds. The military series is faster at 50 nanoseconds maximum.

AC Electrical Characteristics
$T_A = 25°C$, $C_L = 15\,pF$, and input rise and fall times = 20 ns.
Typical temperature coefficient for all values of $V_{DD} = 0.3\%/°C$

	Parameter	Conditions	Min.	Typ.	Max.	Units
	CD4001M					
t_{PHL}	Propagation Delay Time High to Low Level	$V_{DD} = 5.0V$		35	50	ns
		$V_{DD} = 10V$		25	40	ns
t_{PLH}	Propagation Delay Time Low to High Level	$V_{DD} = 5.0V$		35	65	ns
		$V_{DD} = 10V$		25	40	ns
t_{THL}	Transition Time High to Low Level	$V_{DD} = 5.0V$		65	125	ns
		$V_{DD} = 10V$		35	70	ns
t_{TLH}	Transition Time Low to High Level	$V_{DD} = 5.0V$		65	175	ns
		$V_{DD} = 10V$		35	75	ns
C_{IN}	Input Capacitance	Any Input		5.0		pF
	CD4001C					
t_{PHL}	Propagation Delay Time High to Low Level	$V_{DD} = 5.0V$		35	80	ns
		$V_{DD} = 10V$		25	55	ns
t_{PLH}	Propagation Delay Time Low to High Level	$V_{DD} = 5.0V$		35	120	ns
		$V_{DD} = 10V$		25	65	ns
t_{THL}	Transition Time High to Low Level	$V_{DD} = 5.0V$		65	200	ns
		$V_{DD} = 10V$		35	115	ns
t_{TLH}	Transition Time Low to High Level	$V_{DD} = 5.0V$		65	300	ns
		$V_{DD} = 10V$		35	125	ns
C_{IN}	Input Capacitance	Any Input		5.0		pF

AC Electrical Characteristics
$T_A = 25°C$, $C_L = 15\,pF$, and input rise and fall times = 20 ns.
Typical temperature coefficient for all values of $V_{DD} = 0.3\%/°C$

	Parameter	Conditions	Min.	Typ.	Max.	Units
	CD4011M					
t_{PHL}	Propagation Delay Time High to Low Level	$V_{DD} = 5.0V$		50	75	ns
		$V_{DD} = 10V$		25	40	ns
t_{PLH}	Propagation Delay Time Low to High Level	$V_{DD} = 5.0V$		50	75	ns
		$V_{DD} = 10V$		25	40	ns
t_{THL}	Transition Time High to Low Level	$V_{DD} = 5.0V$		75	125	ns
		$V_{DD} = 10V$		50	75	ns
t_{TLH}	Transition Time Low to High Level	$V_{DD} = 5.0V$		75	100	ns
		$V_{DD} = 10V$		40	60	ns
C_{IN}	Input Capacitance	Any Input		5.0		pF
	CD4011C					
t_{PHL}	Propagation Delay Time High to Low Level	$V_{DD} = 5.0V$		50	100	ns
		$V_{DD} = 10V$		25	50	ns
t_{PLH}	Propagation Delay Time Low to High Level	$V_{DD} = 5.0V$		50	100	ns
		$V_{DD} = 10V$		25	50	ns
t_{THL}	Transition Time High to Low Level	$V_{DD} = 5.0V$		75	150	ns
		$V_{DD} = 10V$		50	100	ns
t_{TLH}	Transition Time Low to High Level	$V_{DD} = 5.0V$		75	125	ns
		$V_{DD} = 10V$		40	75	ns
C_{IN}	Input Capacitance	Any Input		5.0		pF

FIGURE 6-31   Alternating current electrical characteristics

Due to the capacitor formed by the gate and channel, it takes time for the input voltage to rise and fall. The transition time HIGH-to-LOW, fall time, is 125 nanoseconds maximum for a 4001M at 5 V and 200 nanoseconds for a 4001C. These long transition times limit the operation of CMOS to less than 2 MHz.

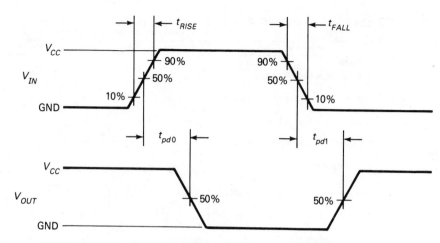

FIGURE 6-32   CMOS rise time, fall time and propagation delays

6.10   54C/74C SERIES

The C in the device number of a 74CXX series component stands for CMOS. These ICs have the same pinout and function as the corresponding 74XX TTL device. For example, the 74C30 is a CMOS 8-input NAND and the 7430 is a TTL 8-input NAND. Their pinouts are the same.

The 54C series can operate between $-55°C$ and $125°C$, while the 74C series can operate between $-40°C$ and $+85°C$. In general, the 54C/74C series is 50% faster and can sink 50% more current than the corresponding IC in the 4000A series.

The 74HC series is composed of high-speed CMOS ICs that are pin-compatible with TTL 74LS devices. For these ICs V_{DD} can range from 2 V to 6 V. A LOW input can range from 0 V to 30% of the supply voltage, and a HIGH input can range from 70% of V_{CC} up to V_{CC}. Operating at 5 V, a LOW input can range from 0 V to 1.5 V, and a HIGH input can range from 3.5 V to V_{CC}.

The 74HCT series is composed of high-speed CMOS ICs with logic levels compatible to TTL 74LS devices. A LOW input can range from 0 V to 0.8 V, and a HIGH input can range from 2 V to V_{CC}. The power supply is limited to 4.5 to 5.5 V.

Figure 6-33 compares three of TTL's subfamilies, LS, ALS, and FAST, to the CMOS families 4000 and HC. Note that TTL FAST is the fastest with a maximum propagation delay of 5 nanoseconds compared to 250 nanoseconds for the 4000 family.

CMOS HC and TTL LS have propagation delays of the same magnitude. Of the families listed in Figure 6-33 FAST can operate the fastest, with clock frequencies of 125 MHz, and CMOS is by far the slowest, with 4 MHz. FAST draws 1.1 mA per gate from the supply, while CMOS draws only 0.1 nanoamp at rest.

The HC subfamily has enough current drive capability to drive 10 TTL LS loads.

	TTL			CMOS		
	LS	ALS	FAST	4000	HC	Units
Maximum propagation delay	15	10	5	250	15	NS
Quiescent supply current per gate	0.4	0.2	1.1	0.0001	0.0005	mA
Maximum clock frequency	33	35	124	4	40	MHz

FIGURE 6-33   Speed-power comparison

6.11   INTERFACING TTL TO CMOS

To interface TTL to CMOS operating at 5 V, care must be taken that high level TTL outputs are high enough to be recognized by the following CMOS ICs. The TTL output should be pulled up through a 10-kΩ external resistor as shown in Figure 6-34.

To interface TTL with CMOS that is operating at high voltage levels, one of the high-voltage open-collector gates can be used as shown in Figure 6-35. The open-collector output is pulled up to the operating voltage of the CMOS gate.

In either case, TTL is capable of sinking sufficient current to drive an unlimited number of CMOS gates at low frequency.

Figure 6-36 shows a CMOS NAND gate that could be operated from 5 V to 18 V. The higher voltages on its output are not compatible with TTL inputs. These levels present no problem to the 4049 hex inverting buffer. The 4049 can operate at a supply voltage of 5 V and handle input voltages up to 15 V. The outputs are TTL compatible. The 4049 sinks sufficient current to drive two standard TTL gates; in

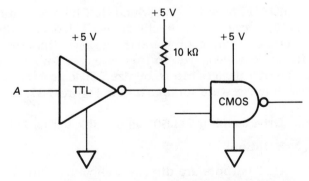

FIGURE 6-34   TTL to 5-volt CMOS

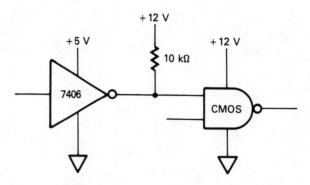

FIGURE 6-35   TTL to higher voltage CMOS

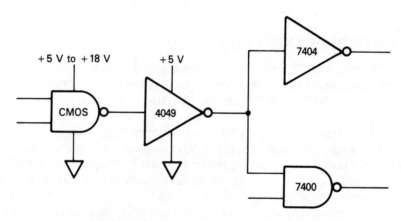

FIGURE 6-36   CMOS to TTL interface

this case a 7404 and a 7400. The 4050 hex noninverting buffer can be used if inversion is not required.

Since CMOS HC operating at 5 V and HCT have outputs that are TTL compatible, HC and HCT devices can directly drive a TTL device. However, TTL HIGH level outputs can range down to 2.5 V and this is lower than the acceptable 3.5 V input that the HC devices recognize. In this case a pull-up resistor is required for the interface. The HCT devices can recognize TTL logic levels and interface directly to them. These interfaces are summarized in Figure 6-37.

Example: Drive as many 74LS00 as possible with a 74HCT00.

Solution:

HCT outputs are directly compatible with TTL devices. No pull-up resistor is needed. A 74HCT00 has a fan-out of 10 LS gates.

HCT TO LSTTL INTERFACING

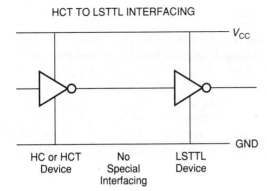

HC or HCT Device	No Special Interfacing	LSTTL Device

LSTTL TO HC INTERFACING

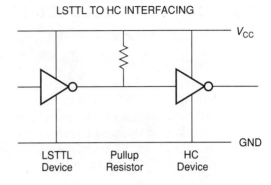

LSTTL Device	Pullup Resistor	HC Device

LSTTL TO HCT INTERFACING

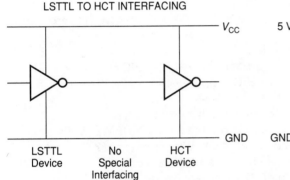

LSTTL Device	No Special Interfacing	HCT Device

LSTTL TO LOW-VOLTAGE HSCMOS

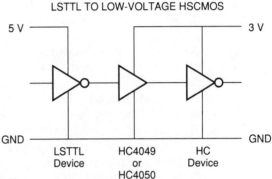

LSTTL Device	HC4049 or HC4050	HC Device

FIGURE 6-37 Interfacing TTL to HC and HCT devices (Courtesy of Motorola, Inc.)

Example: Drive as many 74HC00 gates operating at 5 V as possible with a 74LS00.

Solution:

A 74LS00 high level output voltage V_{OH} can range down to 2.4 V. A 74HC00 high level input V_{IH} has minimum of 3.15 V. The output of the 74LS00 needs to be "pulled-up" with an external pull-up resistor. The maximum input current drawn by the 74HC00 is 1 μA. The low level output current, I_{OL}, for 74LS00 is 8 mA, but the high level output current is only 0.4 mA. Using 0.4 mA to calculate fan-out still allows us to drive

$$\text{fan-out} = \frac{I_{OH}}{I_{IH}} = \frac{0.4 \text{ mA}}{1 \text{ μA}} = 400$$

By using a pull-up resistor (1 kΩ or so), 400 or more 74HC00s may be driven by a 74LS00.

6.12 EMITTER COUPLED LOGIC (ECL)

ECL circuits are designed so that the transistors do not saturate when they turn on. Switching times are decreased and propagation delays of a couple of nanoseconds result. ECL can be operated at frequencies up to 200 MHz and higher. ECL consumes more than twice as much power as TTL. ECL operates with supply voltages of −5.2 V and ground. The maximum low level output is −1.63 V and the minimum high level output is −0.980 V. In ECL, the low level is usually represented by a 1. Therefore, ECL is usually used as negative logic. Since ECL is not compatible with TTL levels, interface gates such as the 10124 and 10125 must be used to interface ECL to TTL or CMOS.

Exercise 6

1. What is the minimum 1 level output on a 74LS08?
2. What is the maximum current that a 74LS08 can sink in the zero output state?
3. How many 5475s can a 7404 drive?
4. What is the maximum current drawn by a 74LS83A IC?
5. What is the maximum LOW-to-HIGH propagation delay time of a 5421? (Assume a load of 50 pF, 2 kΩ.)
6. At 15 V supply, how much current can a 4006BM sink?
7. What is the minimum 1 level output voltage for a 4006BM? (5-V supply)
8. Can a 4006BM (5-V supply) drive a 7400? Why or why not?
9. How much current can a 4049 sink on a LOW level output? (V_{DD} = 5 V)
10. Interface a 7432 OR gate to a 4081 AND gate operating at 5 V.
11. Interface a 7404 inverter to a 4071 OR gate operating at 5 V.
12. Use a 7406 to interface a 7408 AND gate to a 4081 AND gate operating at 12 V.
13. Use a 7406 to interface a 7408 AND gate to a 74C08 AND gate operating at 12 V.
14. Use a 4049 to interface a 4081 AND gate operating at 12 V to a 7404 inverter.
15. Use a 4050 to interface a 74C00 NAND gate operating at 15 V to a 7400 NAND gate.
16. Use a 4049 to interface a 4001 operating at +12 V to as many 7404s as possible.
17. Use a 4049 to interface a 74LS00 to a 74HC06 operating at 3 V.
18. What is the acceptable range of supply voltages for an HC device?
19. What is the acceptable range of power supply voltages for HCT devices?
20. Can a TTL device directly drive an HCT device?
21. Can a TTL device directly drive an HC device operating at 5 V?
22. What does HCT stand for?
23. How many LS devices can a CMOS HC gate drive?
24. Rank these families according to power consumption from least power to most power consumed: CMOS HC, CMOS, TTL, LS, ALS, FAST, S.

25. Rank these families according to speed, in descending order: CMOS HC, CMOS, TTL, LS, ALS, FAST, S.
26. What is the maximum propagation delay for a 4001 NOR gate?
27. At quiescence, how much current does a 4001 NOR gate draw?
28. What is the typical low level drive current for a 4011 NAND gate?
29. What is the maximum low level output voltage for a 4001 quad 2-input NOR gate?
30. Draw the schematic of an open-collector inverter driving a relay coil connected to +20 V.
31. Draw the schematic diagram of three open-collector inverters, 7406s, tied to a pull-up resistor of 5 kΩ and a power supply of 20 V.
32. How does a totem-pole gate differ from an open-collector gate?

6 Specifications and Open-Collector Gates

OBJECTIVES

After completing this lab, you should be able to:

- measure and graph input current versus input voltage for a TTL gate.

- measure and graph output current versus output voltage for a TTL gate.

- measure and graph source current versus applied frequency for a CMOS IC.

- construct a 6-input wired NOR circuit.

COMPONENTS NEEDED

1	7404 IC
1	7406 IC
1	74C14 IC
1	4001 IC
1	1-kΩ resistor
1	1-kΩ potentiometer
1	100-Ω resistor

PROCEDURE

1. Set up the circuit shown and find the unknown values in the table. Then draw a graph of input voltage versus input current. Does the maximum input current measured fall within the IC specifications?

Input Voltage	Current
5.0 V	
4.5 V	
4.0 V	
3.5 V	
3.0 V	
2.5 V	
2.0 V	
1.5 V	
1.0 V	
0.5 V	
0.0 V	

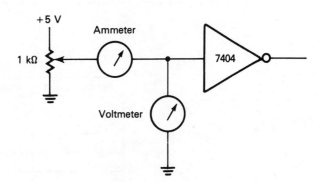

2. Connect the circuit shown and find the unknown values in the table. Graph the output voltage versus the output current. Does the output voltage fall within the IC specifications?

Output Current	Ouput Voltage
4 mA	
6 mA	
8 mA	
10 mA	
12 mA	
14 mA	
16 mA	
18 mA	
20 mA	
25 mA	
30 mA	
40 mA	

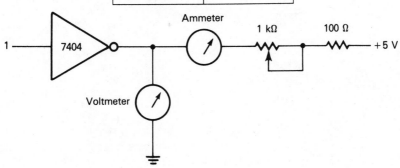

3. Construct the circuit shown and find the unknown values in the table. Graph the output current versus the output voltage.

Output Current	Ouput Voltage
	5.0 V
	4.5 V
	4.0 V
	3.5 V
	3.0 V
	2.4 V
	2.0 V

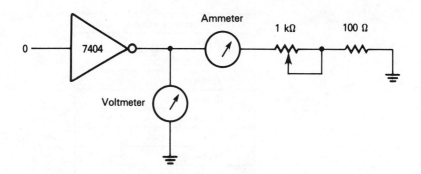

4. Construct the circuit shown and find the unknown values in the table. Graph the input frequency versus V_{DD} current. Use the scope to measure the input frequency.

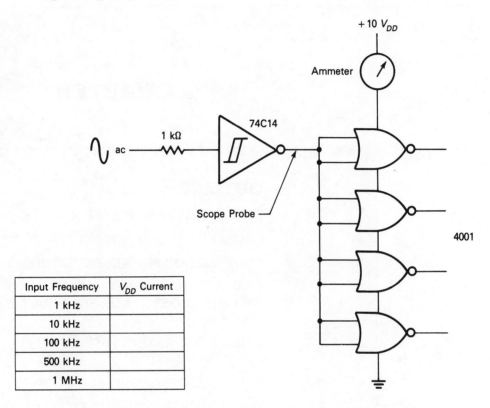

Input Frequency	V_{DD} Current
1 kHz	
10 kHz	
100 kHz	
500 kHz	
1 MHz	

5. Use a 7406 open-collector gate to construct a 6-input wired NOR gate.

CHAPTER 7

Flip-Flops

OBJECTIVES

After completing this chapter, you should be able to:

- explain the operation of a SET-RESET flip-flop.

- use a crossed NAND or crossed NOR flip-flop as a debounce switch.

- explain the operation of a gated SET-RESET flip-flop.

- explain the operation of a transparent D flip-flop.

- explain the operation of a D flip-flop used as a latch.

- explain the operation of a master-slave D flip-flop.

7.1 INTRODUCTION TO FLIP-FLOPS

A *flip-flop* is a digital circuit that has two outputs Q and $\overline{Q}$, which are always in the opposite states. If Q is 1 then $\overline{Q}$ is 0 and the flip-flop is said to be set, on or preset. If Q is 0 then $\overline{Q}$ is 1 and the flip-flop is said to be reset, off or cleared. There are several types of flip-flops and the control inputs vary with each type. The logic levels on the flip-flop's inputs will determine the state of the Q and $\overline{Q}$ outputs according to the truth table for that type of flip-flop.

Unlike the gates studied up to this point, the flip-flop can in some states maintain its output state (on or off) after the input signals which produced the output state change. Thus, the flip-flop can store a bit of information or one place of a larger binary number. There are many other uses for flip-flops as we will see in the next few chapters.

7.2 CROSSED NAND SET-RESET FLIP-FLOPS

A *SET-RESET flip-flop* is a digital circuit whose output is set by the $\overline{\text{SET}}$ input, but can only be reset by the $\overline{\text{RESET}}$ input. The two crossed NAND gates in Figure 7-1 form a $\overline{\text{SET}}$-$\overline{\text{RESET}}$ flip-flop.

The inputs $\overline{\text{SET}}$ and $\overline{\text{RESET}}$ are active LOW. The $\overline{\text{SET}}$ input must be a 0 to set the Q output to a 1. Notice the complement bar over the $\overline{\text{SET}}$ and $\overline{\text{RESET}}$ inputs. This means they are active LOW inputs. The outputs of a flip-flop are usually labeled Q and $\overline{Q}$ meaning that if Q is a 1, $\overline{Q}$ is a 0 and vice versa.

When the $\overline{\text{SET}}$ input goes to a 0 and the $\overline{\text{RESET}}$ input is held at 1, the output of the crossed NAND flip-flop will have the configuration shown in Figure 7-2, because any 0 into a NAND gate makes its output a 1. This will set the Q output to 1 and the $\overline{Q}$ output to a 0.

If the $\overline{\text{SET}}$ input goes to a 1 and the $\overline{\text{RESET}}$ remains at 1, the output does not change, as shown in Figure 7-3. This is because the outputs are fed back to the input of the opposite gate, which makes them retain their original output configuration.

To reset the flip-flop, you must bring the $\overline{\text{RESET}}$ input to a 0 and keep the $\overline{\text{SET}}$ at 1 as shown in Figure 7-4. As can be seen, the $\overline{\text{SET}}$ input cannot reset the Q output to a 0. This can only be done by bringing the $\overline{\text{RESET}}$ input to a 0, while keeping the $\overline{\text{SET}}$ input at 1. The same thing is true for the $\overline{\text{RESET}}$ input. It cannot set or bring the Q output to a 1, only reset it or bring the Q output to a 0.

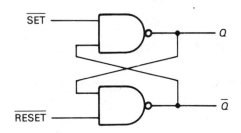

**FIGURE 7-1   Crossed NAND
SET-RESET flip-flop**

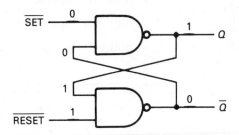

**FIGURE 7-2   Setting the _Q_ output for
a crossed NAND SET-RESET flip-flop**

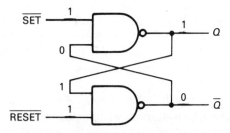

**FIGURE 7-3   The <u>unchanged</u> state for
a crossed NAND SET-RESET flip-flop**

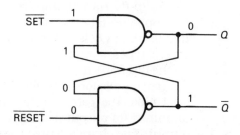

**FIGURE 7-4   Resetting the _Q_ output for
a crossed NAND SET-RESET flip-flop**

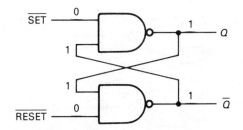

FIGURE 7-5 The unused state for a crossed NAND SET-RESET flip-flop

SET	RESET	$\overline{\text{SET}}$	$\overline{\text{RESET}}$	Q	$\overline{Q}$	
1	1	0	0	1	1	Unused State
1	0	0	1	1	0	
0	1	1	0	0	1	
0	0	1	1	Q	$\overline{Q}$	Unchanged State

FIGURE 7-6 Truth table for a crossed NAND $\overline{\text{SET}}$-$\overline{\text{RESET}}$ flip-flop

The only other possible input state not yet covered for the crossed NAND $\overline{\text{SET}}$-$\overline{\text{RESET}}$ flip-flop is 0 on both inputs, as shown in Figure 7-5. This is an unused state. We never want a flip-flop to have Q and $\overline{Q}$ with the same value.

The input that returns to 1 first will determine the resulting state of the flip-flop. The truth table for a $\overline{\text{SET}}$-$\overline{\text{RESET}}$ crossed NAND flip-flop is as shown in Figure 7-6.

7.3 CROSSED NOR SET-RESET FLIP-FLOPS

Figure 7-7 shows a *crossed NOR SET-RESET flip-flop*. Note that the inputs are not complemented; therefore they are active HIGH.

When the SET input goes to a 1, and the RESET remains at 0, then the Q output goes to a 1 state, as shown in Figure 7-8. Any 1 into a NOR gate will produce a 0 output.

When the SET input returns to 0, and the RESET is also 0, the outputs Q and $\overline{Q}$ do not change, as shown in Figure 7-9. This is because the outputs of the NOR gates are tied back to the opposite gates input. This keeps the gates from changing states.

To bring Q back to a 0, the RESET input must be made 1 while the SET input is held at 0. This is shown in Figure 7-10.

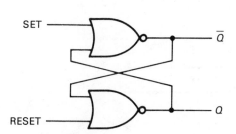

FIGURE 7-7 Crossed NOR SET-RESET flip-flop

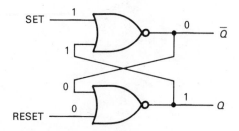

FIGURE 7-8 Setting the Q output for a crossed NOR SET-RESET flip-flop

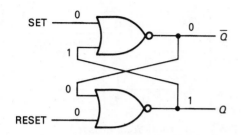

FIGURE 7-9   Unchanged state for a crossed NOR SET-RESET flip-flop

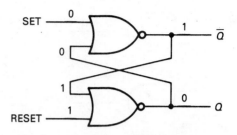

FIGURE 7-10   Resetting the Q output for a crossed NOR SET-RESET flip-flop

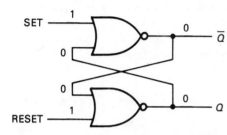

FIGURE 7-11   The unused state for a crossed NOR SET-RESET flip-flop

SET	RESET	$\overline{SET}$	$\overline{RESET}$	Q	$\overline{Q}$	
1	1	0	0	0	0	Unused State
1	0	0	1	1	0	
0	1	1	0	0	1	
0	0	1	1	Q	$\overline{Q}$	Unchanged State

FIGURE 7-12   Truth table for a crossed NOR SET-RESET flip-flop

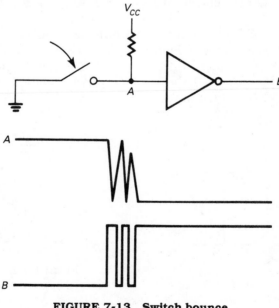

FIGURE 7-13   Switch bounce

The unused state for the crossed NOR SET-RESET flip-flop is where the SET is 1 and the RESET is 1. This is shown in Figure 7-11.

The first input to return to the 0 state will determine the output state of the Q and $\overline{Q}$ outputs. The truth table for the crossed NOR SET-RESET flip-flop is shown in Figure 7-12.

7.4 COMPARISON OF THE CROSSED NAND AND THE CROSSED NOR SET-RESET FLIP-FLOPS

When you compare the truth tables in Figure 7-6 for the crossed NAND $\overline{SET}$-$\overline{RESET}$ flip-flop, and Figure 7-12 for the crossed NOR SET-RESET flip-flop, one difference is the values of Q and $\overline{Q}$ in the unused state.

Another major difference between the two flip-flops is the SET and RESET inputs. The crossed NOR inputs are active HIGH and the $\overline{SET}$, $\overline{RESET}$ inputs on the crossed NAND gates are active LOW. This means that the crossed NOR flip-flop will change state when an input goes HIGH or 1 and the crossed NAND will change state when an input goes LOW or 0.

7.5 USING A $\overline{SET}$-$\overline{RESET}$ FLIP-FLOP AS A DEBOUNCE SWITCH

When a normal metal contact on a single-pole switch closes or opens, the contacts do not make and break the circuit smoothly. Instead they bounce, making and breaking contact many times before they finally come to rest. This happens very fast and can cause havoc with digital circuits which may be counting the number of switch closings that occur. Figure 7-13 shows a typical circuit with a single-pole switch and the corresponding waveforms generated when the switch is closed. If a counter had been clocked by the output of the inverter in Figure 7-13, it would have counted 3 pulses instead of 1.

Figure 7-14 shows how a single-pole double-throw switch can be used in conjunction with a $\overline{SET}$-$\overline{RESET}$ flip-flop to prevent switch bounce.

These circuits work because when the switch is moved the flip-flop outputs will remain unchanged until the center pole of the switch contacts the opposite pole of the switch, at which time the flip-flop changes state and remains there even though the switch bounces. Figure 7-15 shows the switch movement using a crossed NAND $\overline{SET}$-$\overline{RESET}$ flip-flop.

7.6 THE GATED SET-RESET FLIP-FLOP

Figure 7-16 shows a *crossed NAND gated SET-RESET flip-flop* and its truth table. There are two NAND gates which are used to gate the SET-RESET inputs to the $\overline{SET}$-$\overline{RESET}$ flip-flop. The clock input is used to enable or inhibit the two gates. If a 0 is put on the clock input, the output of the two NAND gates will be forced to a 1.

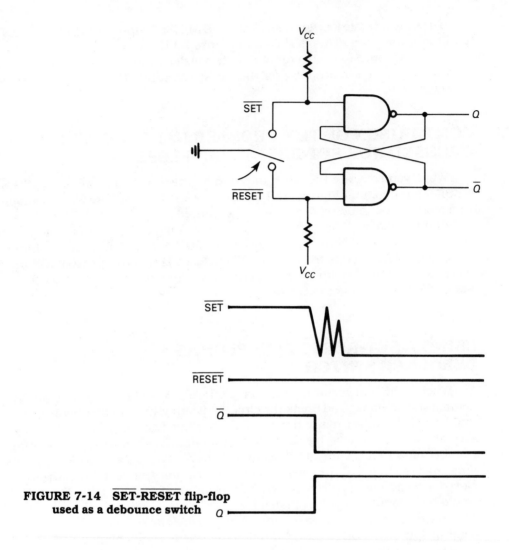

**FIGURE 7-14 $\overline{\text{SET}}$-$\overline{\text{RESET}}$ flip-flop
used as a debounce switch**

This places the crossed NAND $\overline{\text{SET}}$-$\overline{\text{RESET}}$ flip-flop in its remembering or un-changed state. Therefore, when the clock is 0, the flip-flop outputs cannot be changed. When the clock is made 1 the gates are enabled, or turned on, and the values of the inputs are passed through as their complements. This is shown in Figure 7-17.

Because the NAND gates invert the inputs, when the SET is 1 and RESET 0, Q is 1 and $\overline{Q}$ is 0. Also, when the SET is 0 and RESET is 1, Q is 0 and $\overline{Q}$ is 1. This means that when the clock is 1, the Q and $\overline{Q}$ outputs follow the values of the SET and RESET respectively.

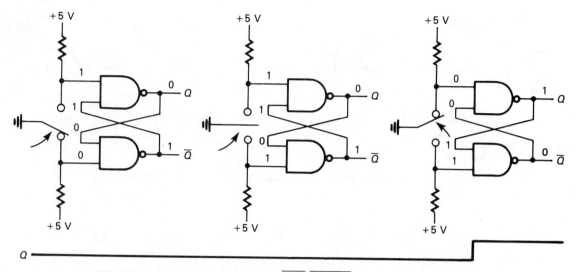

FIGURE 7-15   A crossed NAND $\overline{\text{SET}}$-$\overline{\text{RESET}}$ flip-flop changing states

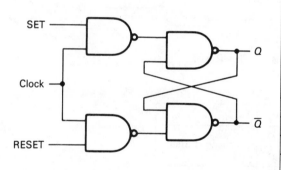

FIGURE 7-16   Gated SET-RESET flip-flop

Clock	SET	RESET	Q	$\overline{Q}$	
0	0	0	Q	$\overline{Q}$	
0	0	1	Q	$\overline{Q}$	
0	1	0	Q	$\overline{Q}$	Unchanged State
0	1	1	Q	$\overline{Q}$	
1	0	0	Q	$\overline{Q}$	
1	0	1	0	1	
1	1	0	1	0	
1	1	1	1	1	Unused State

7.7   THE TRANSPARENT *D* FLIP-FLOP

One problem with the gated NAND SET-RESET flip-flop is there can be 1 on the *Q* and a 1 on the $\overline{Q}$ when the SET and RESET inputs are both 1. This is the unused state, which should be avoided if possible. Also, it would be much more convenient if one input could SET and RESET the flip-flop. Both of these problems can be alleviated by placing an inverter between the SET and RESET inputs as shown in Figure 7-19. This makes a new input which we call the *D* input. Notice that the SET and RESET inputs can never be the same value because of the inverter. This means that the unused state could never exist. Also there is one *D* or data input to SET or RESET the flip-flop.

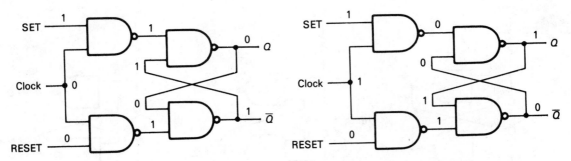

FIGURE 7-17   Enabling a gated SET-RESET flip-flop

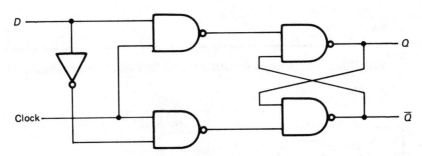

FIGURE 7-18   D flip-flop

D	Clock	Q	$\overline{Q}$	
0	0	Q	$\overline{Q}$	Unchanged State
1	0	Q	$\overline{Q}$	
0	1	0	1	
1	1	1	0	

FIGURE 7-19   Truth table for transparent D flip-flop

It can be readily seen that when the clock is 1, which enables the gates to the $\overline{\text{SET-RESET}}$ flip-flop, the value of D (1 or 0) is transferred to the Q output. When the clock is 0, the Q and $\overline{Q}$ outputs cannot be changed by the D input.

This type of D flip-flop is called a transparent D flip-flop because when the clock is 1, Q changes when D changes. The flip-flop appears transparent until the clock falls to 0, at which time the flip-flop becomes opaque. Figure 7-19 shows the truth table for the transparent D flip-flop.

The D flip-flop is used to store bits of binary numbers. Because it can be turned on or off by the clock, it is also used to catch or latch a binary number present on the D input for a short time and store it on the Q and $\overline{Q}$ outputs. A D flip-flop can be used as the output port of a microcomputer, as shown in Figure 7-20. When the computer wants to output an 8-bit binary number to the printer, it places the binary number on the data bus and then strobes the clocks of the D flip-flops which causes the Q outputs to take on the value of the data bus. The number is now latched on the Q outputs and will not change even though the data bus changes. A typical TTL IC, which contains 4 transparent D flip-flops, is the 7475 quad latch.

The symbol for the D flip-flop is also shown in Figure 7-20.

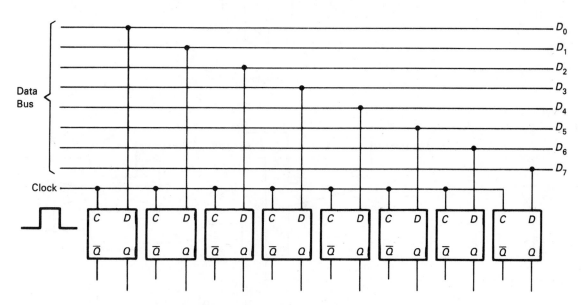

FIGURE 7-20   Computer-output port

7.8   THE MASTER-SLAVE D FLIP-FLOP

Figure 7-21 shows a master-slave D flip-flop made from NAND gates. The master section is a transparent D flip-flop while the slave section is a gated SET-RESET flip-flop. The clock is fed to an inverter which is connected to the slave clock.

This type of master-slave D flip-flop is called a negative edge-triggered D flip-flop because the Q outputs will take on the value of the D input only on the falling edge of the clock pulse.

As shown in Figure 7-22, when the clock is 1, the master part of the flip-flop, which is a transparent D flip-flop, is turned on. The Q' output will follow the D input. The slave part, which is a gated SET-RESET flip-flop, is turned off because the inverter on the clock made the clock a 0. Because the slave is turned off, the Q outputs cannot change.

When the clock falls from 1 to 0, the master is turned off and cannot change; but the slave transfers the values of Q' and $\overline{Q}'$ to Q and $\overline{Q}$ because the slave clock goes to 1. The slave will not change if the D input changes because the master is turned off

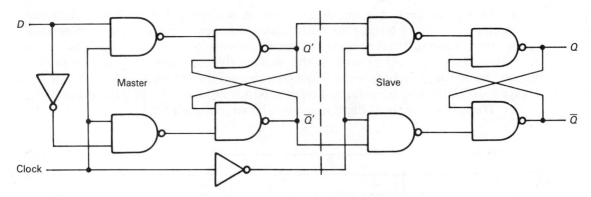

FIGURE 7-21   Master-slave D flip-flop

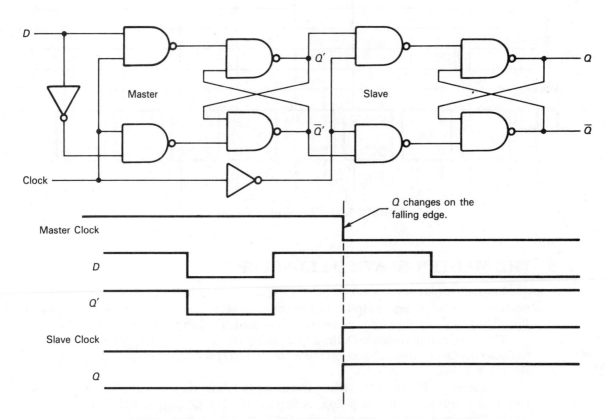

FIGURE 7-22   Negative edge-triggered master-slave D flip-flop

by the 0 on the clock. Therefore, the Q outputs can only change on the falling edge of the clock and will take on the value of the D input at the instant the falling edge happens.

If the inverter is reversed as shown in Figure 7-23, the flip-flop will change states on the positive or rising edge of the clock. Figure 7-23 also shows $\overline{CLEAR}$ and $\overline{PRESET}$ inputs, which can be used to force the output of the flip-flop to 0 or 1, regardless of the clock or D input values. When the $\overline{PRESET}$ goes LOW the Q output is set or forced to 1. When the $\overline{CLEAR}$ goes LOW the Q output is cleared or forced to 0. Notice that an invalid state will result if both $\overline{PRESET}$ and $\overline{CLEAR}$ are LOW or 0 at the same time. The $\overline{PRESET}$ and $\overline{CLEAR}$ inputs could be labeled $\overline{SET}$ and $\overline{RESET}$ also, because they make the flip-flop behave like a simple $\overline{SET}$-$\overline{RESET}$ flip-flop.

Figure 7-24 shows the symbol for a positive edge-triggered D flip-flop and a negative edge-triggered flip-flop. Notice the bubbled input on the active LOW inputs and the $>$ mark which indicates the flip-flop is edge-triggered.

A typical positive edge-triggered flip-flop is the TTL 7474, which has two flip-flops on one IC. The 4013 is a CMOS dual, edge-triggered flip-flop with active HIGH, SET and RESET inputs. Figure 7-25 shows a truth table for a positive edge-triggered flip-flop such as the 7474 IC. Two commonly used symbols for rising edge are the $\_\Gamma$ or $\uparrow$.

As will be seen in the next chapter, master-slave flip-flops are used for many different types of digital circuits, such as shift registers, counters and frequency dividers, because they are edge-triggered.

Figure 7-26 shows some of the commonly used D flip-flops which are manufactured by several integrated circuit manufacturers. The 7475 is a quad transparent D flip-flop which can be used as a latch or storage register. The 74LS74 is a positive edge-triggered flip-flop with active LOW presets and clears. The 74LS174 and 74LS175 are often used as an output port latch for microcomputers because they have an active LOW clear, but the 74LS273 has eight D flip-flops on one 20-pin chip, which means it can store 1 byte. That makes it a better choice for a microcomputer output port.

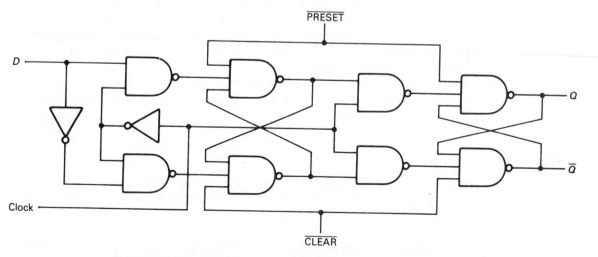

FIGURE 7-23 Positive edge-triggered master-slave D flip-flop

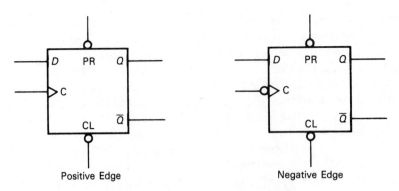

FIGURE 7-24   Flip-flop symbols

CLEAR	PRESET	Clock	D	Q	$\overline{Q}$	
0	1	X	X	0	1	
1	0	X	X	1	0	
0	0	X	X	1	1	Unused State
1	1	⌐⌐	1	1	0	
1	1	⌐⌐	0	0	1	

X = 1 or 0

FIGURE 7-25   Truth table for a positive edge-triggered D flip-flop

7.9   THE PULSE EDGE-TRIGGERED *D* FLIP-FLOP

The master-slave *D* flip-flop is not the only way to make an edge-triggered *D* flip-flop. Figure 7-27 shows an edge-triggered *D* flip-flop which uses a pulse generator on the clock input to enable and inhibit the clock of the transparent *D* flip-flop very quickly. Because this short pulse happens only on the rising edge of the input clock, the *D* flip-flop is an edge-triggered *D* flip-flop.

The circuit which produces the short pulse on the rising edge of the clock is called an edge-triggered one-shot. We will study one-shots more in a later chapter.

Exercise 7

1. Draw a $\overline{\text{SET}}$-$\overline{\text{RESET}}$ flip-flop logic diagram using NAND gates.
2. Draw the logic diagram of a gated $\overline{\text{SET}}$-$\overline{\text{RESET}}$ flip-flop using NAND gates.
3. Draw a logic diagram of a transparent *D* flip-flop using NAND gates or NOR gates.
4. Draw a logic diagram of a negative edge-triggered master-slave *D* flip-flop using NAND gates or NOR gates.

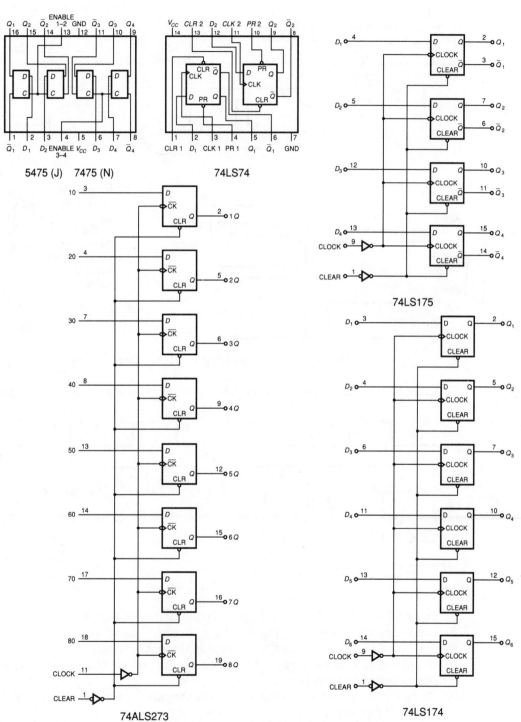

FIGURE 7-26 Common *D* flip-flops

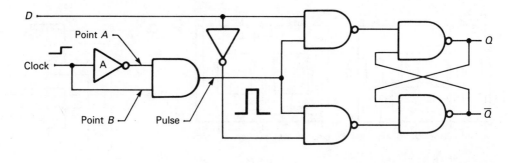

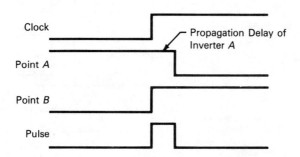

NOTE: When this circuit is made from individual gates, Inverter A should be 3 or 5 7404 inverters "daisy chained" together to produce a sufficient delay for this circuit to work.

FIGURE 7-27   Edge-triggered D flip-flop

5. Complete the waveform for the negative edge-triggered master-slave flip-flop.

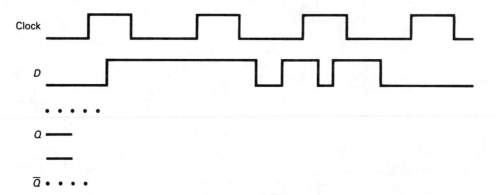

6. Draw a SET-RESET flip-flop logic diagram using NOR gates.
7. Draw a gated SET-RESET flip-flop logic diagram using NOR gates.
8. List two CMOS D flip-flop ICs. Draw the pinouts for these ICs.
9. Using the waveforms in number 5, draw the output waveforms for a positive edge-triggered D flip-flop.
10. Using the waveforms in number 5, draw the output waveforms for a transparent D flip-flop.

11. Draw the logic diagram for a positive edge-triggered D flip-flop using NOR gates.

12. Draw two commonly used symbols for indicating positive edge triggering in flip-flop truth tables.

13. Draw the logic symbol for a negative edge-triggered D flip-flop with active LOW, $\overline{\text{CLEAR}}$ and $\overline{\text{PRESET}}$.

14. What would be the value (one or zero) of the Q output of a $\overline{\text{SET}}$-$\overline{\text{RESET}}$ flip-flop made from NAND gates if the $\overline{\text{SET}}$ input and the $\overline{\text{RESET}}$ input were LOW or zero?

15. Draw the logic diagram for a debounce switch using NOR gates.

LAB **Flip-Flops**

OBJECTIVES

After completing this lab, you should be able to:

■ construct a debounce switch.

■ explain the operation of a gated SET-RESET flip-flop.

■ explain the operation of a master-slave D flip-flop.

COMPONENTS NEEDED

3	7400 quad NAND gate ICs
1	7408 quad AND gate IC
1	single-pole, double-throw switch

PROCEDURE

1. Use a 7400 quad NAND IC and a single-pole double-throw switch to make a debounce switch.
2. Draw the logic diagram for a negative edge-triggered D master-slave flip-flop with a $\overline{CLEAR}$ and $\overline{PRESET}$. Use one 7408 IC and three 7400 ICs.
3. Construct the flip-flop in number 2 and have the instructor check the operation.
4. Write the truth table for the flip-flop in number 3 and verify its operation.
5. When does the flip-flop change states?

If your circuit does not work properly, consider these points:

1. Test all power-supply connectors to the ICs.
2. Check all input and output for proper voltage levels.
3. Disconnect the outputs of the first two NAND gates from the master part of the flip-flop. Now test their inputs and outputs for proper operation.
4. If the operation in step 3 of the circuit check does not correct the problem, then remove the output of the crossed NAND $\overline{SET}$-$\overline{RESET}$ flip-flop from the input NAND gate of the slave and test it for proper operation.

242

5. If step 4 of the circuit check does not correct the problem, then repeat steps 3 and 4 for the slave part of the flip-flop.
6. Bring both $\overline{\text{CLEAR}}$ and $\overline{\text{PRESET}}$ HIGH and place a wire from $\overline{Q}$ to the D input. Next put a 1-kHz TTL square wave signal on the clock of the master-slave D flip-flop and compare it to the Q output with the scope. What is the frequency of the Q output? The toggling of a master-slave flip-flop is the subject of the next chapter. You may want to read the next few pages of Chapter 8 to better understand the flip-flop's action.

CHAPTER 8

OUTLINE

Master-Slave
D and *JK* Flip-Flops

OBJECTIVES

After completing this chapter, you should be able to:

■ make a master-slave *D* and *JK* flip-flop toggle.

■ explain the operation of a *JK* flip-flop.

■ explain the operation of a 2-phase nonoverlapping clock.

■ explain the operation of a shift counter.

8.1   TOGGLING A MASTER-SLAVE *D* FLIP-FLOP

Figure 8-1 shows a master-slave *D* flip-flop constructed from NAND gates that is wired to toggle. A flip-flop is said to *toggle* when the Q outputs change states on each clock pulse. This means if Q is 1, after the next clock pulse, Q would toggle or change to 0 and after the next clock pulse, Q would toggle back to 1. This would continue as long as the clock pulses continued. Notice the $\overline{Q}$ output is tied back to the D input. This is why the flip-flop toggles.

Figure 8-2 shows the waveforms for a toggling of the master-slave *D* flip-flop in Figure 8-1. When the clock goes HIGH, the master part of the flip-flop, which is a transparent *D* flip-flop, transfers the value of the D input, which is 0, to the Q' outputs of the master. The slave part of the flip-flop cannot change its output yet, because the inverter between the master and slave makes its clock 0, and turns off the slave. When the clock falls to 0, the master is turned off first, preventing Q' and $\overline{Q}'$ from changing. A few nanoseconds later, the Q and $\overline{Q}$ outputs change states to match the values of Q' and $\overline{Q}'$. This changes the value of the D input to 1. The

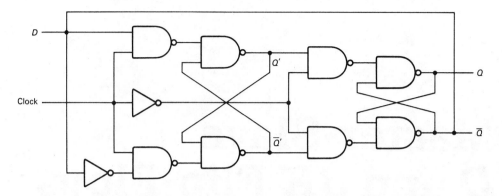

FIGURE 8-1   Master-slave *D* flip-flop wired to toggle

master is turned off by the LOW clock. Q' and $\overline{Q}'$ cannot change until the clock returns to a 1.

When the clock returns to a 1, the slave is turned off by the 0 on its clock and a few nanoseconds later the master turns on and passes the 1 on the *D* input to the Q' outputs and the whole cycle starts over again.

This circuit will work because of the propagation delay of the NAND gates used. The slave will turn off before the outputs of the master change states and the master will turn off before the outputs of the slave can change states. On the falling edge of every clock pulse to the flip-flop, the *Q* output switches states. Notice in Figure 8-2 that the *Q* output has one positive-going pulse for two positive-going pulses of the clock.

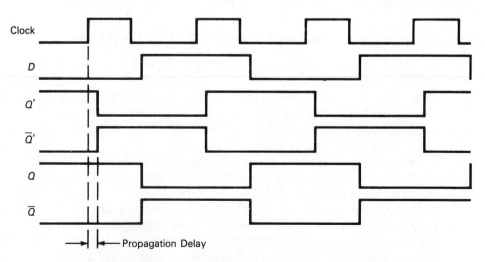

FIGURE 8-2   Toggling a master-slave *D* flip-flop

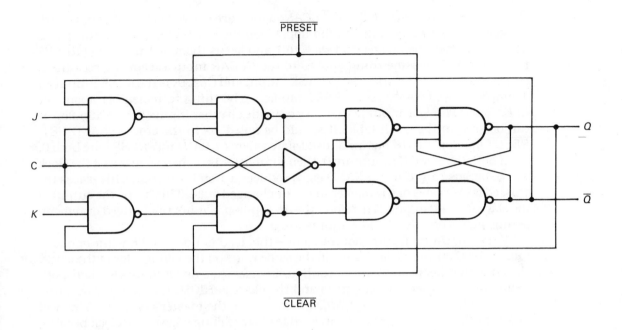

PRESET

CLEAR

PRESET	$\overline{CLEAR}$	J	K	C	Q	$\overline{Q}$	
0	1	X	X	X	1	0	
1	0	X	X	X	0	1	
0	0	X	X	X	1	1	Unused State
1	1	0	1	⊐_	0	1	
1	1	1	0	⊐_	1	0	
1	1	0	0	X	Q	$\overline{Q}$	Unchanged State
1	1	1	1	_⊓_	Toggle		

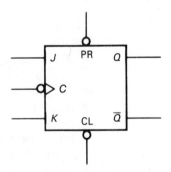

FIGURE 8-3   *JK* flip-flop

8.2   THE *JK* FLIP-FLOP

The *JK* flip-flop is a special kind of master-slave flip-flop. Figure 8-3 shows a *JK* flip-flop constructed with NAND gates, the logic symbol used for a negative edge-triggered *JK* flip-flop and its truth table.

This type of flip-flop can be wired or programmed to do the job of any type of flip-flop. The *Q* and $\overline{Q}$ outputs are wired back to *K* and *J* gates respectively. This will allow the flip-flop to toggle when the *J* and *K* inputs are 1s. The *J* and *K* inputs are used to steer the *Q* outputs. There are 2 inputs, called $\overline{PRESET}$ and $\overline{CLEAR}$, which force the *Q* outputs to 1 and 0, respectively, when they are brought LOW.

Notice that the $\overline{PRESET}$ and $\overline{CLEAR}$, which are active LOW inputs, go to the master and slave parts of the JK flip-flop. When the $\overline{PRESET}$ goes LOW, this forces the Q outputs of the master and slave to 1 and keeps them at 1 until the $\overline{PRESET}$ returns to 1. The same thing applies to the $\overline{CLEAR}$ input except it forces the Q output to 0. $\overline{PRESET}$ works just like a $\overline{SET}$ input and $\overline{CLEAR}$ is the $\overline{RESET}$ input for a simple crossed NAND $\overline{SET}$-$\overline{RESET}$ flip-flop. The thing to remember about the $\overline{CLEAR}$ and $\overline{PRESET}$ input is that they override all other inputs to a JK flip-flop as can be seen in the truth table. It should be noted, as in the crossed NAND $\overline{SET}$-$\overline{RESET}$ flip-flop, the JK has an unused state where $\overline{CLEAR}$ and $\overline{PRESET}$ are both 0.

If both the J and K inputs are 0, then the master is turned off just as if the clock was 0, because any 0 into a NAND gate produces a 1 on its output. This places the crossed NAND gates of the master in the unchanged state. This keeps the output of the slave from changing. Therefore, the JK flip-flop is in its unchanged or remembering state when the J and K inputs are 0.

When the JK inputs are not the same—that is, J is 1 or 0 and K is the opposite value—the Q outputs will change to the same value on the falling edge of the clock.

The only other possible input configuration for the J and K inputs is where both inputs are 1s. When this happens and the clock is HIGH, the Q and $\overline{Q}$ outputs, which are tied back to input NAND gates, control the master's outputs. This will cause the flip-flop to toggle or change states every falling edge of the clock because the Q output is tied back to the opposite K gate and the $\overline{Q}$ is tied back to the opposite J gate.

Facts to remember about negative edge-triggered JK flip-flops include the following:

- The Q output only changes on the falling edge of the clock except when the $\overline{CLEAR}$ or $\overline{PRESET}$ go LOW.
- $\overline{CLEAR}$ and $\overline{PRESET}$ override all other inputs on the JK flip-flop.
- When the J and K are both 1, the flip-flop will toggle on the falling edge of the clock.
- When J and K are not equal the output will follow the J and K on the falling edge of the clock.
- When J and K are both 0, the Q outputs will not change their values.

Figure 8-4 shows the waveform for the Q and $\overline{Q}$ outputs of a negative edge-triggered JK flip-flop for a given set of input waveforms on the J, K, $\overline{CLEAR}$, and $\overline{PRESET}$. Notice that except for when the $\overline{CLEAR}$ or $\overline{PRESET}$ go active the Q and $\overline{Q}$ outputs do not change states except on the falling edge of the CLK (clock) signal. The value of the Q and $\overline{Q}$ outputs is determined by the value of the J and K inputs before the falling edge of the CLK input to the flip-flop. They follow the truth table in Figure 8-3.

8.3 THE NONOVERLAPPING CLOCK

Figure 8-5 shows a JK flip-flop used to make a nonoverlapping clock. Notice that CP and CP' are one-half the frequency of the clock and are 180° out-of-phase.

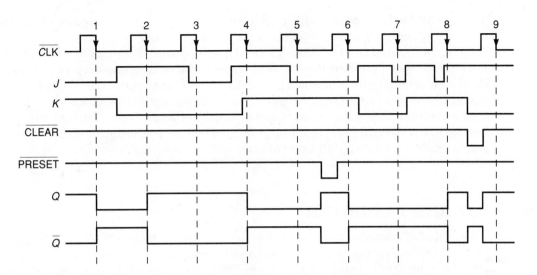

FIGURE 8-4   Output waveforms for a *JK* flip-flop

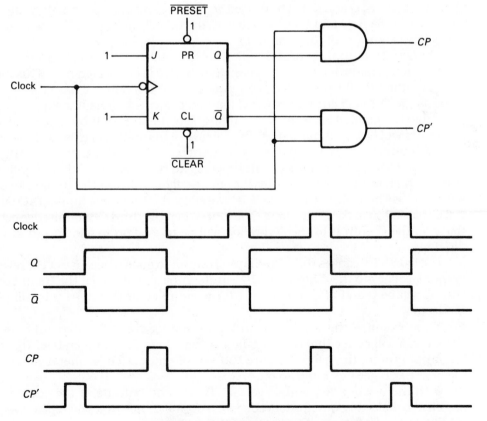

FIGURE 8-5   Nonoverlapping clock

They are said to be nonoverlapping because the leading or rising edges and trailing or falling edges of *CP* and *CP'* will never occur at the same time.

The *JK* flip-flop is wired to toggle on the falling edge of the clock. This enables the *CP* AND gate and then the *CP'* AND gate on the next falling edge of the clock. Each time an AND gate is enabled by the flip-flop's *Q* or $\overline{Q}$ output, the next positive clock pulse is passed through the gate. On the falling edge of that clock, the flip-flop toggles, inhibiting one AND gate and enabling the other AND gate. This process can be seen in the wave diagram in Figure 8-5. This type of nonoverlapping clock is used for generating strobe signals and various waveforms for digital devices.

8.4   THE SHIFT COUNTER

Figure 8-6 is a shift counter made from three *JK* flip-flops. The *Q* and $\overline{Q}$ outputs of the *A* flip-flop are connected to the *J* and *K* inputs of the *B* flip-flop, respectively, and the *B* flip-flop's *Q* and $\overline{Q}$ outputs are connected to the *J* and *K* inputs of the *C* flip-flop in the same way.

Because all the clocks of the three flip-flops are tied together when the *CP* falls, the value of the *A*, *Q* and $\overline{Q}$ will be passed on to the *Q* and $\overline{Q}$ outputs of the *B* flip-flop. The value of the *B*, *Q* and $\overline{Q}$ outputs before the *CP* falling edge will be passed on to the *Q* and $\overline{Q}$ outputs of the *C* flip-flop.

The *Q* and $\overline{Q}$ outputs of the *C* flip-flop are tied back to the *J* and *K* inputs of the *A* flip-flop in reverse order. *Q* is connected to *K*, and $\overline{Q}$ is connected to *J*. This means that after the *CP* falls, the *Q* and $\overline{Q}$ outputs of the *A* flip-flop will have the opposite value as the *Q* and $\overline{Q}$ outputs of the *C* flip-flop before the *CP* falling edge.

If the shift counter is cleared by bringing all of the $\overline{CLEAR}$ inputs LOW, all of the *Q* outputs will be 0 and the $\overline{Q}$ outputs will be 1. This state will exist as long as the $\overline{CLEAR}$ inputs are LOW. When they are brought back HIGH and after the falling edge of the next *CP*, the *A* flip-flop's *Q* output will go to the opposite of the *C* flip-flop's *Q* output which was 0. Therefore after the falling edge of the first *CP*, the output of the *A* flip-flop will be 1. After the next *CP* falling edge, the 1 on the *A* flip-flop's *Q* output will be passed on to the *B* flip-flop's *Q* output. When the third *CP* falling edge occurs, the 1 on the *Q* output of the *B* flip-flop will be passed on to the output of the *C* flip-flop.

The fourth *CP* causes the *A* flip-flop's *Q* output to go to 0 because of the reversed output feedback to the *A* flip-flop's *JK* inputs. After the fifth *CP*, the *B* flip-flop's *Q* output will be 0 and after the sixth *CP* the entire counter will be 0 on all of the *Q* outputs.

If you examine the waveforms in Figure 8-6 for the 3 flip-flop shift counter shown, you will notice that each flip-flop is 1 for three *CP* and then 0 for three *CP*. The outputs of the three flip-flops are 120° out-of-phase; that is, after *A* goes to 1 on the next *CP*, *B* goes to 1 and so on.

With the use of a few simple gates, *CP*, *CP'* and outputs of the counter it is possible to make any waveform needed for a digital device that will repeat every six *CP*. Figure 8-7 shows some examples of this.

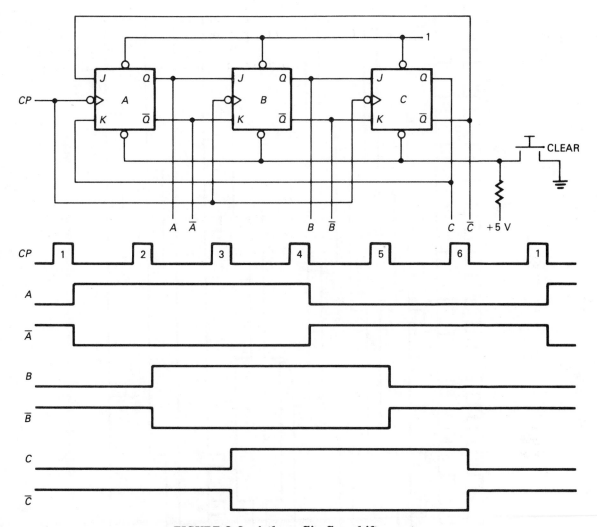

FIGURE 8-6   A three-flip-flop shift counter

Figure 8-8 shows four commonly used *JK* flip-flops and their pinouts. The 74LS73 is a negative edge-triggered master-slave *JK* flip-flop with only $\overline{\text{CLEAR}}$ inputs. The 74LS76 and 7476 are full-blown negative edge-triggered master-slave *JK* flip-flops with both $\overline{\text{CLEAR}}$ and $\overline{\text{PRESET}}$ inputs. This costs two more pins on the IC package, though. If the designer wishes to have a 14-pin IC package and to retain both $\overline{\text{CLEAR}}$ and $\overline{\text{PRESET}}$, then he or she can choose the 74LS78, which has them but also has a common $\overline{\text{CLOCK}}$ and $\overline{\text{CLEAR}}$ pin for both flip-flops. A typical positive edge-triggered *JK* flip-flop is the 74LS109.

All these flip-flops will work for the examples in this textbook as well as others not mentioned here. The student should refer to a manufacturer's IC specification manual to get a feel for the number and diversity of *JK* flip-flop ICs available today.

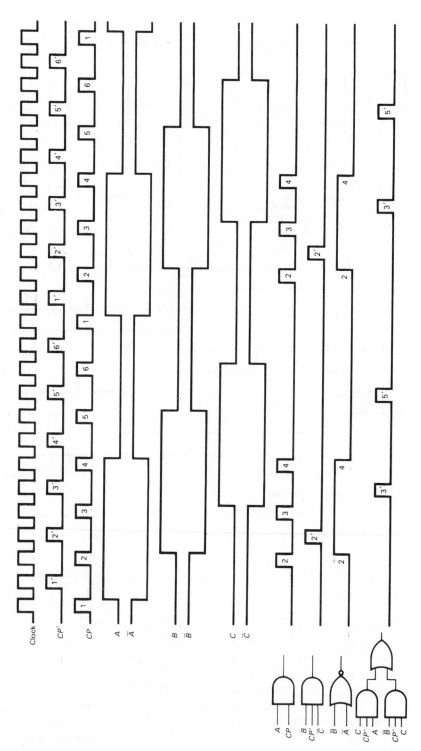

FIGURE 8-7 Nonoverlapping clock and three-flip-flop shift-counter waveforms

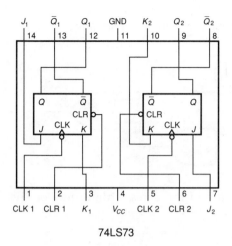

74LS73

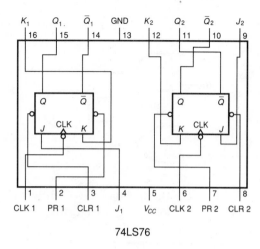

74LS76

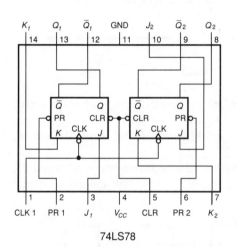

74LS78

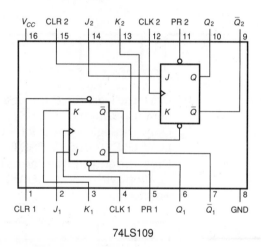

74LS109

FIGURE 8-8   Commonly used *JK* flip-flops

Exercise 8

1. Draw a logic diagram of a delayed clock and shift counter which will repeat every 10 *CP*. Include the IC pinout and pin number on the logic diagram.
2. Using the TTL and CMOS data manuals, locate two ICs which have negative clocks and two ICs which have positive clocks for both TTL and CMOS. Draw their pinouts.
3. Draw the logic diagram for a 2-phase nonoverlapping clock using one *JK* flip-flop and three NOR gates. Label the IC used and put pin numbers on the drawing.

4. Using Figure 8-7, draw the waveforms for the following gates.

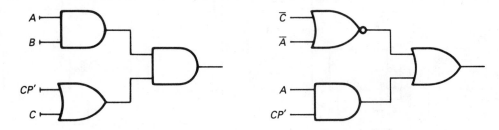

5. Using Figure 8-7, draw the logic diagram of the gates which produce the following waveforms.

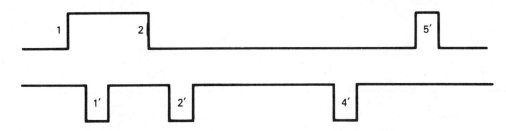

6. Draw the logic diagram for a negative edge-triggered *JK* flip-flop with active LOW, $\overline{CLEAR}$ and $\overline{PRESET}$ using NOR gates.

7. Draw the logic symbol for a negative edge-triggered *JK* flip-flop with active, LOW, CLEAR and PRESET.

8. Draw the output waveforms for the *Q* output of a negative edge-triggered *JK* flip-flop with active LOW, CLEAR and PRESET. Use the following input waveforms.

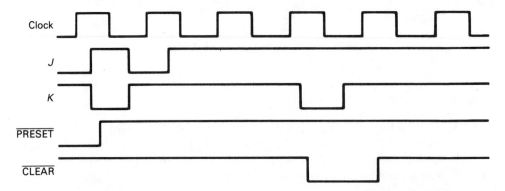

9. Repeat number 8 with a positive edge-triggered flip-flop with active HIGH, CLEAR and PRESET.

10. What would be the value of the Q output of a negative edge-triggered *JK* flip-flop after the falling edge of the clock if the *J* input was 0 and the *K* input was 0 before the falling edge of the clock?
11. Draw the logic symbol for a flip-flop and put the correct values on the inputs to cause it to toggle.
12. Draw the waveforms for the flip-flop shown using the waveforms in Figure 8-7 as the inputs.

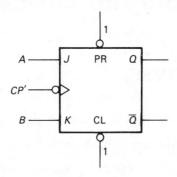

13. Using the waveforms in Figure 8-7 and a single *JK* flip-flop, draw the logic diagram of a circuit which will produce an even-duty-clock which has the same frequency as *CP*.
14. What would have to be put on the inputs of the flip-flop in Figure 8-3 for Q and $\overline{Q}$ to be HIGH?
15. What would be the value of *CP* in Figure 8-5 if the $\overline{PRESET}$ input were tied LOW on the *JK* flip-flop?

8 Shift Counter and Delayed Clock

OBJECTIVES

After completing this lab, you should be able to:

- use a *JK* flip-flop to make a nonoverlapping clock.
- use *JK* flip-flops to make a shift counter.
- use simple gates to make various waveforms.

COMPONENTS NEEDED

2	7476 *JK* flip-flop ICs
1	7408 quad AND gate IC
1	10-kΩ, ¼-W resistor
1	7404 hex inverter IC
1	7432 quad OR gate IC
1	7410 triple three-input NAND gate IC
1	7427 triple three-input NOR gate IC

PROCEDURE

1. Use a *JK* flip-flop and two AND gates to generate a clock, *CP*, and delayed clock, *CP'*.

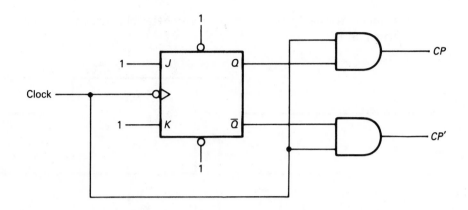

2. On a sheet of graph paper, plot the waveforms that you would expect to see for clock, Q, $\overline{Q}$, CP, CP'.
3. Use an oscilloscope to observe the actual waveforms of the following: clock, Q, $\overline{Q}$, CP, CP'. Graph the actual waveforms.
4. Construct the shift counter shown and connect to it the delayed clock from number 1.

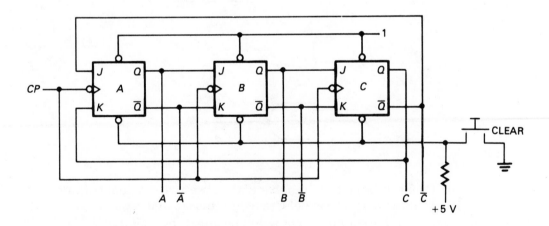

5. Plot the waveforms that you would expect to see for CP, CP', A, $\overline{A}$, B, $\overline{B}$, C, and $\overline{C}$.
6. Use an oscilloscope to observe the actual waveforms CP, CP', A, $\overline{A}$, B, $\overline{B}$, C, and $\overline{C}$. Graph the actual waveforms.

7. Connect the gates necessary to produce the following waveforms and observe them on the scope.

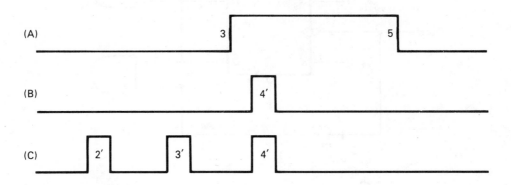

8. Predict the outputs of these gates.

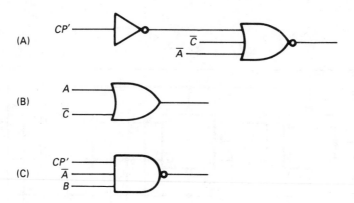

9. Connect the gates in number 8 and observe the actual outputs.
10. Make up a waveform not covered in this lab and create it.

If your circuit does not work properly, consider these points:

1. Check the power supply connection to each IC.
2. Check all the inputs and outputs of each gate for proper voltage levels.
3. The nonoverlapping clock:
   A. Make sure that you have tied the J, K, $\overline{CLEAR}$ and $\overline{PRESET}$ to $+5$ V. JK flip-flops do not always consider an unconnected input as a logic 1.

B. Use your scope to trace the clock signal to the *JK* flip-flop's clock input pin and to the inputs of the two NAND gates.

C. Use one channel of the scope to display the input clock and the other channel to display the Q output of the flip-flop. This will tell you if the flip-flop is toggling or not.

4. Shift counter:

A. Check to see if you left a $\overline{PRESET}$ or $\overline{CLEAR}$ unconnected.

B. Be sure you have cleared all the flip-flops to properly sequence them.

C. Use the scope to trace the *CP* signal to each flip-flop's clock input pin.

CHAPTER 9

Shift Registers

OBJECTIVES

After completing this chapter, you should be able to:

■ explain the operation of a shift register.

■ describe how to parallel load a shift register and serially shift the data out and serial-in, parallel-out.

■ describe typical serial digital data transition methods.

■ describe typical IC shift registers.

9.1 SHIFT REGISTER CONSTRUCTED FROM *JK* FLIP-FLOPS

Figure 9-1 shows a 4-bit serial-in, parallel-out shift register made from *JK* flip-flops. Notice the inverter between the *J* and *K* inputs of the *A* or first flip-flop. This means that the *J* and *K* inputs can never be the same. When the *A* flip-flop is clocked with the falling edge of the clock, the *Q* and $\overline{Q}$ outputs will take on the values of the *J* and *K* inputs. The *Q* outputs of the *A* flip-flop are connected to the *JK* inputs of the *B* flip-flop. The *B* flip-flop is connected to the *C* flip-flop in the same way. This method of connecting the flip-flops would continue on until you obtained the number of binary bits you want your shift register to have.

Because the clock inputs of all the flip-flops are tied together, the flip-flops will all change states at the same time and their *Q* outputs will reflect the *J* input before the falling edge of the clock.

At the falling edge of the first clock pulse in Figure 9-1, the *Q* output of the *A* flip-flop (Q_A) is set to 1 because the serial-in was 1 before the falling edge of the clock. The *Q* output of the *B* flip-flop (Q_B) is still 0 after the falling edge of clock pulse 1 because Q_A was 0 before the falling edge of the first clock pulse. The outputs of the *C* and *D* flip-flops (Q_C and Q_D) are 0 for the same reason.

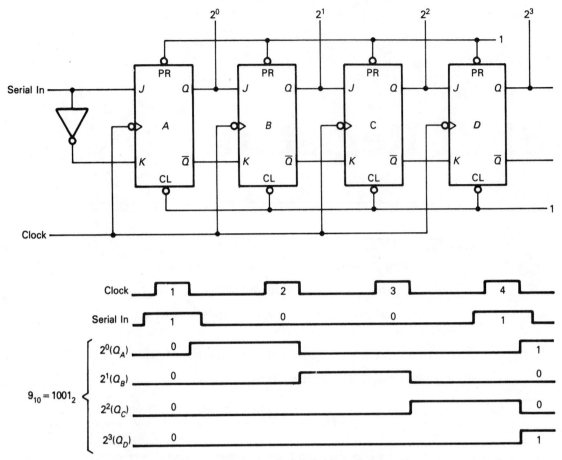

FIGURE 9-1   A 4-bit shift register

At the falling edge of the second clock pulse, Q_A changes to 0 because the serial-in input was 0 before the falling edge. Q_B now changes to 1 because Q_A was 1 before the falling edge of the second clock pulse.

On the third falling edge, the 1 on the output of the B flip-flop would be transferred to the output of the C flip-flop and after the fourth clock pulse the register would be full. In Figure 9-1, the binary number equivalent to 9 was clocked into the shift register in serial form and was converted to parallel form after the fourth clock pulse.

9.2   PARALLEL AND SERIAL DATA

Data in serial form is fed one bit at a time over only one line or wire at a rate which is constant and in phase with a clock reference. Parallel data on the other hand has one line or wire for each bit in the binary number or data word and does not have to be referenced to a clock to transfer it from one register to another.

The frequency of the reference clock or bits per second of the serial-in input is usually called *baud rate* of the serial transfer. A typical baud rate for a teletype machine is 110 bits per second. At this rate you can send ten 11-bit binary numbers in one second.

If the same reference clock is used to clock a parallel register with 1 line for each bit of the 11-bit binary number then you can transfer one hundred ten 11-bit numbers in one second.

As can be seen, parallel data transmission is much faster but takes many more lines or wires than serial transmission. When digital data is transferred over long distances or is placed on a long magnetic tape for storage, serial methods are used because it only takes one line or wire to do the job.

9.3   PARALLEL-IN SERIAL-OUT

Figure 9-2 shows a parallel-in serial-out shift register and the waveform for loading the binary number equivalent to 9 and shifting it out to the right. The A flip-flop has a 0 on the J input and a 1 on the K input. This will cause the Q output of the A flip-flop to go to 0 after the falling edge of the clock input. If the shift register was clocked 4 times, the Q outputs of all the flip-flops would be 0 and stay 0 until the shift register was broadside loaded or parallel loaded with a new binary number.

This is done by placing a binary number on the parallel inputs and raising the parallel load input or control to 1. This enables the NAND gates which feed the $\overline{\text{PRESET}}$ and $\overline{\text{CLEAR}}$ inputs of each flip-flop. Because of the inverter between the two NAND gate inputs, one NAND gate output will be 1 while the other will be 0. They can never be the same value while the NAND gates are enabled. This will cause the Q output of the flip-flop to be set or reset depending on the value of the parallel input to the NAND gates feeding the flip-flop.

Because $\overline{\text{CLEAR}}$ and $\overline{\text{PRESET}}$ take precedence over all other inputs of the JK flip-flop, the Q outputs of the flip-flops will not change as long as the parallel load input is 1 since this enables the NAND gates. When the parallel load input falls to 0, the NAND gates are inhibited and their outputs go to 1 because any 0 into a NAND gate produces a 1 on the output.

The shift register is now loaded with the binary number desired. In the case of Figure 9-2, the binary number is equivalent to 9. The $\overline{\text{PRESET}}$ and $\overline{\text{CLEAR}}$ are 1, which means when the falling edge of the clock occurs, the shift register will shift each bit right one place, shift in a 0 on the left and shift out a 1 on the right. After four clock pulses the number will be shifted out to the right, the shift register will be empty or 0 and ready for a new number.

The Q output of the D flip-flop could be connected to a serial-in parallel-out shift register which was clocked by the same clock and the binary number equivalent to 9 would have been transferred to the other shift register in 4 clock pulses. This idea is represented in Figure 9-3.

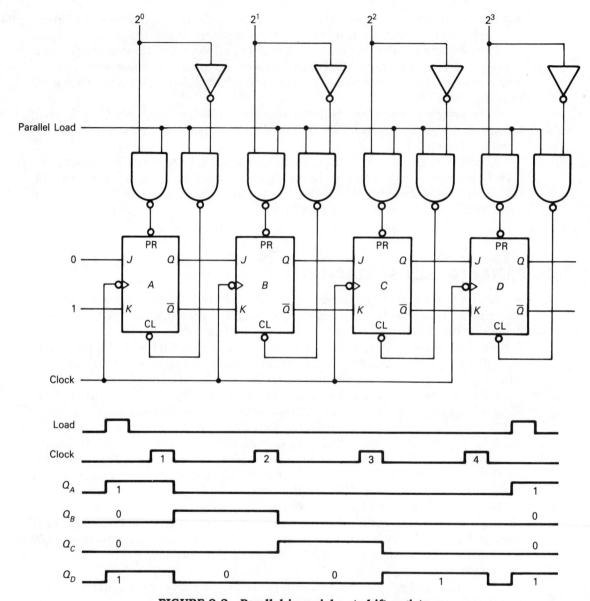

FIGURE 9-2   Parallel-in serial-out shift register

9.4   SERIAL DATA TRANSMISSION FORMATS

There are several serial data transmission formats which are standardized. Two of these are the RS232 serial interface used in computers and the 20 mA current loop used on older teletypes. Figure 9-4 shows a typical word format for these serial interfaces. This is called asynchronous serial transfer.

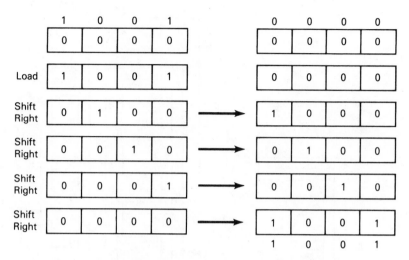

FIGURE 9-3   Serial data transfer

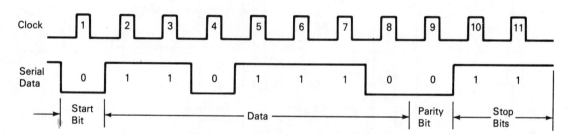

FIGURE 9-4   Asynchronous serial format

Each word is started with a LOW-going start bit which starts the shift register clocking in the data. The next 7 bits are data usually in the form of ASCII code for letters of the alphabet. These are followed by a parity bit and two stop bits which are 1s. Notice that it takes 11 clock pulses to shift in one word. A typical teletype runs at a 110 baud rate. This means that if it used the format in Figure 9-4, it could send or receive 10 words or characters per second. This is quite slow compared to today's computer, which runs at 9600 baud rate or higher.

The start bit and the stop bits are called framing bits and are used to start and stop the serial shift register which is receiving the data. This type of serial data transfer is called asynchronous serial data transfer because the data comes in these neat little packages and at any time interval between these packages. Synchronous serial data does not use framing bits and the data must come one word after the other at a constant rate. Synchronous serial data transfer is usually done in blocks of binary numbers or words. A good example of synchronous serial data transfer is the way most disk drives store data. The data is stored on the disk in sectors. These sectors are usually 128, 256, or 512 bytes long. A *byte* is an 8-bit binary number.

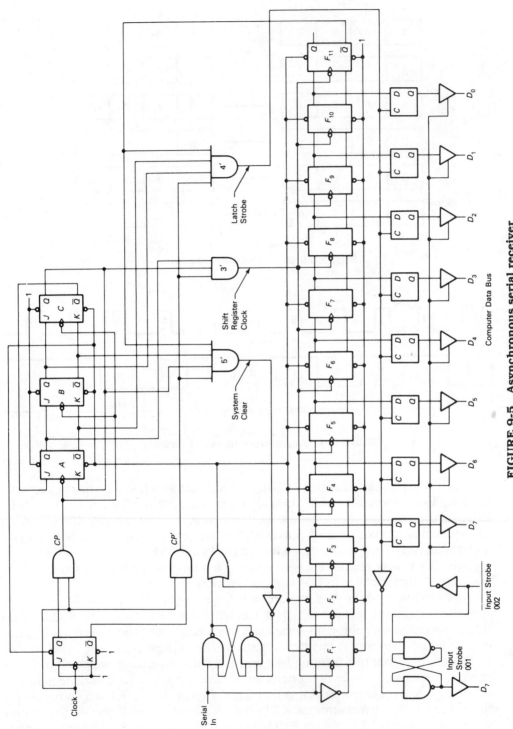

FIGURE 9-5 Asynchronous serial receiver

To store one byte on a disk using asynchronous serial methods would take a minimum of 10 bits, one start bit, eight data bits, and one stop bit. To store one byte on a disk using the synchronous methods would take only 8 bits for the actual data. Therefore, the synchronous method would let us store 20% more data on the disk than the asynchronous method. This is good for disk storage but for serial data transfer over long distances synchronous methods are not easy to use. The clock must be carried with the data stream, in some method, to run the serial shift register which converts the data from serial form to parallel form. Because the clock is started with the start bit and stopped with the stop bit, in asynchronous serial data transfer, the clock can be derived from a very accurate crystal clock at the receiver end of the serial transfer and does not need to be carried with the data stream.

Figure 9-5 shows a serial input logic system designed to input serial data in the format of Figure 9-4. This is a typical system which would input serial data and latch it out into a set of D flip-flops for a microprocessor to input into its memory. Such logic circuits are not usually constructed from basic gates and flip-flops today because they are put on LSI circuits as one integrated circuit. The LSI circuits are called UART ICs and include a parallel-to-serial transmitter logic system and all the necessary logic to check the parity, stop bits and framing of the incoming data. Most of the LSI UART ICs are programmable by the microprocessor they are feeding. The word UART is an acronym for universal asynchronous receiver transmitter. Figure 9-5 is an example of how the receiver of a typical UART might be constructed.

Figure 9-6 shows the waveforms for the 11th clock pulse of the shift register clock. The shift register clock is obtained by decoding the 3' pulse from the delayed clock and shift counter. This means the clock which is driving the delayed clock must be 12 times the baud rate of the incoming serial data.

When there is no data being sent to the serial input, the Q output of the crossed NAND gates is 0. This causes the flip-flops of the shift counter to be cleared, and the Q outputs will be 0 as long as the Q of the crossed NAND gates is 0. Also the outputs of the shift register will be 1 because the Q of the crossed NAND gates is tied through an OR gate to the $\overline{\text{PRESET}}$s of the 11 flip-flops of the shift register.

This condition will exist until the serial input goes LOW. This causes the crossed NAND gates to change states producing a 1 on the Q output of the crossed NAND gates. Three clock pulses of CP' later, the shift register will be clocked. This makes F_1 go LOW and shifts a 1 out of the end of the shift register.

The shifting of the shift register will continue for eleven 3' clocks after which the output $\overline{Q}_{F_{11}}$ will go to a 1 enabling the 4' and 5' AND gates. When the 4' pulse goes HIGH, it latches the value of the 8 bits of data into the D latches. When the 5' pulse goes HIGH, the shift counter is cleared and the shift register is preset because the crossed NAND gates are reset by the 5' pulse. The receiver is now ready to shift in another serial word when another HIGH-to-LOW transition of the serial input occurs. When the 4' pulse goes HIGH, latching the data into the D flip-flops, it also sets a crossed NAND $\overline{\text{SET-RESET}}$ flip-flop, which the computer will scan to see if any data has been received.

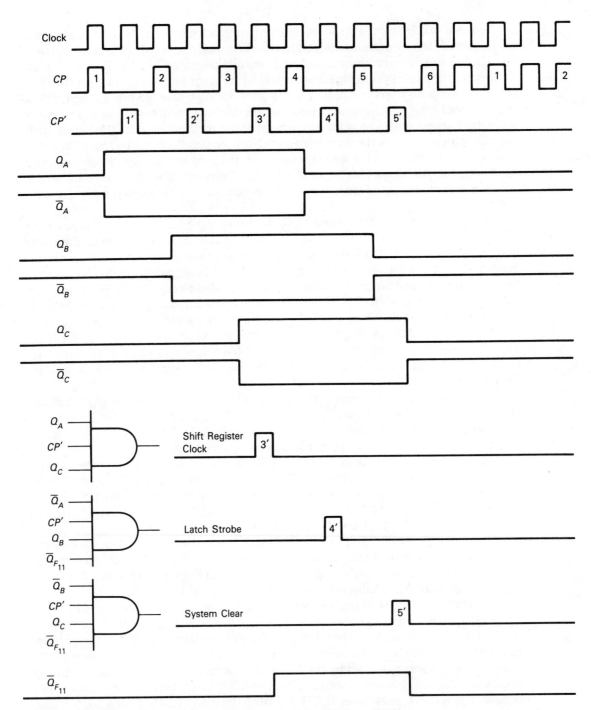

FIGURE 9-6 The last clock cycle of the asynchronous serial receiver shown in Figure 9-5

9.5   IC SHIFT REGISTERS

Figure 9-7 shows a logic diagram of a 7495 4-bit shift register. This shift register can be wired to shift right or left, parallel-in, serial-out or serial-in, parallel-out. Each of the JK inputs on the 4 flip-flops is controlled by a set of two AND gates fed into a NOR gate. The AND gates are enabled or inhibited by the mode control (pin 6). When the mode control goes HIGH, the AND gates which have the A, B, C, and D parallel inputs are enabled and the outputs are inhibited. This means when the mode is HIGH the value of the JK inputs of the flip-flops will be controlled by the $ABCD$ parallel inputs. This is a parallel load or broadside load.

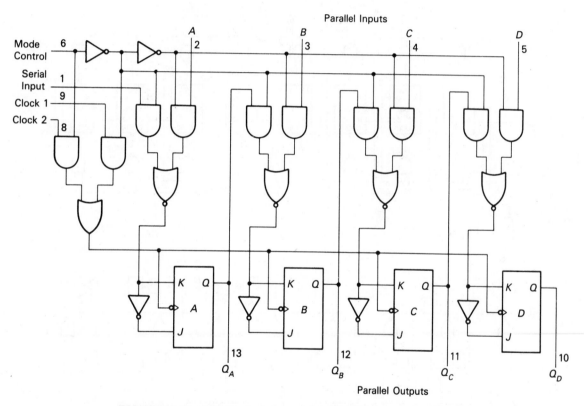

FIGURE 9-7   A logic drawing of the 7495 4-bit shift register

When the mode control is made LOW, the parallel inputs are inhibited and the serial inputs to the JK flip-flops are enabled. Notice the A flip-flop is fed from the serial input (pin 1) and the B flip-flop is fed from Q_A and so on. In this configuration, when clock 1 is brought LOW the flip-flops will shift right one place. When the mode control is HIGH, the shift register can be broadside loaded and when the mode control is LOW the shift register can be shifted right.

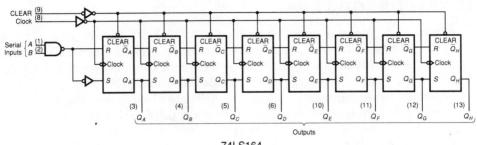

74LS164

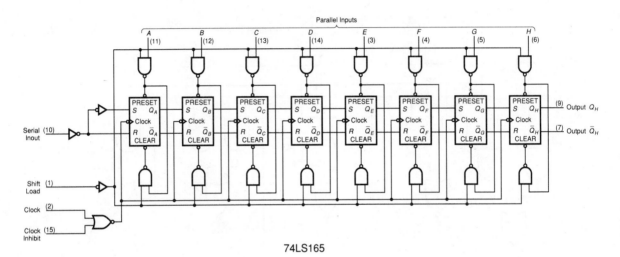

74LS165

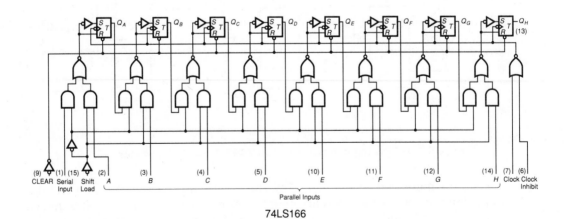

74LS166

FIGURE 9-8   Common shift registers

To get the 7495 to shift left, you must wire Q_D back to the C input, Q_C back to the B input and Q_B back to the A input. Then place the mode control HIGH to enable the $ABCD$ inputs.

The 7495 shift register is only 4 bits wide but is quite versatile. Figure 9-8 shows three 8-bit shift registers which are not as versatile but have the advantage of shifting a whole byte in one IC.

The 74LS164 is an 8-bit serial-in, parallel-out shift register with an active LOW CLEAR. The counterpart to this IC is the 74LS165, which is a parallel-in, serial-out IC with 8 bits also. The 8-bit IC which most resembles the old 7495 is the 74LS166. As you can see, it has the same logic configuration as the 7495, but the parallel outputs are missing and not brought out of the IC.

It is recommended that the student refer to a good IC manufacturer's specification manual for other shift register ICs.

Exercise 9

1. Draw the logic diagram for a 5-bit shift register using JK flip-flops.
2. Draw a logic diagram for a 4-bit shift register using 4027 CMOS JK flip-flops. Show pin numbers on the drawing.
3. Draw the wave diagram for Figure 9-2 if the binary number equivalent to 6 is on the parallel inputs.
4. What is the purpose of the start bit in the serial format of Figure 9-4?
5. Draw a logic diagram for an 8-bit shift register using two 7495 ICs.
6. Draw the waveforms for the 4-bit shift register in Figure 9-1 if the binary number equivalent to 6 is shifted into the shift register.
7. Draw the waveforms for the shift register in Figure 9-2 if the binary number equivalent to 7 is on the parallel inputs.
8. Using the clock pulse given in Figure 9-4, draw the waveform for an asynchronous serial word which has no parity bit, one stop bit and data of the binary number equivalent to 62.
9. Using Figure 9-5 as an example, design an asynchronous serial receiver that uses a clock 16 times the incoming baud rate of the serial data.
10. Draw the logic diagram of a 7495 4-bit shift register wired to shift left.
11. Draw the logic diagram of a 7495 4-bit shift register wired to be a parallel-in, serial-out shift register.
12. Which serial method of storing data on a magnetic device will store the most data?
13. Write the code for the words DIGITAL ELECTRONICS in ASCII code. Use the ASCII chart at the end of Lab 9.
14. Look up the 74164 IC in your data book and write a description of it.
15. Draw the logic diagram of a 16-bit shift register using the 74164. Show pin numbers.

LAB **9** **Shift Registers**

OBJECTIVES

After completing this lab, you should be able to:

- use the 7495 shift register.

- use the ASCII code.

- use a team approach to solving a problem.

COMPONENTS NEEDED

Part 1

1	7476 *JK* flip-flop IC
1	7414 hex Schmitt trigger input inverter IC
1	7408 quad AND gate IC

Part 2

2	7476 *JK* flip-flop ICs
1	7420 dual four-input NAND gate IC
1	7410 triple three-input NAND gate IC

Part 3

3	7495 shift register ICs
1	7400 quad NAND gate IC
1	7402 quad NOR gate IC

Part 4

2	7475 quad transparent *D* latch ICs
8	red LEDs
8	330-Ω, ¼-W resistors

PROCEDURE

1. The instructor will assign each lab group one part of the serial receiver to construct.
2. Each lab group will construct their part of the serial receiver and test it to be sure it works.
*3. After construction, the separate lab groups will combine their parts of the serial receiver to make a complete serial receiver and connect the input to the output of an ASCII terminal which the instructor will provide.
4. There will be a message sent in serial ASCII form. The characters will be spaced about 5 seconds apart to allow you to read them from the LEDs on your serial receiver.

Part I

A. Construct the nonoverlapping clock shown using one 7476 *JK* flip-flop, one 7408 AND gate IC and one 7414 IC.
**B. Feed the clock with the signal from the sine wave frequency generator which has been squared up by the 7414 Schmitt-trigger inverter.
C. Set the signal generator at 1320 Hz and measure the frequency on the scope to get it correct. This is very important or the serial receiver will not work.
D. Put the inputs and outputs of the clock on a blank spot of the protoboard to make it easy to connect to the next part of the serial receiver.

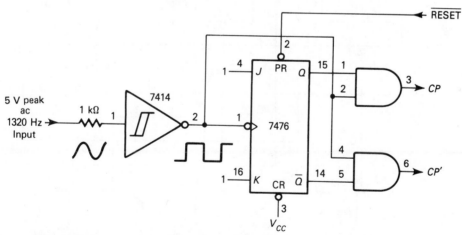

If your circuit does not work properly, consider these points:

1. Check all power supply connections.
2. Check to be sure there are no unconnected inputs on the *JK* flip-flop.
3. Make sure you have connected the signal generator ground to the circuit ground.

---

* If an RS232C equipped terminal is used for the serial output device, be sure to use a buffer IC such as LM1489 to convert the RS232C voltage levels to standard TTL voltage levels.
** A TTL level frequency generator could be used in the place of the 7414 IC.

4. Use the scope to trace the input clock signal from the signal generator to all points in the circuit.

Part II

A. Construct the shift counter shown and the 3′, 4′ and 5′ outputs. Use two 7476 *JK* flip-flops, one 7420 4-input NAND gate and one 7410 3-input NAND gate.

B. Feed the shift counter with the clock from your trainer and check the output waveforms on the scope to be sure they are correct.

C. Place the inputs and outputs of the shift counter on a blank spot on your protoboard to make it easy to connect to the next part of the serial receiver.

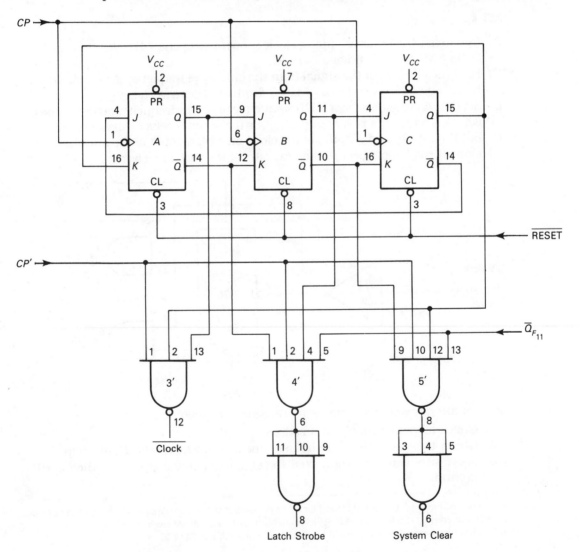

If your circuit does not work properly, consider these points:

1. Check all the power supply connections to the ICs.
2. Connect a clock of about 1 kHz to the CP input and RESET the shift counter by momentarily bringing the $\overline{\text{RESET}}$ input LOW.
   A. Make sure that the PRESET inputs to the JK flip-flops are HIGH.
   B. Use the scope to test the Q outputs for their proper signal. Q_A should be one clock pulse out-of-phase with Q_B and Q_C should be two clocks out-of-phase with Q_A. If these are not the waveforms you see, then check your wiring.
   C. Check your wiring by putting channel one of the scope on a pin and channel two on a pin which should be tied to the first pin. If the signals are not the same, then they are not connected.
3. Tie CP' and $\overline{Q}_{F_{11}}$ HIGH.
   A. Use the scope to test the "clock," "latch strobe" and "system clear" for proper operation. These should be one clock pulse out-of-phase.
   B. Bringing $\overline{Q}_{F_{11}}$ LOW and CP' HIGH should cause only "latch strobe" and "systems clear" to be LOW.
   C. If the circuit does not work in the manner described, then use the method described in 2C to trace your wiring.

Part III

A. Construct the shift register shown and the $\overline{\text{SET}}$-$\overline{\text{RESET}}$ starting flip-flop shown. Use three 7495 shift register ICs, one 7400 two-input NAND gate IC and one 7402 NOR gate IC.
B. Hold the system clear input LOW, put the clock input on a debounce button of your trainer, put a zero on the serial input and clock the shift register full of zeros. Repeat the procedure with the serial input HIGH.
C. Bring the system clear HIGH. This should preset all the Q outputs to one.
D. Put all the inputs and outputs of the shift register on a blank spot of your protoboard to make it easy to connect to the next part of the serial receiver.

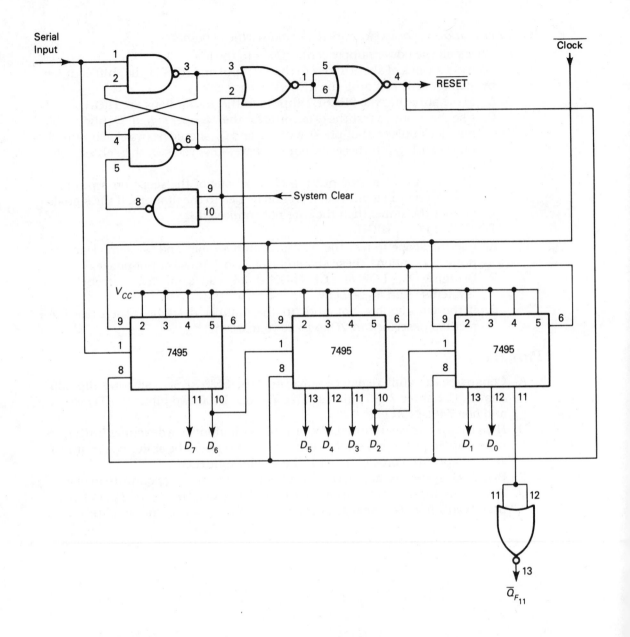

If your circuit does not work properly, consider these points:

1. Check all power supply connections.
2. Check all inputs and outputs for proper voltage levels.
3. Supply a 1 Hz or slower $\overline{\text{Clock}}$ to the clock input, tie the serial input HIGH and pulse system clear, HIGH, and then LOW.

A. Pin 3 of the 7408 AND gate should be LOW. If it is HIGH, then there is something wrong with the crossed NAND $\overline{\text{SET}}$-$\overline{\text{RESET}}$ flip-flop or the system clear input.

B. Using the scope or voltmeter, check all the pins for their expected logic levels. The D_0 through D_7 should be HIGH.

4. Put channel one of the scope on the slow 1 Hz clock input to the circuit. Put channel two of the scope on pin 11 of the 7493, which supplies the signal for $\overline{Q}_{F_1}$. Momentarily bring the serial input LOW for one clock pulse of the 1 Hz clock.

A. If the shift registers are working correctly, pin 11 should go LOW 11 clock pulses later.

B. If pin 11 never goes HIGH, then back up to D_2 and repeat the procedure.

C. If D_2 never goes LOW, then repeat the procedure with D_1.

D. Use the scope to trace wiring to find any faults discovered.

Part IV

A. Construct the parallel register shown. Use two 7475 D latches and eight LEDs.

B. Test your register to make sure it works properly.

C. Put the inputs to your register on the protoboard in a logical order to make it easy to connect to the next part of the serial receiver.

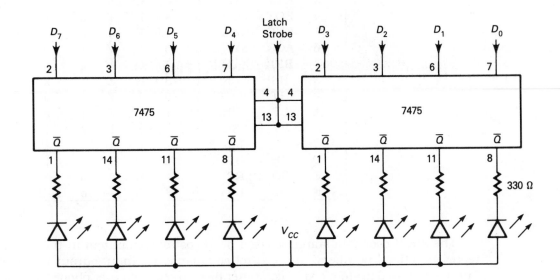

If your circuit does not work properly, consider these points:

1. Check all power supply connections.
2. Bring D_0 through D_7 HIGH and pulse the latch strobe HIGH. This should turn on all the LEDs.
3. Bring D_0 through D_7 LOW while the latch strobe is LOW. This should not change the state of the LEDs until you pulse the clock HIGH again at which time they should turn off.
4. If the current limiting resistors are too low, then the 7475 IC will not work properly because of the heavy I_{CC}.

Part V

If the total circuit does not work properly after you have combined all the parts, consider these points:

1. Is each circuit connected to a common ground?
2. Put a scope probe on the serial input to detect the incoming data word from the terminal.
3. Use the other scope channel to trace the signal back from the serial input.

ASCII CODE CONVERSION TABLE

BITS 4 thru 6	—	0	1	2	3	4	5	6	7
	0	NUL	DLE	SP	0	@	P	`	p
	1	SOH	DC1	!	1	A	Q	a	q
	2	STX	DC2	"	2	B	R	b	r
	3	ETX	DC3	#	3	C	S	c	s
	4	EOT	DC4	$	4	D	T	d	t
	5	ENQ	NAK	%	5	E	U	e	u
	6	ACK	SYN	&	6	F	V	f	v
BITS 0 thru 3	7	BEL	ETB	'	7	G	W	g	w
	8	BS	CAN	(	8	H	X	h	x
	9	HT	EM	)	9	I	Y	i	y
	A	LF	SUB	*	:	J	Z	j	z
	B	VT	ESC	+	;	K	[	k	{
	C	FF	FS	,	<	L	\	l	¦
	D	CR	GS	-	=	M	]	m	}
	E	SO	RS	.	>	N	∧	n	~
	F	SI	US	/	?	O	—	o	DEL

The ASCII code is a standard used by almost all computers in the world to represent letters, numbers, and control commands for input/output terminals. This code is used to store text in computer files, to send text over phone wires, and similar tasks.

The first 32 ASCII codes are commands for the computer terminal, and the rest with the exception of 7F hex are printable characters.

The binary numbers that your instructor will send to your newly constructed serial receiver will be in ASCII code. You must write the binary code in hex form and use the ASCII conversion table to translate the message to letters and numbers. Do this by using the most significant hex digit to find the column and the least significant hex digit for the row in the table. For instance the hex number 4A would be the ASCII code for the capital letter J.

If the ASCII characters being sent to your serial receiver are coming from a standard computer terminal with an RS-232C interface, hold the control key down and press the letters starting with A. You will find that this will give you the first 32 control codes in the ASCII table.

CHAPTER 10

OUTLINE

Counters

OBJECTIVES

After completing this chapter, you should be able to:

- explain the operation of a ripple counter.
- describe the decode and clear method of making a divide-by-N counter.
- explain the use of a presettable counter.
- explain how to design a divide-by-N synchronous counter.
- use typical MSI counter ICs.
- describe the up-down counter.

10.1 THE RIPPLE COUNTER

Figure 10-1 shows a 4-bit ripple counter and the waveform it generates. The negative-edge-triggered JK flip-flops are set to toggle. The clock is tied to the Q output of the previous flip-flop. This means the first flip-flop (A) must change states from a Q of HIGH-to-LOW in order for the next flip-flop (B) to toggle. Notice that flip-flop A changes states on every trailing edge of the input clock, flip-flop B changes states on the trailing edge of flip-flop A and flip-flop C changes states on the trailing edge of flip-flop B. This procedure continues for as many flip-flops as are in the counter.

Notice also that the output frequency of each flip-flop is one-half the frequency of the previous flip-flop. This means the output frequency of flip-flop A is one-half the clock frequency and the output frequency of flip-flop B is one-half the frequency of flip-flop A or one-fourth the clock frequency.

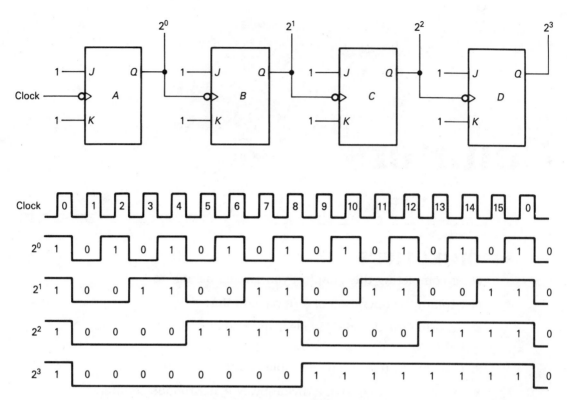

FIGURE 10-1   A 4-bit ripple counter

This divide-by-2 for each succeeding flip-flop continues for as many flip-flops as are in the counter.

Because the counter divides the frequency of the previous flip-flop by 2, the value of the outputs for each flip-flop at any count of the input clock will be the binary number of that clock pulse. This means the counter in Figure 10-1 will count in binary from 0000 to 1111, and then start over again with 0000. The largest number the counter will display is a function of the number of flip-flops in the counter, because each flip-flop produces 1 bit in the binary number. Therefore, the formula for the largest binary number that can be displayed for a given number of bits will determine the largest number that can be displayed for a given counter of N flip-flops.

$$\text{Largest binary number} = 2^N - 1$$

For the counter in Figure 10-1, the largest number that can be displayed is $2^4 - 1$ or 15. Since the D output divides the input frequency by 2^4 or 16, we would call this counter a divide-by-16 ripple counter. If you were to add one more flip-flop to the counter in Figure 10-1 it would be a divide-by-2^5 or divide-by-32 counter.

The last flip-flop in a ripple counter must wait for the input signal to ripple down through each preceding flip-flop before it can change. Because of this cumulative propagation delay of a ripple counter, the larger it is, the slower it is. If each flip-flop in Figure 10-1 has a propagation delay of 25 nanoseconds it would take 4×25 nanoseconds for the counter to change from 0111 to 1000. The output does not change together, but one after the other.

10.2 THE DECODE AND CLEAR METHOD OF MAKING A DIVIDE-BY-N RIPPLE COUNTER

The divide-by-5 counter in Figure 10-2 is a ripple counter which uses the decode and clear method of resetting the counter after the 5th clock pulse. The number 5 is decoded by the NAND gate which produces a LOW output when the number 5 appears on the output of the counter. As the waveforms show, the number 5 is not present for very long and then the counter is reset to 000 causing the LOW output of the NAND gate to go back HIGH and starting the counter over again at 000 for the next pulse of the input clock. This effectively resets the counter at the number 5 but it produces a small spike at the 5 count on the A output. Also the C output is a bit longer than it should be. These spikes can cause problems if the output which contains them is used for input clocks to other counters because they are counted as an extra pulse. A counter of divide-by-any number you wish can be constructed by this method.

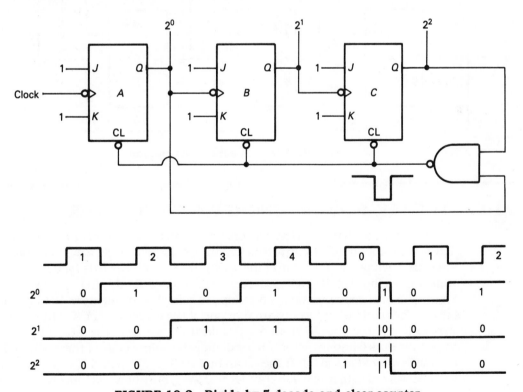

FIGURE 10-2   Divide-by-5 decode-and-clear counter

The divide-by-*N* ripple counter has many problems which have been previously discussed. To correct these problems you can use a synchronous counter. A synchronous counter has all the clock inputs tied together so that each flip-flop changes state at the same time. This means that the propagation for the whole counter is the same as one flip-flop, no matter how many flip-flops are in the counter. To do this, each flip-flop must be steered by using the *JK* input so it will change to its proper state when the next clock comes along. The counter in Figure 10-2 does this by allowing the flip-flop to toggle or not to toggle by using AND gates to produce a 1 on the JK input only when there is 1 on all the outputs before the flip-flop. This counter will produce the same waveforms as the ripple counter in Figure 10-1, but it is much faster.

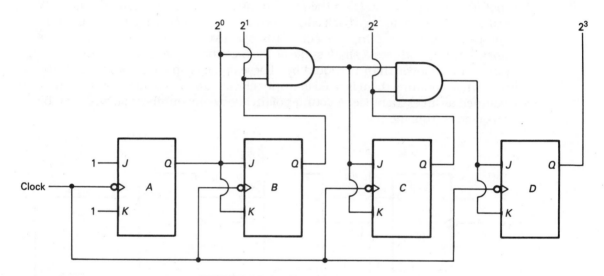

FIGURE 10-3   Synchronous counter

10.3   THE DIVIDE-BY-*N* SYNCHRONOUS COUNTER

To design a divide-by-*N* synchronous counter, the first thing you must do is define the values of the *JK* inputs before the clock pulse to get a desired change in *Q* after the clock pulse. The truth table in Figure 10-4 shows this for a negative edge-triggered *JK* flip-flop. If you have trouble with the truth table in Figure 10-4, refer to Figure 8-3 in the chapter on *JK* flip-flops to refresh your memory. Notice that in all lines of the truth table there is an *X* for one of the values of *J* or *K*. This means that the *J* or *K* input can be either a 1 or a 0. Take the first line in the table, for instance. If the *Q* output is 0 and we want it to be 0 after the falling edge of the clock to the flip-flop, then *J* must equal 0 but the *K* can be 1 or 0. If *J* = 0 and *K* = 1, the *Q* output would be forced to 0, and if *J* = 0 and *K* = 0 the *Q* output would not change from the 0 state it was already in.

Before Clock	After Clock	Before Clock	
Q	Q	J	K
0	0	0	X
0	1	1	X
1	0	X	1
1	1	X	0

X = 1 or 0

FIGURE 10-4   Truth table for a negative edge-triggered *JK* flip-flop

The second step in the design of a divide-by-*N* synchronous counter is to define the *JK* input for each flip-flop in the counter before the clock to obtain the desired *Q* output after the clock. Figure 10-5 shows this for a divide-by-5 synchronous counter. The desired *Q* output in this case is a binary count from 0 to 101. The output for the first *Q* values before the first clock would be 000. The desired *Q* output after the LOW-going edge of the first clock would be 001. This output can be achieved by making *J* = 0 and *K* = *X* (0 or 1) on the two most significant flip-flops, *C* and *B*, and by making *J* = 1 and *K* = *X* (1 or 0) on the *A* flip-flop. After the first clock the *Q* output will be 001 and will be ready to be set up for the second LOW-going clock, which will change it to 010.

Q Before the Clock			Q After the Clock			C		B		A	
C	B	A	C	B	A	J	K	J	K	J	K
0	0	0	0	0	1	0	X	0	X	1	X
0	0	1	0	1	0	0	X	1	X	X	1
0	1	0	0	1	1	0	X	X	0	1	X
0	1	1	1	0	0	1	X	X	1	X	1
1	0	0	0	0	0	X	1	0	X	0	X

X = 1 or 0

FIGURE 10-5   *JK* input for a divide-by-5 synchronous counter

$$K_A = A\,\overline{B}\,\overline{C} + A\,B\,\overline{C}$$
$$K_A = A\,\overline{C}\,(\overline{B} + B)$$
$$K_A = A\,\overline{C}$$

FIGURE 10-6   Reducing to minimal terms

Once this is done, a Boolean expression needs to be made for each *J* and *K* input which will express the *J* or *K* input with respect to the *Q* outputs before the clock. This can be done by expressing each 1 in the *J* and *K* columns of the truth table as an AND gate whose inputs are the *Q* values before the clock, then OR each AND expression together as shown in Figure 10-6. Reduce the expression to its lowest Boolean terms.

This method will give a correct Boolean statement, but it will not always be the simplest form which will work for the truth table because it presumes that all the other combinations of the truth table must be 0. This is not always the case as can be seen if you examine the K_A column of the truth table in Figure 10-5. Notice that two input combinations must be a 1. All the other input combinations can be 1 or 0 as shown by the X. Therefore, K_A could be made 1 and fulfill the requirements for the truth table.

Figure 10-7 shows the simplest Boolean expressions for the truth table in Figure 10-5. Examine these expressions and confirm that there is no simpler form.

All that is needed to complete the design of our divide-by-5 synchronous counter is to draw the logic diagram from the Boolean expressions, as shown in Figure 10-8.

$$
\begin{array}{ccc}
J_A = \overline{C} & J_B = A & J_C = AB \\
K_A = 1 & K_B = A & K_C = 1
\end{array}
$$

FIGURE 10-7   Minimal terms for divide-by-5 synchronous counter

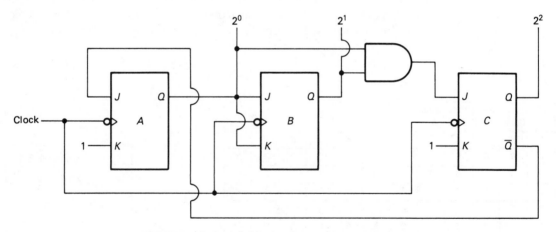

FIGURE 10-8   Divide-by-5 synchronous counter

This design method could be used to design a string of JK flip-flops that would step through any sequence of output you desired.

10.4   PRESETTABLE COUNTERS

The presettable counter shown in Figure 10-9 uses a set of NAND gates to supply a $\overline{\text{CLEAR}}$ or $\overline{\text{PRESET}}$ signal to each flip-flop in the ripple counter. A 1 on the PRESET control will enable the NAND gates allowing the data on the PRESET input to pass, thus setting the counter's outputs to the value on the PRESET inputs.

While the PRESET control is HIGH the counter will hold the value of the PRESET input because the $\overline{\text{CLEAR}}$ or $\overline{\text{PRESET}}$ of a JK flip-flop overrides the clock of

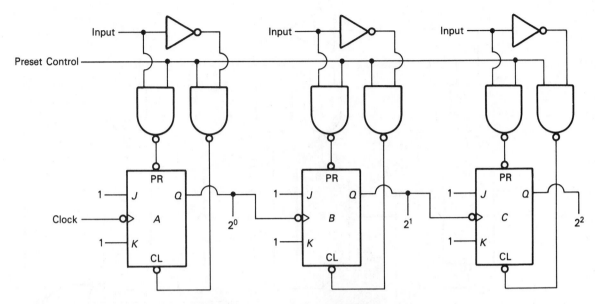

FIGURE 10-9   3-bit presettable counter

the flip-flop, but when the PRESET control falls LOW the $\overline{\text{PRESET}}$ and $\overline{\text{CLEAR}}$ will both be HIGH allowing the counter to start counting from the PRESET value on the next falling edge of the clock.

The ability of this counter to preset to a predetermined value before counting begins allows us to use it as a programmable divide-by-N counter as shown in Figure 10-10. Notice the decode 0 and preset routine produces a short spike just as in the decode and clear method used for the divide-by-N counter. The advantage here is we can control the divisor by the binary number put on the PRESET inputs. The counter in Figure 10-10, PRESET to the binary number equivalent to 5, starts counting up to the binary number equivalent to 7 and then is RESET on the next falling edge of the clock to 5 again. Therefore, this is a divide-by-3 counter.

10.5   TYPICAL MSI COUNTER ICS

Figure 10-11 shows the logic diagram for three medium-scale-integration TTL counters. The 7490 is a combination divide-by-2 and divide-by-5 counter which can be configured to be a divide-by-10 counter by connecting the output of the divide-by-2 counter to the input of the divide-by-5 counter. Notice that the divide-by-5 part of the 7490 is part ripple and part synchronous.

The 7492 is a divide-by-2 and divide-by-6 counter. A divide-by-12 counter can be made by connecting these two counters. Also, the divide-by-6 part of the 7492 is part ripple and part synchronous.

The 7493 is a straight divide-by-2 and divide-by-8 ripple counter which can be connected to produce a divide-by-16 counter.

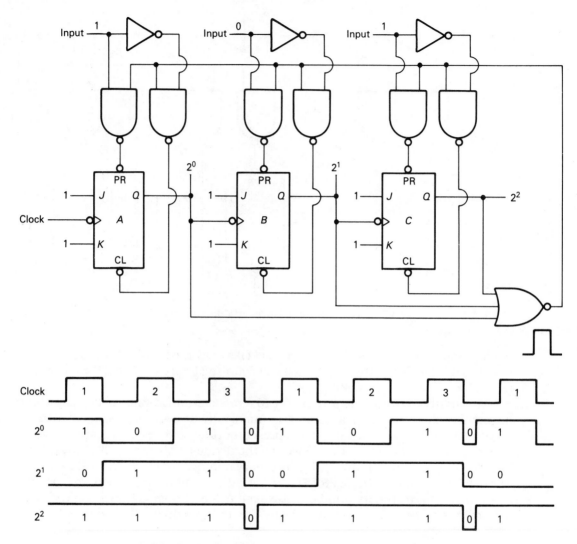

FIGURE 10-10   Presettable counter set to divide by 3

Figure 10-12 shows these counters used to make a digital clock. The 7447 IC is a binary-coded decimal to 7-segment decoder/driver, used to display the time on the common-anode LEDs. We will study these ICs and LEDs in a later chapter in this text.

10.6   THE UP-DOWN COUNTER

If you construct a ripple counter by using the $\overline{Q}$ of each flip-flop as the clock to the next flip-flop, the counter will count down from its maximum count to 0 and then start over again. Figure 10-13 shows this for a divide-by-8 countdown counter.

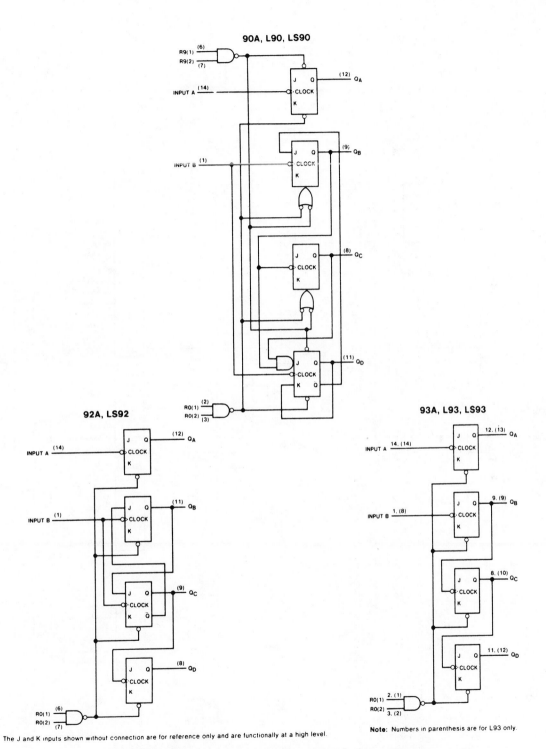

The J and K inputs shown without connection are for reference only and are functionally at a high level.

Note: Numbers in parenthesis are for L93 only.

FIGURE 10-11   Medium-scale-integration TTL counters

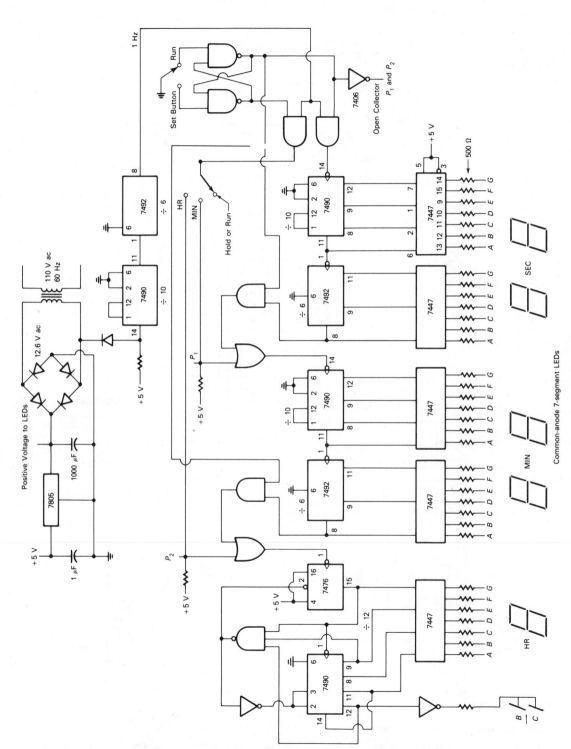

FIGURE 10-12 Digital clock

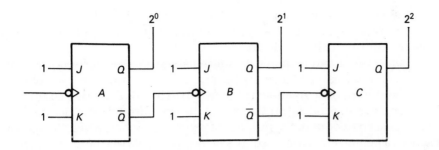

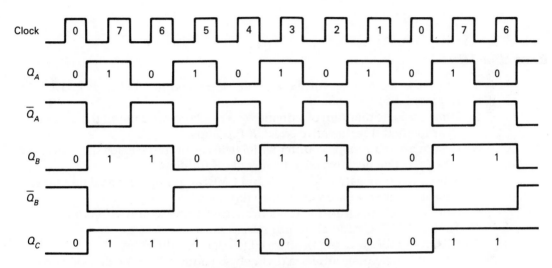

FIGURE 10-13   Ripple down counter

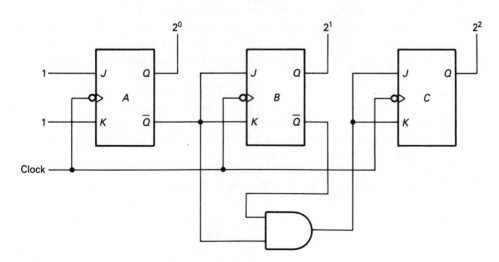

FIGURE 10-14   Synchronous down counter

Figure 10-14 shows a synchronous countdown counter. Notice that the steering logic which controls the flip-flops is derived from the $\overline{Q}$ outputs of the flip-flops. This causes the counter to count down.

An up-down counter can be made by adding steering logic derived from the Q outputs to the counter as in Figure 10-14, and a control input to control which steering logic will be enabled. As shown in Figure 10-15, this gives us a counter that will count up or down depending on the level of the up-down input. Such a counter could be used to keep a count of the number of people in a room. When the sensor at the door senses a person entering, it counts up one pulse. When the sensor senses a person leaving the room, it counts down one count.

Exercise 10

1. Draw the logic diagram for a 5-bit ripple counter using negative edge-triggered *JK* flip-flops.
2. Draw the logic diagram of a divide-by-9 ripple counter using the decode and clear method. Use negative edge *JK* flip-flops.
3. Design a synchronous counter which will count in the following order: 000, 100, 001. Use negative edge-triggered *JK* flip-flops.
4. Make a list of negative-edge TTL and CMOS flip-flops in the data manuals.
5. Repeat number 4 for positive-edge flip-flops.
6. Draw the logic diagram for a synchronous counter which will count from 0 to 12. Use 7476 ICs and show pin numbers.
7. Repeat number 3 using positive edge-triggered flip-flops.
8. Draw the logic diagram for a divide-by-6 counter. Use the decode and clear method and one 7490 IC.
9. Draw the waveforms for the counter in number 8.
10. Draw the logic diagram for a ripple-down counter which will count from 15 to 0 and start over again at 15. Use 7476 ICs and show pin numbers.
11. Draw the waveforms for the down counter in number 10.
12. Draw the logic diagram for a presettable counter which will count from 3 to 10 and start over again at 3. Use 7476 ICs and show pinouts.
13. Draw the waveforms for the counter in number 12.
14. Draw the logic diagram for a synchronous up-down counter which will count up from 0 to 15 or down from 15 to 0. Use 7476, 7432, 7408 and 7404 ICs and show pinouts.
15. Draw the waveforms for the up-down counter in number 14 with the following count: 0, 1, 2, 3, 4, 5, 4, 3, 2, 3, 4.

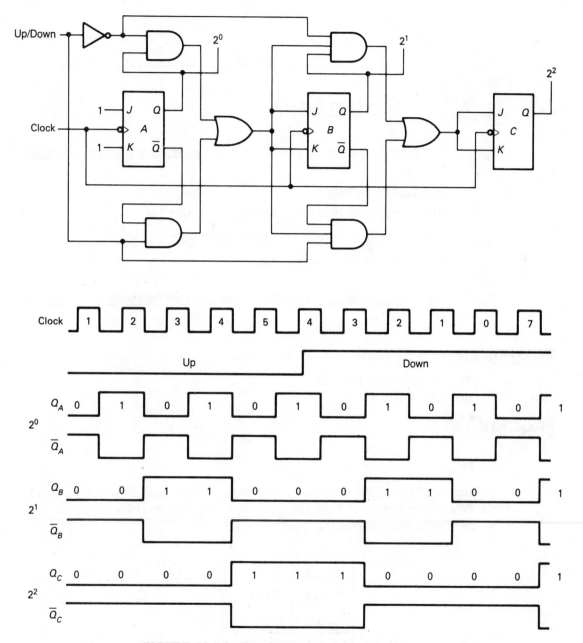

FIGURE 10-15   Synchronous up-down counter

10 Counters

OBJECTIVES

After completing this lab, you should be able to:

■ design a divide-by-*N* synchronous counter.

■ use typical TTL counter ICs.

COMPONENTS NEEDED

2	7476 dual *JK* flip-flop ICs
1	quad AND gate IC

PROCEDURE

Part I

Using negative edge-triggered *JK* flip-flops, design and construct a divide-by-10 synchronous counter. Find the missing values in the tables shown.

A. Define the operation of the flip-flop to be used.

Before Clock	After Clock		Before Clock	
Q	Q	$\rightarrow$	J	K
0	0	$\rightarrow$		
0	1	$\rightarrow$		
1	0	$\rightarrow$		
1	1	$\rightarrow$		

B. Define the *JK* for each flip-flop in the counter for each count required.

Before Clock				After Clock				JK state before Clock							
D	C	B	A	D	C	B	A	J_D	K_D	J_C	K_C	J_B	K_B	J_A	K_A
0	0	0	0	0	0	0	1								
0	0	0	1	0	0	1	0								
0	0	1	0	0	0	1	1								
0	0	1	1	0	1	0	0								
0	1	0	0	0	1	0	1								
0	1	0	1	0	1	1	0								
0	1	1	0	0	1	1	1								
0	1	1	1	1	0	0	0								
1	0	0	0	1	0	0	1								
1	0	0	1	0	0	0	0								

C. Write the Boolean statement for each *J* and *K* input in the counter.

$$J_A = \qquad J_B = \qquad J_C = \qquad J_D =$$
$$K_A = \qquad K_B = \qquad K_C = \qquad K_D =$$

D. Convert the Boolean statements for the *JK* inputs to a logic diagram.

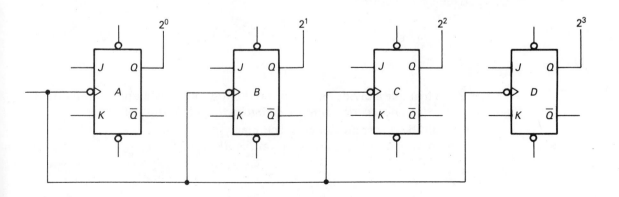

E. Construct the circuit you designed and use it to draw the following wave-forms. Have your instructor check the operation.

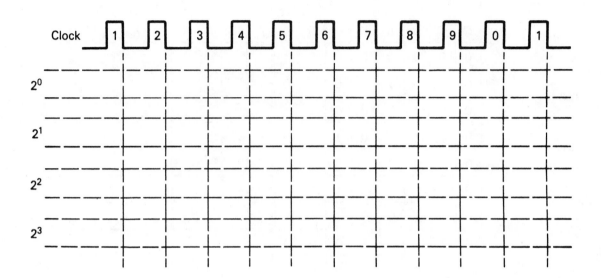

Part II

Use a 7490 BCD counter to make the following counters. Have your instructor check the operation.

A. A divide-by-5 counter.
B. A divide-by-2 counter.
C. A divide-by-10 counter.

Part III

Use a 7493 to make a divide-by-10 counter using the decode-and-clear method. *If your counter does not work properly, consider the following points:*

1. Check all the power supply connections.
2. Check for any unconnected input on the *JK* flip-flop.

3. Use the scope to trace the signal from the input clock to the last flip-flop. They should match the expected waveform for a BCD counter.
4. If the waveforms do not match the expected BCD waveform either your wiring is at fault or your design is incorrect.

CHAPTER 11

Schmitt-Trigger Inputs and Clocks

OBJECTIVES

After completing this chapter, you should be able to:

- explain the operation of a Schmitt-trigger input.

- use a Schmitt-trigger input to square up a sine wave.

- use the Schmitt trigger in the construction of a clock.

- describe how the 555 timer works and how it is used as a clock.

- use a 4001 CMOS IC to construct a crystal oscillator.

11.1 THE SCHMITT-TRIGGER INPUT

Figure 11-1 shows the graph of the input voltage versus the output voltage of a typical TTL Schmitt-trigger input. As the input voltage rises, the output stays at a LOW or 0 value until the input voltage reaches about 1.8 V. At this upper threshold, the output snaps to a logic 1 value. When the input voltage drops, the output does not return to logic 0 until the input voltage drops below the lower threshold of about 0.8 V. The difference in the upper and lower thresholds is called the hysteresis of the Schmitt-trigger and is typically 1 V for a TTL Schmitt-trigger input. The symbol for a Schmitt-trigger input is the graph in Figure 11-1 as shown in the noninverting gate on the right side of the figure.

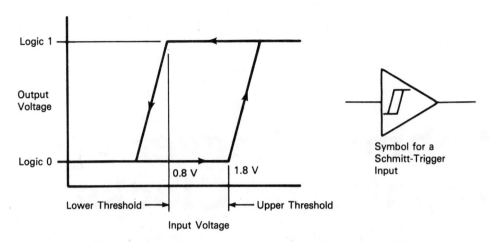

FIGURE 11-1 Output voltage versus input voltage of a Schmitt trigger

11.2 USING A SCHMITT TRIGGER TO SQUARE UP AN IRREGULAR WAVE

The fact that the Schmitt trigger has hysteresis is the reason it can be used to square up a wave such as a sine wave. As the input voltage rises and passes the upper threshold, the output voltage changes state. It will not change state again until the input voltage drops below the lower threshold. This is shown in Figure 11-2 which uses a 7414 Schmitt-trigger input to an inverter.

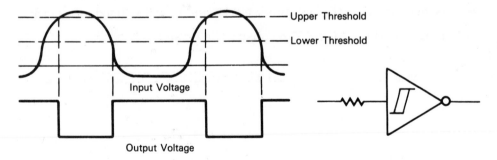

FIGURE 11-2 Using a Schmitt-trigger input to square up a sine wave

Notice the input voltage only goes 0.7 V below ground. This is because the lower part of the sine wave has been clipped off by clamping diodes in the input of the 7414 IC. To protect this clamping diode, a current-limiting resistor is used to feed the sine wave.

11.3 A SCHMITT-TRIGGER CLOCK

A *clock* is an oscillator or sometimes called an astable multivibrator and is used in a digital circuit. Figure 11-3 shows a simple clock made from a 7414 Schmitt trigger. When point *A* (which is the inverter output) is HIGH or at logic 1, the capacitor will charge through the 1-kΩ resistor and the TTL input as shown in Figure 11-4A. When the capacitor voltage reaches the upper threshold of the Schmitt trigger, the output of the inverter drops to 0 voltage or logic 0. This causes the capacitor to discharge through the 1-kΩ resistor as shown in Figure 11-4B. When the capacitor voltage has dropped to the lower threshold, the inverter output will change back to a logic 1, thus completing one cycle of the clock as shown in Figure 11-4. Notice that the capacitor charges much faster than it discharges because it can charge through the 1-kΩ resistor and the TTL input of the inverter, but can only discharge through the 1-kΩ resistor which takes longer.

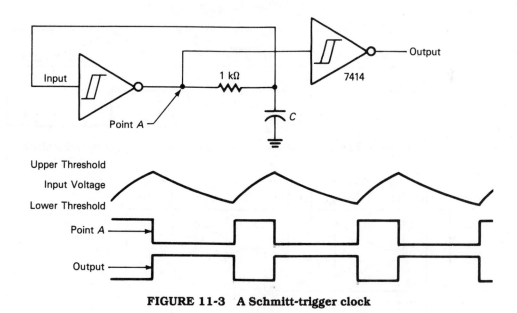

FIGURE 11-3 A Schmitt-trigger clock

The frequency at which the clock will run is dependent on the *RC* time constant of charge and discharge. Because the TTL inverter input is a source load and helps to charge the capacitor, the resistor cannot be much larger than 1 kΩ or the discharge voltage will never fall below the lower threshold of the Schmitt-trigger input. Therefore, we must keep the resistor *R* at about 1 kΩ, but we can change the capacitor, which will change the *RC* time constant for charge and discharge. The formula for the frequency of the clock is shown as a function of the capacitors assuming that *R* = 1 kΩ.

$$F \approx \frac{6.79 \times 10^{1-4}}{C}$$

The second inverter is used as a current buffer to drive other circuits without affecting the operation of the clock circuit.

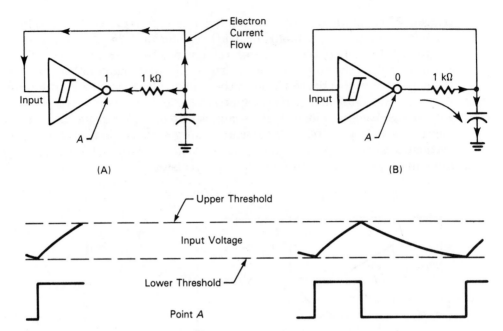

(A)   (B)

FIGURE 11-4   One clock cycle of the Schmitt-trigger inverter clock

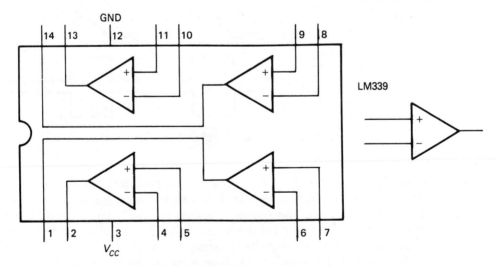

FIGURE 11-5   The LM339 voltage comparator

11.4   THE 555 TIMER USED AS A CLOCK

The 555 timer is a general purpose timer IC which can be used for many applications. To understand its operation, we must first understand the operation of a voltage comparator made from an operational amplifier IC. Figure 11-5 shows an LM339 voltage comparator IC. On one IC there are four comparators which have two inputs each, one marked + and the other marked − . Each comparator also has one open-collector output which is not typical of most operational amplifiers. This is because this operational amplifier is designed to be used as a voltage comparator in a digital circuit where the output will be ground or V_{CC} only.

The supply voltage for the IC can range from 3 V to 15 V, and the inputs have very high impedance. This means it can be used with CMOS circuits and that the inputs will not have any effect on a circuit to which they are connected.

If a reference voltage is placed on the negative input as shown in Figure 11-6A and the positive input is raised to a voltage greater than the negative input, the output will go to HiZ which will produce a logic 1 because of the external pull-up resistor. If the positive input is now lowered to a voltage below the negative input voltage, the output will go to ground which produces a logic zero as shown in Figure 11-6B. In short, if the positive input is greater than the negative input, the output will be HiZ. If the positive input is less than the negative input, the output will be 0 or ground.

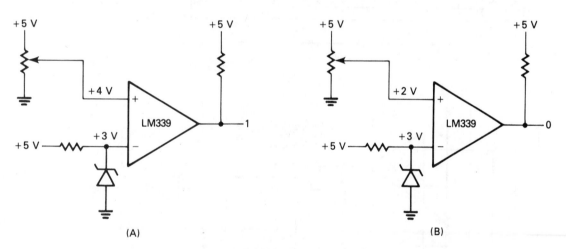

FIGURE 11-6   Operation of the LM339 voltage comparator

The 555 uses two of this type of comparator to set and reset a flip-flop. Figure 11-7 shows a 555 IC set up to make a clock. The reference voltage for the comparators is set by a voltage divider made up of three 5-kΩ resistors, which is where the chip got the name 555. This voltage divider places ⅓ of the supply voltage on the positive input of the lower comparator and ⅔ of the supply voltage on the negative input of the upper comparator. The negative input of the lower comparator is called the trigger and the positive input of the upper comparator is called the threshold.

If the threshold and trigger are tied together, they can be used to SET and RESET the flip-flop inside the 555 by raising them above ⅔V_{CC} and lowering them below ⅓V_{CC}.

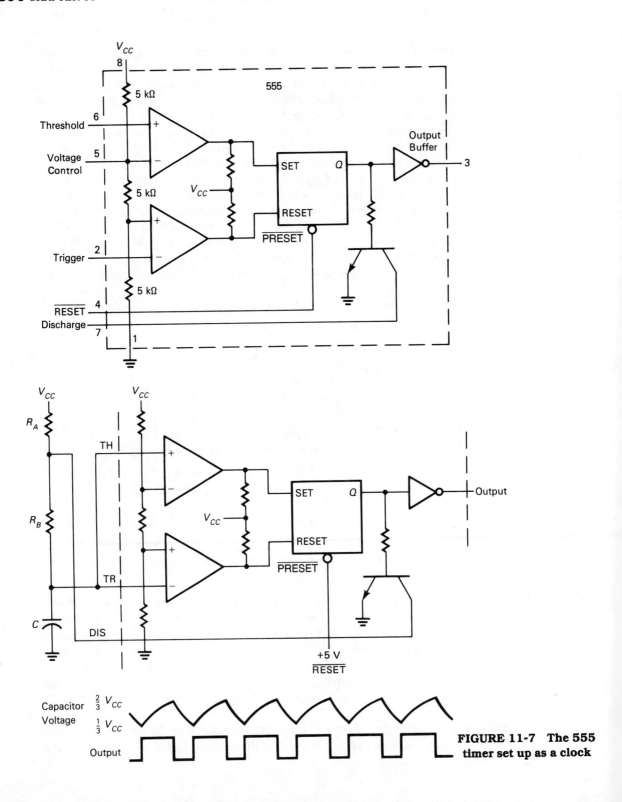

FIGURE 11-7   The 555
timer set up as a clock

When the trigger and threshold are above $\frac{2}{3}V_{CC}$ then the top comparator is on or at logic 1 because the positive input is above the negative input which is at $\frac{2}{3}V_{CC}$ at all times. The bottom comparator is at logic 0 because the positive input is set at $\frac{1}{3}$ V_{CC} by the three 5-kΩ resistor voltage dividers and the negative input is greater than that voltage. This sets the flip-flop to a one state and the ICs output goes LOW because of the inverting buffer.

As the voltage on the trigger and threshold decreases below $\frac{2}{3}V_{CC}$ but greater than $\frac{1}{3}V_{CC}$ the top comparator goes to 0 and the lower comparator stays at 0. This is the no-change state of the internal flip-flop. Therefore, it stays set to logic 1. When the input voltage on the inputs drops further to below $\frac{1}{3}V_{CC}$, the bottom comparator turns on or goes to logic 1. This causes the flip-flop to be RESET. Therefore, we can SET or RESET the internal flip-flop of the 555 by raising the two inputs' threshold and trigger above $\frac{2}{3}V_{CC}$ and below $\frac{1}{3}V_{CC}$.

The output of the internal flip-flop is also connected to the base of the discharge transistor inside the IC. When the Q output is at logic 1, it turns on the transistor connecting the discharge pin to ground. When the Q output is 0, the transistor is off and the discharge pin is at HiZ or not connected to anything.

By connecting a resistor capacitor circuit to the threshold, trigger and discharge inputs as shown in Figure 11-7, the 555 will oscillate or become a clock. When the flip-flop is RESET to a 0 state, the discharge is in the HiZ state which allows the capacitor to charge through R_A and R_B. When the capacitor voltage reaches a voltage just above $\frac{2}{3}V_{CC}$, the flip-flop will change states to a logic 1. This causes the discharge transistor to be turned on and the capacitor starts to discharge through R_B and the discharge transistor until the voltage drops to a little below $\frac{1}{3}V_{CC}$ at which time the flip-flop is RESET and the whole cycle starts over. Figure 11-7 also shows the waveform for the 555 clock. Remember that the output of the 555 is the complement of the Q output of the flip-flop.

A formula for the frequency of the output can be computed by using the formula for the voltage charge of a capacitor as a function of the RC time constant and the time of charge.

$$V_C = V_S (1 - e^{\frac{-T}{RC}})$$

where V_C = capacitor voltage
V_S = supply voltage
T = time of charge
RC = resistance $\times$ capacitor, i.e., RC time constant

First, we need to solve the equation for the time of charge:

$$V_C = V_S (1 - e^{\frac{-T}{RC}})$$

$$\frac{V_C}{V_S} = 1 - e^{\frac{-T}{RC}}$$

$$\frac{V_C}{V_S} - 1 = -e^{\frac{-T}{RC}}$$

$$\frac{-V_C}{V_S} + 1 = e^{\frac{-T}{RC}}$$

$$\ln\left(\frac{-V_C}{V_S} + 1\right) = \frac{-T}{RC}$$

$$-T = (RC)\ln\left(\frac{-V_C}{V_S} + 1\right)$$

$$T = -(RC)\ln\left(\frac{-V_C}{V_S} + 1\right)$$

The time needed to change from $\frac{1}{3}V_S$ to $\frac{2}{3}V_S$ is equal to:

$$T = \left[-(RC)\ln\left(\frac{-\frac{2}{3}V_S}{V_S} + 1\right)\right] - \left[-(RC)\ln\left(\frac{-\frac{1}{3}V_S}{V_S} + 1\right)\right]$$

$$T = -RC\left[\ln\left(-\frac{2}{3} + 1\right) - \ln\left(-\frac{1}{3} + 1\right)\right]$$

$$T = -RC\left[\ln\left(\frac{1}{3}\right) - \ln\left(\frac{2}{3}\right)\right]$$

$$T = -RC\left[-1.10 - (-0.41)\right]$$

$$T = 0.69(RC)$$

Now we have an equation which will give us the time it will take to charge and discharge the capacitor in the middle $\frac{1}{3}$ of the voltage supply as a function of the RC time constant.

One cycle consists of one charge time and one discharge time. The RC time constant for the charge is $C(R_B + R_A)$ because the capacitor charges through both R_A and R_B but the discharge RC time constant is CR_B because the capacitor discharges through R_B only. Therefore, the charge time will be longer than the discharge time because of the difference in the RC time constants. With this knowledge in hand, we can construct an equation for the total time of one cycle of the clock.

$$T_C = 0.69(R_B + R_A)C$$

and

$$T_{DC} = 0.69CR_B$$

where T_C = time to charge the middle $\frac{1}{3}$ of V_{CC}
T_{DC} = time to discharge the middle $\frac{1}{3}$ of V_{CC}

Therefore the time for one cycle (P) is

$$P = T_C + T_{DC}$$
$$P = 0.69(R_B + R_A)C + 0.69CR_B$$
$$P = 0.69CR_B + 0.69CR_A + 0.69CR_B$$
$$P = 0.69C(R_B + R_A + R_B)$$
$$P = 0.69C(2R_B + R_A)$$

The frequency of a clock is equal to the reciprocal of the period P.

$$F = \frac{1}{P}$$

Therefore:

$$F = \frac{1}{0.69C(2R_B + R_A)}$$

$$F = \frac{1.44}{C(2R_B + R_A)}$$

This is the frequency formula for the 555 timer clock in Figure 11-7.

The 555 will produce a fairly stable output frequency from very long periods to about 0.5 MHz. It can run on 5 V to 18 V for a supply voltage and draws about 3 mA to 10 mA of current with no output load. One nice feature is its current output ability. The 555 can sink or source up to 200 mA of current, which means you can drive a heavy load with it.

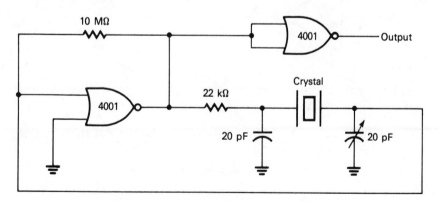

FIGURE 11-8   CMOS crystal oscillator

11.5   CRYSTAL OSCILLATORS

When very precise and stable clocks are needed, a quartz crystal is used to generate the frequency. Figure 11-8 shows a crystal oscillator made from a 4001 CMOS NOR gate. This circuit works well up to the limits placed on it due to the

propagation delays of the CMOS NOR gate. The frequency of oscillation is determined by the vibrating frequency of the quartz crystal.

By placing a 10-MΩ resistor from the output to the input of the NOR gate, the
input is biased at $\frac{1}{2}V_{DD}$. This basically makes a high-gain amplifier out of the gate.
By placing a crystal and a PI network of capacitors from output to input, we can
cause the amplifier to ring or oscillate at the frequency of the crystal. The variable
20-pF capacitor will vary the output frequency slightly to fine-adjust the output
frequency. The buffer gate is used to keep the circuit being connected to the clock
from affecting the operation of the clock.

Exercise 11

1. Using the Data Manual, find the upper and lower thresholds for a 74C14
   when V_{DD} is 10 V.
2. Find the operating frequency of a 555 timer wired as in Figure 11-7. Use the
   following values for the components.

	R_B	R_A	C
a.	1 kΩ	1 kΩ	0.01 μF
b.	3 kΩ	1 kΩ	0.1 μF
c.	1 kΩ	5 kΩ	10 μF

3. Draw the waveforms for the circuit in Figure 11-9.

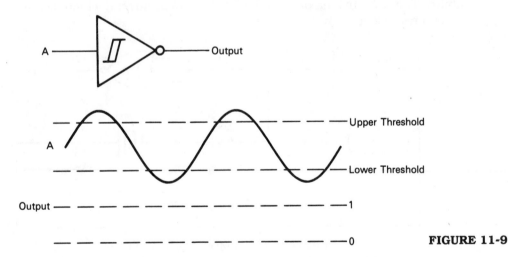

FIGURE 11-9

4. How much time will it take for the capacitor shown in Figure 11-10 to charge
   to $\frac{1}{4}V_S$, $\frac{1}{2}V_S$ and $\frac{3}{4}V_S$ from no charge when the switch is closed?

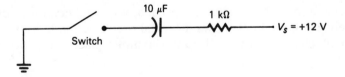

FIGURE 11-10

5. Using the approach used to compute the formula for the frequency of the 555 timer clock, compute the formula for the CMOS Schmitt clock shown in Figure 11-11. Remember the CMOS input is very HiZ.

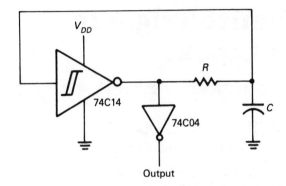

V_{DD}

R

74C14

74C04

C

Output

FIGURE 11-11

6. Draw the logic diagram for a Schmitt-trigger clock, like the one shown in Figure 11-3, which will oscillate at a frequency of 5 kHz. Use a 7414 IC and show pinouts and component values.
7. Why does it take longer to discharge the capacitor in Figure 11-3 than to charge it?
8. What is the output value of an LM339 voltage comparator if the positive input is greater in voltage than the negative input?
9. Draw the logic diagram for a clock which will produce a 2 kHz even duty cycle TTL square wave. Use a 555 timer and a 7476 IC and show pinouts.
10. Draw the waveforms for the clock in number 9. Show the capacitor voltage of the 555 timer, the output of the 555 timer and the output of the JK flip-flop.
11. Compute the formula for the frequency of the 555 time clock in Figure 11-7 if a 5-kΩ resistor is placed from the voltage control to ground.
12. Repeat number 11 but place the resistor from the voltage control to V_{CC}.
13. Draw the logic diagram for a Schmitt-trigger circuit to square up a 5 V peak-to-peak ac sine wave and produce a 10 V HIGH-to-LOW dc wave. Use a 7414 and a 7407.
14. Draw the waveforms for the circuit in number 13.
15. Compute the formula for the frequency of the clock in Figure 11-3 if a 74C14 IC is used and the V_{DD} is 5 V.

11 Schmitt Triggers and Clocks

OBJECTIVES

After completing this lab, you should be able to:

- explain the operation of a Schmitt trigger and measure the upper and lower thresholds.

- explain the operation of an oscillator made from a Schmitt-trigger inverter.

- explain the operation of a 555 timer used as an oscillator.

COMPONENTS NEEDED

1	7414 hex Schmitt-trigger inverter IC
2	1-kΩ ¼-W resistors
2	capacitors (value to be computed)
1	555 timer IC
1	0.01-μF capacitor
2	20-pF capacitor
2	crystals of different frequency below 1 MHz
1	10-MΩ, ¼-W resistor
1	22-kΩ, ¼-W resistor
1	4001 CMOS quad NOR gate IC

PROCEDURE

Part I

A. Construct the circuit for the figure shown and supply the input with a 4 V ac to 5 V peak ac input waveform at about 1 kHz frequency.

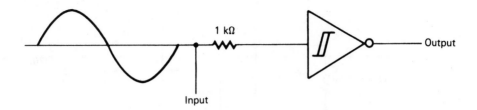

Input

B. Use the scope to measure the upper and lower thresholds. Using graph paper, draw the input and output waveforms.

Part II

A. Construct the oscillator shown.

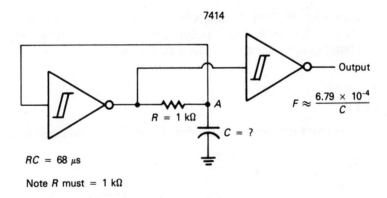

7414

Output

$$F \approx \frac{6.79 \times 10^{-4}}{C}$$

$R = 1\ k\Omega$

$C = ?$

$RC = 68\ \mu s$

Note R must $= 1\ k\Omega$

B. Compute the expected frequency for the oscillator.
C. Display the output wave on the scope and measure the frequency.
D. Display the output wave and point A on the scope. Using graph paper, draw the two waveforms.

Part III

A. Construct the 555 astable multivibrator shown.

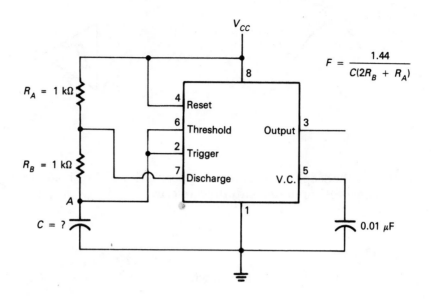

$$F = \frac{1.44}{C(2R_B + R_A)}$$

B. Compute the value C that will produce a frequency of 9.6 kHz.

C. Display the output and point A on the scope and measure the frequency. Using graph paper, draw the waveforms.

Part IV

A. Construct the crystal oscillator shown using the crystal provided.

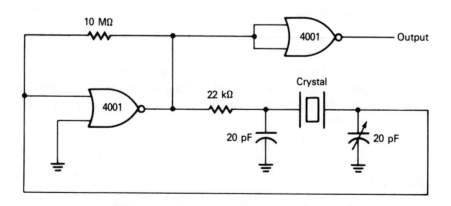

B. Measure the output frequency of the oscillator on the scope. Does it match the frequency of the crystal?

C. Change the crystal to a different crystal with a different frequency and measure the output frequency again. Does the output frequency match the crystal?

CHAPTER **12**

OUTLINE

One-Shots

OBJECTIVES

After completing this chapter, you should be able to:

- describe how to use an *RC* network to debounce a switch.

- describe how to make a pulse stretcher.

- describe how to condition the input of the pulse stretcher to make a retriggerable one-shot.

- use the 555 timer as a one-shot.

- use the 74121 and 74122 one-shots.

- make a data separator from one-shots.

12.1 A ONE-SHOT DEBOUNCE SWITCH

The circuit shown in Figure 12-1 uses an *RC* time constant and a Schmitt trigger to debounce a momentary switch or button. When the button is pushed, the capacitor is discharged very rapidly. When the button is released, the bounce of the metal contacts makes and breaks the circuit in a random manner. The open switch allows the capacitor to start to charge through the resistor connected to V_{CC}. The time for the voltage to reach the upper threshold of the Schmitt-trigger input is dependent on the *RC* time constant. Therefore, the switch must remain open for a certain period of time before the output will change states.

This is a retriggerable one-shot, because each time the switch closes, the capacitor is discharged and the timing cycle begins again. The input may bounce,

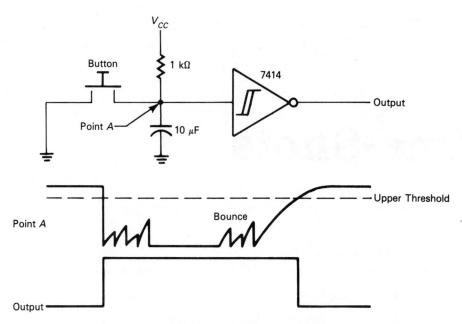

FIGURE 12-1   Debounce switch

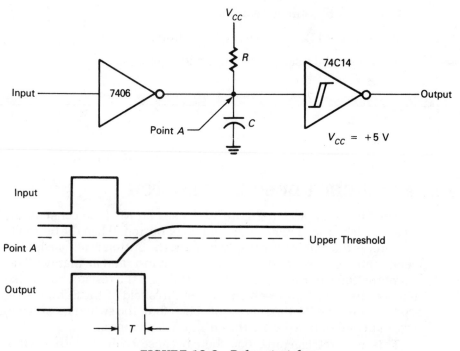

FIGURE 12-2   Pulse stretcher

but the output will not change until the switch has stayed open for a period of time determined by the RC time constant.

12.2 THE PULSE STRETCHER

By adding an open-collector gate such as a 7406 to the input of the debounce switch in Figure 12-1 and changing from a 7414 to a 74C14, a pulse stretcher can be made as shown in Figure 12-2. The input to the 74C14 is held LOW by the output of the 7406 as long as the 7406 input is HIGH. When the input goes back LOW, the capacitor starts to charge through the resistor. When the capacitor voltage reaches the upper threshold of the 74C14 Schmitt trigger, the output will snap LOW. This will stretch the positive going pulse by the amount of time it takes the capacitor to charge to the upper threshold of the 74C14 Schmitt trigger.

If we use a CMOS Schmitt trigger we can ignore the input impedance because a CMOS input has very high impedance; therefore, the only thing which will affect the charge time is the resistor and capacitor used. By using a little algebra, we can devise a formula for the time which this circuit will stretch the input pulse.

$$T = -(RC)\ln\left(\frac{-V_C}{V_S} + 1\right)$$

where V_C = capacitor voltage
V_S = supply voltage
T = time of charge
RC = resistance × capacitor, i.e., RC time constant

Using the CMOS data specification for a 74C14, we find that the upper threshold voltage is 3.6 V at a V_{DD} of 5 V. Substituting this in the equation we get a simpler formula for the time stretch for the circuit in Figure 12-2.

$$T = -(RC)\ln\left(\frac{-3.6\,\text{V}}{5\,\text{V}} + 1\right)$$

$$T = 1.27\,(RC)$$

This formula does not take into account the propagation delays for the two gates used; however this is of no concern except in very short total time pulses.

12.3 THE RETRIGGERABLE ONE-SHOT

In the circuit just discussed, the output could not time out or change states until the input returns LOW. This can be changed by conditioning the input of the 7406 with a capacitor, resistor, diode, edge-triggered circuit as shown in Figure 12-3.

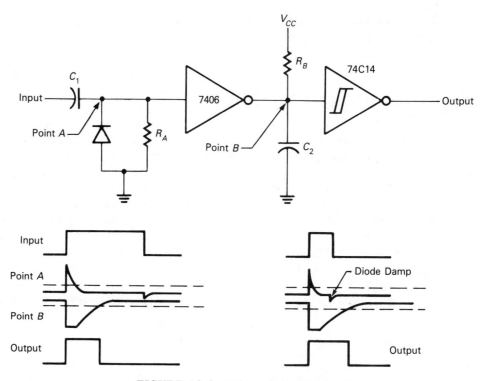

FIGURE 12-3   Edge-triggered one-shot

When the input goes HIGH, capacitor C_1 has no voltage drop and all the voltage drop is across the resistor R_A. This causes the 7406 output to go LOW. After the capacitor charges to a large enough voltage, the input of the 7406 will be LOW causing the 7406 output to go to HiZ.

This means that on the positive edge of the input the 7406 output will put out a very short negative-going pulse. This negative-going pulse discharges capacitor C_2 and starts the timing cycle of the 74C14 Schmitt trigger.

The RC time constant for the 7406 conditioned input should be very short so that output of the 7406 will be a very short negative-going pulse. If this pulse is very short compared to the time of the 74C14, then it can be ignored in the calculations of the time out for the whole circuit. Therefore, the formula for the time out of the one-shot is the same as the pulse stretcher.

The advantage of this circuit is that the output pulse width is independent of the length of the input pulse width. If the input is retriggered by a positive edge before the 74C14 times out, capacitor C_2 is discharged and the timing cycle starts all over again. This is shown in the waveform in Figure 12-4.

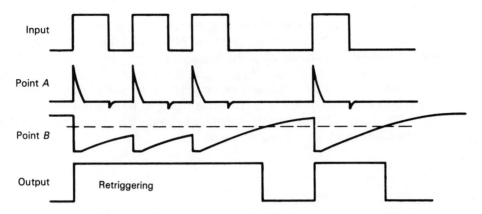

FIGURE 12-4   Triggering the one-shot in Figure 12-3

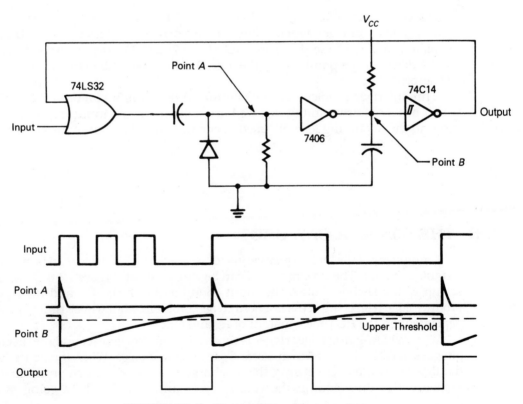

FIGURE 12-5   Nonretriggerable one-shot

Notice that if the one-shot is retriggered quickly enough, the capacitor C_2 never reaches the upper threshold voltage of the 74C14 and the output stays HIGH. This is what is meant whenever we say a one-shot is retriggerable.

The diode in parallel with R_A prevents the discharging capacitor from pulling the input voltage to the 7406 any lower than the forward bias voltage of the diode which is 0.7 V. This diode does the same thing as the clamping diodes inside the 7406. The diode can be left out in many cases where the size of the capacitor is small and the discharge current is also small.

12.4   THE NONRETRIGGERABLE ONE-SHOT

If it is not desired to have the one-shot retriggerable, an OR gate can be used to inhibit the input for the period of the one-shot's timing period. This is shown in Figure 12-5.

Notice that the OR gate is a 74LS32 IC. This is because we used a CMOS Schmitt-trigger inverter which can just barely drive one LS input load of 0.36 mA. Therefore, we must take the output off of the 74LS32 output or we will load the 74C14 down to the point where the 1 or 0 values would be out of the standard TTL levels.

This interface problem can be eliminated by using a standard 7414 TTL IC which has more drive ability, but we have to figure in the input impedance of its input to the RC timing formula used to compute the time out of the one-shot.

12.5   THE 555 AS A ONE-SHOT

Using some of the techniques we just discussed, the 555 timer can be made into a stable one-shot. The timing period can be long or short. Figure 12-6 shows a 555 set up as a one-shot. Notice the input conditioning of the RC and the diode to produce edge triggering. This is a nonretriggerable one-shot whose timing period is dependent on the RC time constant of R_A and C_A.

When a LOW-going edge trigger is input, the flip-flop is reset and the discharge pin goes to HiZ. This allows capacitor C_A to start to charge. When it reaches $\frac{2}{3}V_{CC}$, the flip-flop will be set causing the discharge transistor to turn on discharging capacitor C_A, thus ending the timing cycle until another LOW-going edge is inputted.

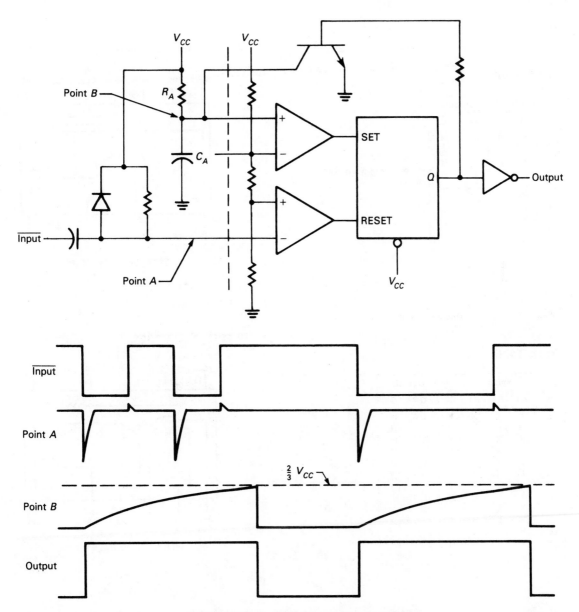

FIGURE 12-6   555 timer as a one-shot

The time of pulse width is the time it takes C_A to change to $\frac{2}{3}V_{CC}$. From previous calculations we know the equation for the time of charge for a resistor and capacitor, and with a little algebra we can compute a formula for the timing period of the 555 one-shot in Figure 12-6.

121 One Shots

Truth Table

Inputs			Outputs	
A1	A2	B	Q	Q̄
L	X	H	L	H
X	L	H	L	H
X	X	L	L	H
H	H	X	L	H
H	↓	H	⊓	⊔
↓	H	H	⊓	⊔
↓	↓	H	⊓	⊔
L	X	↑	⊓	⊔
X	L	↑	⊓	⊔

Pulse Width = 0.7(RC)

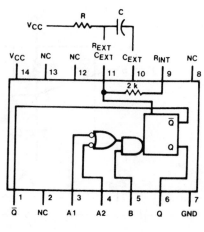

54121 (J,W); 74121 (N)

122 Retriggerable One Shots with Clear

Truth Table

Inputs					Outputs	
Clear	A1	A2	B1	B2	Q	Q̄
L	X	X	X	X	L	H
X	H	H	X	X	L	H
X	X	X	L	X	L	H
X	X	X	X	L	L	H
X	L	X	H	H	L	H
H	L	X	↑	H	⊓	⊔
H	L	X	H	↑	⊓	⊔
H	X	L	H	H	L	H
H	X	L	↑	H	⊓	⊔
H	X	L	H	↑	⊓	⊔
H	H	↓	H	H	⊓	⊔
H	↓	↓	H	H	⊓	⊔
H	↓	H	H	H	⊓	⊔
↑	L	X	H	H	⊓	⊔
↑	X	L	H	H	⊓	⊔

Pulse Width = 0.45(RC)

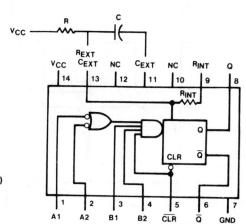

54LS122 (J,W); 74LS122 (N)

Notes: ⊓ = one high-level pulse, ⊔ = low-level pulse.
To use the internal timing resistor of 54121/74121, connect R_{INT} to V_{CC}
An external timing capacitor may be connected between C_{EXT} and R_{EXT}/C_{EXT} (positive)
For accurate repeatable pulse widths, connect an external resistor between R_{EXT}/C_{EXT} and V_{CC} with R_{INT} open-circuited
To obtain variable pulse widths, connect external variable resistance between R_{INT} or R_{EXT}/C_{EXT} and V_{CC}.

FIGURE 12-7   One-shot truth tables

$$T = -(RC) \ln \left(\frac{-V_C}{V_S} + 1 \right)$$

$$T = -(RC) \ln \left(\frac{-\frac{2}{3}V_S}{V_S} + 1 \right)$$

$$T = -(RC) \ln (\frac{1}{3})$$

$$T = 1.1 \, (RC)$$

12.6 THE 74121 AND 74LS122

The 74121 is a nonretriggerable one-shot which has three inputs to the timing circuits. The 74LS122 is a retriggerable one-shot with an active LOW clear. Figure 12-7 shows their pin output and truth table. The pulse width can be controlled by an external resistor and capacitor or by an external capacitor and the internal resistor. These are very handy ICs for many applications. The 74123 is a dual retriggerable one-shot with clear.

12.7 THE DATA SEPARATOR

A data separator is a digital circuit which separates the data from the system clock of stored serial data. This data can be stored on magnetic tape or a magnetic disk and it can be stored by several different methods.

The data separator shown in Figure 12-8 is designed to separate the clock from the data for the input to a UART. This type of serial data transmission was discussed earlier in the chapter on shift registers. The data separator in Figure 12-8 is designed to work with the asynchronous receiver shown in the chapter on shift registers. The data separator produces a clock 12 times the baud rate of the data. This is the clock frequency the receiver needs to input the serial data and transfer it to the parallel register.

To understand the operation of this data separator, we first need to look at the data as it comes off the tape. Data is stored on the tape using two sine wave frequencies. A logic 1 is indicated by a 3.6 kHz frequency and a 0 by one-half of the 3.6 kHz frequency, 1.8 kHz. The baud rate or bit rate is one-twelfth of the one frequency or 3.6 kHz divided by 12, which equals 300 bits per second. This is a commonly used baud rate for serial data stored on tape recorders.

The first 741 op amp is used to amplify the input signal from the tape recorder's output. The second 741 op amp is a zero crossing detector to make a square wave which changes state on the zero transition of the sine wave. This zero detector has a hysteresis of a little over 1 V to help eliminate any noise problem at the transition

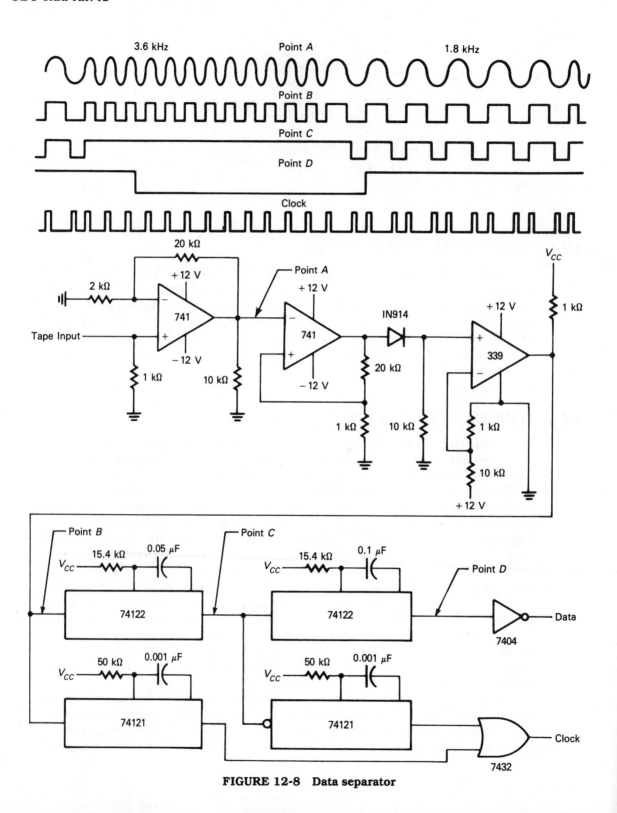

FIGURE 12-8   Data separator

point. A diode is used to prevent the input signal to the 339 op amp from going below ground. The 339 op amp converts the +12 V square wave to a standard TTL square wave of 0 V to +5 V as shown in Figure 12-8 at point B.

After the signal is recovered and converted to a TTL level input, it is fed to a 74122 retriggerable one-shot whose pulse width is set to 1.25 times the period of the 1 frequency. This means that the 74122 one-shot will be retriggered when the frequencies are 3.6 kHz or 1 and the output will not time out and return to a 0 level. When the input frequency to the 74122 one-shot is 1.8 kHz or a 0, the output will time-out producing a square wave whose frequency is that of the 0 frequency or 1.8 kHz. This is shown on the timing diagram in Figure 12-8 at point C.

The output of the first 74122 is fed to the input of the second 74122. The pulse width of this one-shot is 1.25 of one cycle of the 0 frequency or 1.8 kHz. When the first one-shot is HIGH because it is being retriggered, the second one-shot will time-out and go LOW for the period of time a 1 is on the incoming signal to the first one-shot. When the incoming frequency drops to the 0 frequency, the first one-shot passes on the 1.8 kHz frequency to the second one-shot. This causes the second one-shot to be retriggered and the output will go HIGH for the time that a 0 is being fed to the first one-shot. This is shown in Figure 12-8 at point D. Notice the second one-shot produces a 0 when a 1 frequency is fed to the first one-shot and a 1 when a 0 frequency is fed to the first one-shot's input. This is corrected by the inverter on the output of the second one-shot.

The clock is recovered by two nonretriggerable 74121 one-shots which have a short pulse width. Their outputs are ORed together to produce the clock which is 12 times the baud rate. One of the 74121 one-shots is fed from the positive edge of the incoming TTL signal and produces the proper clock when the incoming frequency is at 3.6 kHz or a 1 frequency, but will only produce half that when the incoming frequency drops to a 0 value. At this point the second 74121 supplies the missing clock pulse because it is fed from the falling edge of the output of the first one-shot which will only be there when a 0 frequency is present. This is shown in Figure 12-8. Notice that the duty cycle of the clock wave is not even, but it will still drive the asynchronous receiver quite well.

This method of recording the receiver data and clock on the tape at the same time removes the problem of uneven baud rates due to the uneven mechanical speed of the tape recorder and also allows us to use a device which was really designed for storing analog speech to store digital data.

Exercise 12

1. What is the total pulse width for the circuit shown?

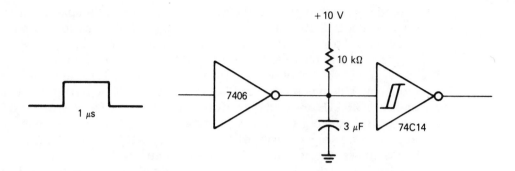

2. What would be the capacitor's value to produce a 1 μs pulse using a 74121 if the resistor used is 10 kΩ?
3. At what frequency will the circuit in Figure 12-3 stop timing out and have a constant 1 for the output? ($R = 1$ kΩ and $C = 0.01$ μF)
4. Complete the waveform for the one-shot described in number 3 for the given input waveform.

5. Draw the logic diagram for a pulse stretcher which will stretch the incoming pulse 5 microseconds. Use a 74C14 IC, a capacitor and resistor and a 7406 IC.
6. Draw the waveform for the input signal, capacitor voltage and the output waveform of the circuit in number 5.
7. Draw the logic diagram for an edge-triggered one-shot which has a 15-millisecond time-out pulse. Use a 74C14 IC, a 7406 IC and two capacitors and resistors.
8. Using the nonretriggerable one-shot circuit in Figure 12-5, draw the logic diagram for a one-shot with a pulse width of 30 microseconds.
9. Draw the logic diagram for a one-shot using a 555 timer with a time-out period of 30 seconds.
10. Use a 74121 to produce a one-shot with a time-out period of 2 microseconds. Show pin numbers.
11. At what input frequency will a retriggerable one-shot, made from a 74122, stop producing a pulse on its output if the capacitor used is 0.1 microfarad and the resistor is 1 kilohm.

12. Design a retriggerable one-shot which will stop producing a pulse on its output when the input frequency is 1.5 kHz.
13. Make a list of CMOS one-shots and draw the pinouts.
14. What is the time-out period for the one-shot in Figure 12-2 if the capacitor is 0.5 microfarad and the resistor is 3.3 kilohms?

LAB 12 One-Shots

OBJECTIVES

After completing this lab, you should be able to:

■ construct a one-shot from a 74C14 and an RC circuit.

■ use a 74121 to shorten a positive pulse.

■ use the oscilloscope to observe the waveforms for the circuits in this lab.

COMPONENTS NEEDED

1	7414 TTL hex Schmitt-trigger input inverter IC
1	74C14 CMOS hex Schmitt-trigger input inverter IC
1	7406 hex open-collector output inverter IC
2	1-kΩ, ¼-W resistor
1	0.01 µF capacitor
1	1N914 diode or equivalent

PROCEDURE

1. Construct the circuit shown and use the ac signal generator to produce a 20-kHz input signal.

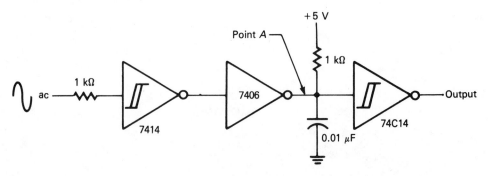

2. Using the figure shown in number 1, draw the expected waveforms and the actual waveform seen on the scope for the output and point *A*. Use graph paper.

3. Make the one-shot edge-triggered by adding the diode, capacitor and resistor as shown.

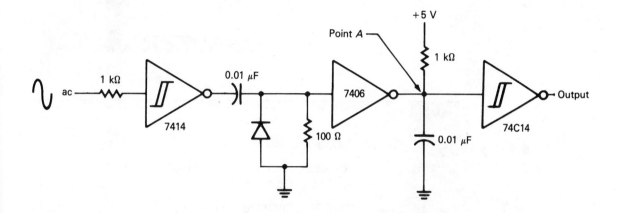

4. Using the figure shown in number 3, draw the expected waveforms and the actual waveform seen on the scope for the output and point *A*. Use graph paper.

5. Draw the logic diagram for a circuit which will take an ac 10 kHz even-duty cycle wave and produce a 10 kHz wave which is HIGH 25% of the time and LOW 75% of the time. Use a 7414 and 74121 to construct the circuit.

CHAPTER **13**

OUTLINE

Digital-to-Analog and Analog-to-Digital Conversions

OBJECTIVES

After completing this chapter, you should be able to:

- use resistor networks for digital-to-analog conversion.

- explain the operation of a TTL digital-to-analog converter.

- use voltage comparators to produce an analog-to-digital converter.

- describe the count-up and compare method of analog-to-digital conversion.

- describe the successive approximation method of analog-to-digital conversion.

13.1 RESISTOR NETWORKS FOR DIGITAL-TO-ANALOG CONVERSION

We will look at two resistor networks to do the job of converting a binary number to a proportional analog voltage. The first is the binary ladder. Figure 13-1 shows the binary ladder made with a switch for each binary bit instead of TTL outputs. This will help to simplify the explanation.

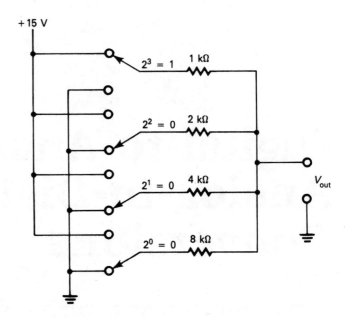

FIGURE 13-1   The binary ladder D-to-A converter

The binary number which is set by the switches is 1000 or decimal number 8. The largest number that the four switches can express is 1111 or decimal number 15. In this case a 1 is + 15 V and a 0 is ground. Therefore, if the binary number 1111 or decimal number 15 is put on the switches, the output of the binary ladder is tied to the + 15 V supply through all the resistors in parallel, as shown in Figure 13-2. This produces a 15 V output voltage. If all the switches are switched to the 0 position, the output is 0 V or ground as shown in Figure 13-3.

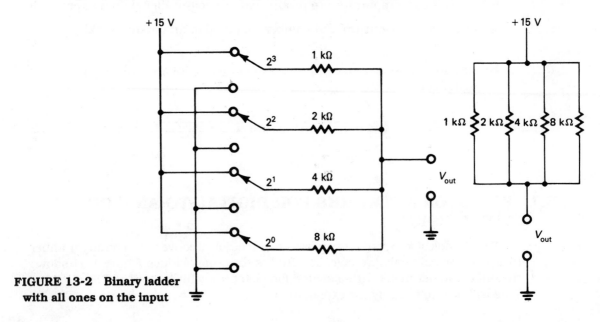

FIGURE 13-2   Binary ladder with all ones on the input

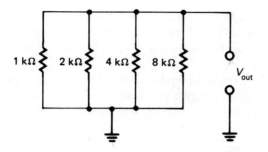

FIGURE 13-3   Binary ladder with all zeros on the input

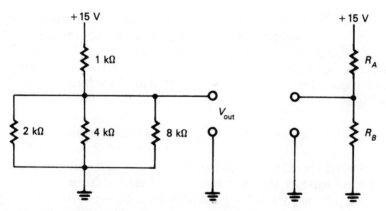

FIGURE 13-4   Equivalent circuit for the binary ladder with 1000 on the input

Now let us analyze the switch configuration in Figure 13-1. The equivalent circuit is shown in Figure 13-4. If we reduce the circuit to two series-equivalent resistors, then the voltage output will be the voltage drop across R_B. Using the voltage divider formula we can find the output voltage for the binary number 1000 or decimal number 8 to be 8 V.

$$R_B = \frac{1}{\dfrac{1}{2\ k\Omega} + \dfrac{1}{4\ k\Omega} + \dfrac{1}{8\ k\Omega}} = 1.1429\ k\Omega$$

$$V_{out} = V_S \left(\frac{R_B}{R_A + R_B} \right)$$

$$V_{out} = 15\ V \left(\frac{1.1429\ k\Omega}{1\ k\Omega + 1.1429\ k\Omega} \right)$$

$$V_{out} = 8\ V$$

The output voltage for the other possible binary number inputs can be computed in a similar manner. You will find the voltage increments to be 1 V for this binary ladder. The binary number equivalent to 10 produces a voltage of 10 V and the binary number equivalent to 7 produces a voltage of 7 V. In other words, the supply voltage is divided into increments equal to the supply voltage divided by the largest binary number which can be input to the resistor network. Therefore, the voltage increment for a binary ladder is found using the following formula.

$$\text{Binary ladder voltage increment} = \frac{V_S}{2^N - 1}$$

where V_S = voltage supply
N = number of bits in the binary number input

When N equals the number of bits in the binary number input to the binary ladder, $2_N - 1$ is the largest number that can be expressed for a binary number of N bits. If this formula is the amount of each increment, then the final output voltage must be equal to the binary number input to the binary ladder times the voltage increment. The following formula is used for the voltage output of the circuit shown in Figure 13-1.

$$\text{Binary ladder: } V_{\text{out}} = \text{binary number input} \left(\frac{V_S}{2^N - 1} \right)$$

The values of the resistors in the binary ladder are divided by 2 for each binary power increase, that is, a 2^0 resistor is 8 kΩ, a 2^1 resistor is 4 kΩ, a 2^2 resistor is 2 kΩ and a 2^3 resistor is 1 kΩ. If a fifth bit is added to the binary ladder, the resistor value is one-half the 2^3 resistor or 500 Ω. You can see that the larger the binary number gets, the smaller the resistor must get. Also it is not easy to get resistors with the exact values that fit this pattern.

These two problems can be eliminated by using another type of resistor network to produce the proportional output voltage. This is the $2R$ network shown in Figure 13-5. Using the same method used in Figure 13-4, we can determine the output voltage for the binary number switched into the $2R$ network in Figure 13-5. This is shown in Figure 13-6.

The $2R$ network is similar to the binary ladder except that the voltage increments are equal to the voltage supply divided by the total number of combinations in the binary number input. The total number of combinations in a binary number of N bits is 2^N. Therefore, the formula for the voltage output of the $2R$ network is equal to the binary number input times the voltage supply divided by 2^N.

$$2R \text{ network: } V_{\text{out}} = \text{binary number} \left(\frac{V_S}{2^N} \right)$$

Both of these networks give very accurate output voltages as long as the output load impedance is very high with respect to the network's impedance. If the load resistance is lowered the linearity of the output is reduced. To reduce this problem a HiZ buffer is usually used such as an operational amplifier to drive the analog load.

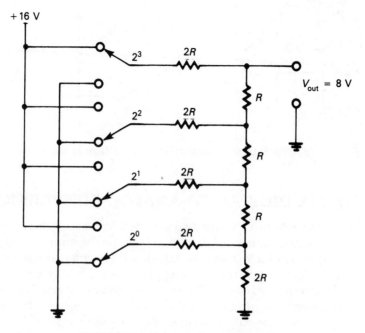

FIGURE 13-5   2R D-to-A converter

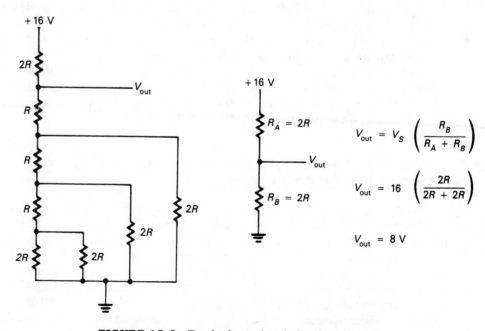

FIGURE 13-6   Equivalent circuit for 2R network

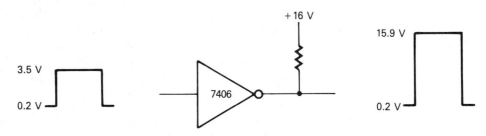

FIGURE 13-7   Converting TTL voltage levels to D-to-A levels

13.2   THE TTL DIGITAL-TO-ANALOG CONVERTER

The binary number in the two previous networks was not a TTL level input. To make the network work, the 1 voltage must be the supply voltage and the 0 voltage must be ground or 0 V. The TTL output voltage will give a good 0 voltage, or in the worst case, 0.4 V, but the 1 voltage is about 3.5 V typically. By using an open-collector output, such as the 7406 inverter has, and a pull-up resistor to V_S of the D-to-A network, we can convert the TTL level voltages to the voltage needed by the D-to-A network. The output voltage of the 7406 is not absolute ground or absolute V_S but it is very close. This is shown in Figure 13-7.

Figure 13-8 shows this type of buffer used to make a TTL D-to-A converter where the top output voltage is 15 V. When the output of the 7406 goes LOW, it will go to 0.1 V or 0.2 V above ground at best which introduces some error into the D-to-A converter and when the 7406 output goes to HiZ the 1-kΩ resistor pulls the 20-kΩ resistor up to + 16 V. In doing so, it adds its 1-kΩ resistor to the 20-kΩ resistor. This means the $2R$ resistor is a little larger (5% in this case) when it is at + 16 V and does not quite reach ground when it is brought LOW.

These errors can be eliminated by other methods but for many uses of D-to-A converters these errors are tolerable. Figure 13-9 shows a D-to-A converter used to control the speed of a small dc motor such as might be used on a robotic arm. Notice the operational amplifier used to buffer the D-to-A converter.

13.3   ANALOG-TO-DIGITAL CONVERSION USING VOLTAGE COMPARATORS

The voltage comparator, which was discussed in the chapter on clocks, can be used to make a very fast analog-to-digital converter. An analog-to-digital converter produces a binary number which is in direct proportion to an analog voltage input.

Figure 13-10 shows a 3-bit A-to-D converter made from seven LM339 voltage comparators. The negative input of each comparator is tied to a resistor voltage divider which divides the 8-V supply into 1-V increments. Each voltage comparator has a reference voltage of 1 V greater than the previous comparator. All of the positive inputs to the comparators are tied together so the input voltage will increase on all comparators at the same time.

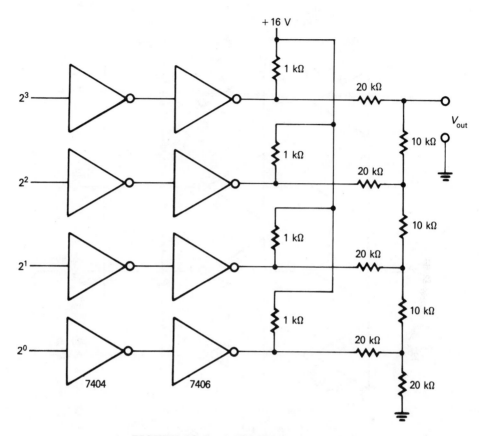

FIGURE 13-8   A TTL 2R D-to-A converter

If the input voltage increases to 2.5 V, then the output of the first two comparators will be + 5 V or logic 1 because the positive input will be greater than the negative input, but the rest of the comparators' outputs will be at ground or logic 0. The output of the LM339 is an open-collector output so by using a pull-up resistor to + 5 V the output will be standard TTL levels even though the input may increase to 8 V. As the voltage increases to 3.5 V the third comparator's output will change to a logic 1. If the analog voltage increases to above 7 V, all the comparators will be at logic 1. The comparators will go to logic 0 when the input voltage goes below the reference voltages set by the voltage divider.

The output of all the comparators is decoded to form a 3-bit binary number by the logic gates shown in Figure 13-10. When all the comparator's outputs are at logic 0, the output of their corresponding NAND gate is 1, because any 0 into a NAND gate produces a 1 output. If the analog input voltage rises to 1.5 V, the first comparator's output goes to logic 1 which is fed to the 2-input NAND gates. The other input to this NAND gate is from the inverter which comes off the output of the second comparator. The output of the second comparator is still 0 because the

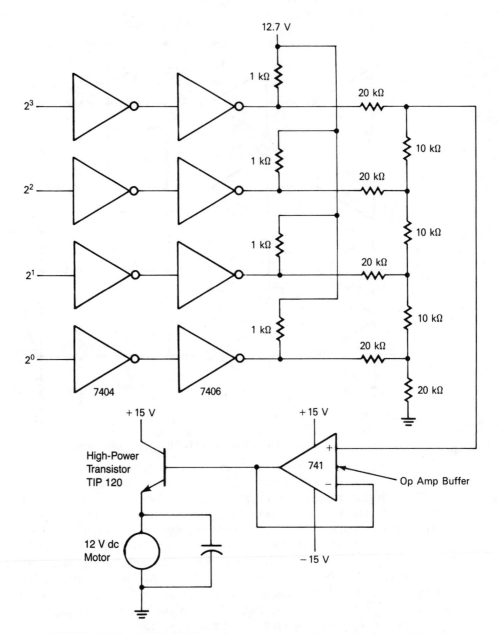

FIGURE 13-9   A TTL 2R D-to-A converter used to control a small motor

analog voltage is still 1.5 V and is not yet large enough to change the state of the second comparator. This 0 is inverted to a 1 and fed to the input of the NAND gate for the first comparator. At this point the first NAND gate has 1 and 1 on its inputs and

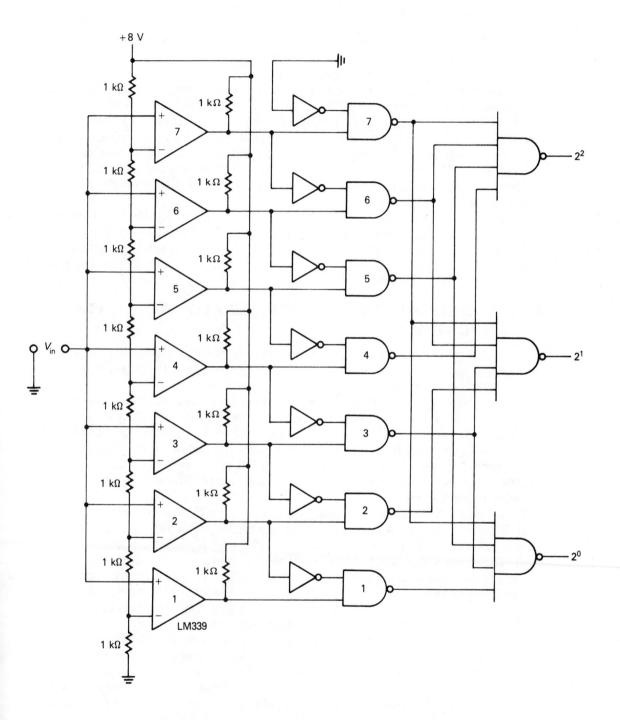

FIGURE 13-10   An A-to-D converter using voltage comparators

0 on its output which is fed to the 2^0 NAND gate. This produces a logic 1 on the output of the 2^0 NAND gate or binary number 1. A binary number 1 on the output means that the analog input voltage lies between 1 V and 2 V.

When the input voltage increases to 2.5 V, the second comparator's output will be logic 1 also. This produces a 0 on the inverter's output which inhibits the first NAND gate, thus removing the 1 on the 2^0 NAND gate's output. The logic 1 on the second comparator's output enables the second NAND gate, producing a 0 on its output which produces a logic 1 on the output of the 2^1 NAND gate or binary number 10. This means that the analog input lies between 2 V and 3 V. It can be seen that as the voltage increases, the binary output of the A-to-D converter will change to reflect the value of the analog input.

In order to increase the accuracy of the comparator in Figure 13-10 you need more comparators and NAND gates. This is a major drawback for this type of A-to-D converter but it operates very fast. The only thing to slow it down is the propagation delay of the comparator and NAND gates, which is in the order of 50 ns to 75 ns.

13.4   THE COUNT-UP AND COMPARE ANALOG-TO-DIGITAL CONVERTER

This type of A-to-D converter uses one voltage comparator and a D-to-A converter. Figure 13-11 shows a count-up and compare A-to-D converter using an LM339 voltage comparator and a 2R resistor network.

The D-to-A converter is driven by a 7493 binary counter which can count from 0000 to 1111 in binary or 0 to 15 in decimal. The output of the D-to-A converter is fed to the negative input of a voltage comparator and the positive input comes from the analog voltage which we want to measure. The output of the LM339 voltage comparator is used to enable or inhibit a NAND gate which feeds the clock to the 7493 counter. If the NAND gate is inhibited, then the counter will not receive a clock pulse and will stop counting.

When the RESET button is pushed, the counter is cleared and all its outputs go to logic 0. This puts a 0 or ground voltage on the negative input of the voltage comparator. Let us say that the analog input to the voltage comparator has 7.75 V on it. This means that the positive input is greater than the negative input of the voltage comparator and its output will be at logic 1. A logic 1 on the input of the NAND gate which is fed from the comparator enables the NAND gate, and the clock is passed on to the 7493 counter. As the counter counts, its display gets larger and the analog voltage output of the 2R resistor network also gets larger.

When the voltage at the negative input of the voltage comparator gets larger than the analog voltage input to the positive input of the comparator, the output of the comparator goes LOW which inhibits the NAND gate and stops the clock and the 7493 counter. The stopped counter will now contain the binary number which produces a voltage one increment of the D-to-A converter larger than the analog input voltage.

The A-to-D converter stays at this point until the RESET button is pushed or until the analog input voltage increases at which time the counter simply counts

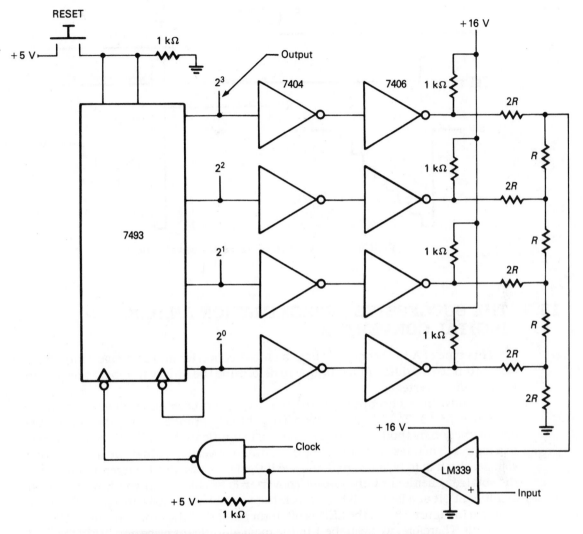

FIGURE 13-11   A count-up and compare A-to-D converter

back up to the new voltage. If the analog input voltage drops after the counter stops, the counter does not change and has to be reset to count back up to the lower voltage. Figure 13-12 shows the waveform for the analog output of the D-to-A converter as the analog input voltage changes.

To increase the accuracy of the count-up and compare A-to-D converter, increase the size of the counter and the D-to-A converter. In the case of the A-to-D converter in Figure 13-11, if you added one more 7493 to the converter the + 16 V would be divided into 256 increments as compared to 16 increments with one 7493 counter. The main drawback to this type of A-to-D converter is speed because of the time it takes to count-up and compare.

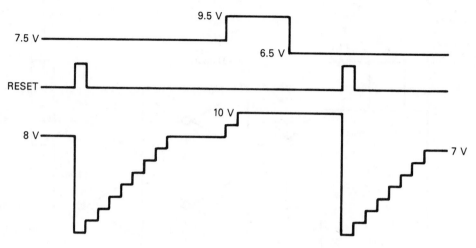

FIGURE 13-12   D-to-A converter waveforms

13.5   THE SUCCESSIVE APPROXIMATION ANALOG-TO-DIGITAL CONVERTER

This type of A-to-D converter uses a D-to-A converter and a voltage comparator also, but it uses a different technique to determine the binary number to feed in to the D-to-A converter.

To study this technique let us use the D-to-A converter and voltage comparator in Figure 13-13. To determine the correct binary number using the successive approximation method, you first set the most significant bit of the D-to-A converter to a logic 1. Then test the output of the D-to-A converter against the input analog voltage which is to be measured and see if the D-to-A output is larger or smaller. If the voltage generated by the D-to-A converter is smaller, then you leave the most significant bit at a logic 1. If it is larger, then you bring it LOW, to a logic 0. In the example in Figure 13-13, the LED is off, indicating that the binary number 1000 is too small. Therefore, we keep the 1 in the most significant place and bring the next place to a logic 1. Now we have the binary number 1100 or decimal number 12 input to the D-to-A converter. The LED is still off because the analog voltage being measured is greater than 12 V; therefore leave the second bit at logic 1 also. Next, we bring the third bit HIGH and we have binary number 1110 or decimal number 14 on the input to the D-to-A converter. The voltage output of the D-to-A converter is now larger than the 12.5 V of the analog input voltage and the LED is on. In this case we remove the 1 on the third bit and put a 1 on the last and least significant bit. This gives us the binary number 1101 input to the D-to-A converter. The output voltage is now 13 V which is still greater than the analog input voltage and the LED is still on. Therefore, remove the 1 and make it a 0. This gives us the final binary number 1100 or decimal number 12, which is one increment of the D-to-A converter less than the actual analog input voltage being measured.

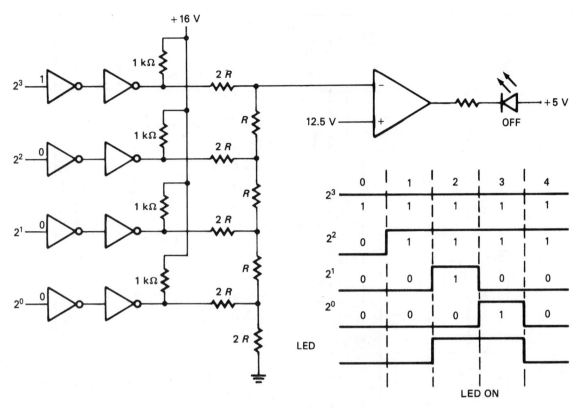

FIGURE 13-13   The successive approximation method of A-to-D conversion

The successive approximation method just described takes only 4 cycles to determine the correct binary number for the given analog input voltage and it takes the same time to determine a large number as a small one. Therefore, the successive approximation method is faster than the count up method but not as fast as the voltage comparator method.

Figure 13-14 shows a successive approximation A-to-D converter which uses a CP and CP' nonoverlapping clock generator, shift register and storage registers. When the RESET button is pushed, the clock generator and A flip-flop are preset to 1 while all the rest of the flip-flops are reset to 0. The A-to-D converter will stay in this configuration until the RESET button is released, thus allowing the clock generator flip-flop to run. After the clock is allowed to run, Q_A is preset to 1 which places a 1 on the 2^3 or most significant bit of the D-to-A converter. The contents of the comparator are fed to an AND gate which is controlled by the CP clock of the nonoverlapping clock generator. When it goes HIGH, the content of the comparator is passed through the AND gate to the A_S storage flip-flop NAND gate. If the comparator is 1, then the flip-flop will be set; if it is 0 then the flip-flop is not set. Next comes the CP' clock which shifts the A, B, C, and D flip-flop shift register 1

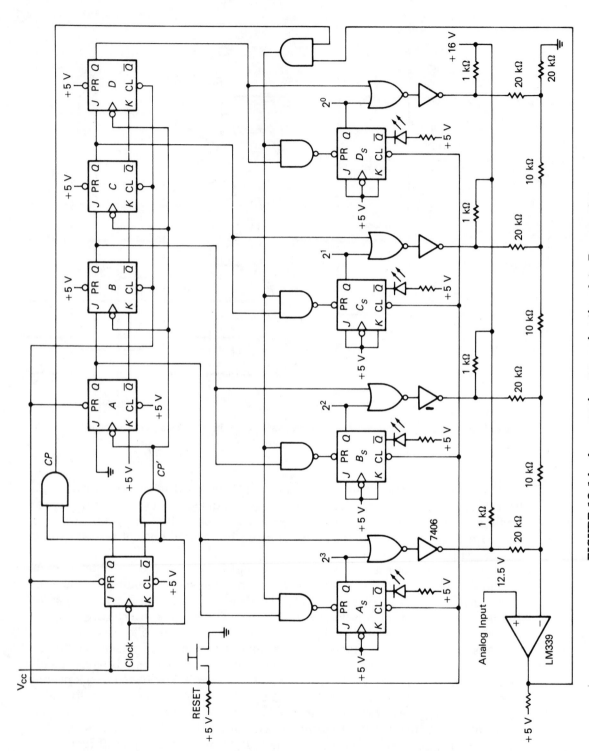

FIGURE 13-14 A successive approximation A-to-D converter

place. This makes Q_B a 1 and the cycle is repeated. After the 1 which was generated in Q_A by the RESET pulse is shifted out of the A, B, C, and D flip-flop shift register the correct binary number is present on the 2^0 and 2^3 outputs.

Notice that it took only 4 CP' pulses to obtain the correct binary number. Figure 13-15 shows the waveforms for the operation of the A-to-D converter in Figure 13-14.

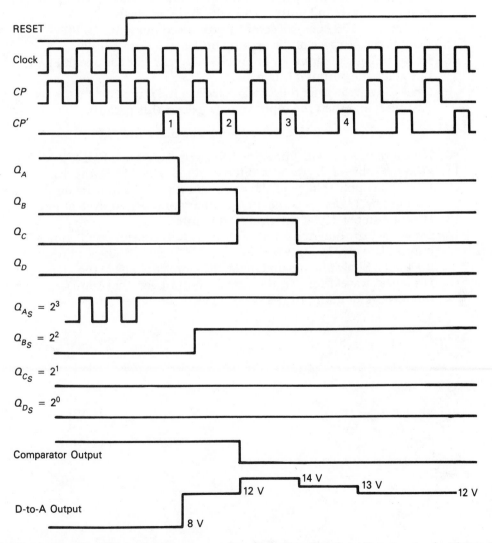

FIGURE 13-15 Waveforms for the successive approximation A-to-D converter in Figure 13-14

Exercise 13

1. Draw the logic diagram for a TTL $2R$ D-to-A converter with 256 increments and 10 V of maximum output.
2. What is the output voltage if the binary number 0110 is put on the TTL inputs of the D-to-A converter in Figure 13-8?
3. What is the voltage increment of the D-to-A converter in Figure 13-8 if the supply voltage to the $2R$ network is changed to 5 V, 10 V and 32 V?
4. What is the binary number output of the A-to-D converter in Figure 13-10 if 6.3 V is put on the analog input?
5. Draw the waveform for the analog output of the successive approximation A-to-D converter in Figure 13-14 with 5.5 V applied.
6. Why is the 7406 open-collector inverter used in Figure 13-8?
7. What is the purpose of the op amp buffer in Figure 13-9?
8. Draw the logic diagram for an A-to-D converter like the one in Figure 13-10 but make the output 4 bits wide.
9. Draw the waveform for the A-to-D converter in Figure 13-11 if the converter is first reset and then the input voltage follows this pattern.

0 V to 3.3 V to 6.4 V to 5.2 V to 7.3 V

10. Repeat number 9 with the A-to-D converter in Figure 13-14.
11. Find the value of the resistor if the 2^4 bit is added to Figure 13-1.
12. Find the supply voltage of the circuit in Figure 13-5 if the output voltage changes by 3 V for a change in the binary number equivalent to 2.
13. Repeat number 12 for the circuit in Figure 13-2.
14. Draw the logic diagram for a count up A-to-D converter which has an 8-bit output. Use two 7493, two 7407, one 7400, one LM339, and resistors for a $2R$ ladder. Show pinouts and make the supply voltage to the $2R$ ladder 12 V.
15. Draw the waveforms for the analog output for the A-to-D converter in number 14 for an input voltage of 1.2 V.

13 Digital-to-Analog and Analog-to-Digital

OBJECTIVES

After completing this lab, you should be able to:

- construct a D-to-A converter.

- use an LM339 voltage comparator to construct an A-to-D converter.

- use the scope to observe the stair-step waveform.

COMPONENTS NEEDED

1	7493 four-bit ripple counter IC
1	7404 hex inverter IC
1	7406 hex open-collector output inverter IC
1	7400 quad NAND gate IC
1	LM339 quad op amp comparator
4	1-kΩ, ¼-W resistors
1	1k or larger pot
4	10-kΩ, ¼-W resistors
5	20-kΩ, ¼-W resistors
4	red LEDs
4	330-Ω, ¼-W resistors

PROCEDURE

1. Construct the A-to-D converter shown in the following order.
   A. Wire up the 7493 counter and get it to run.
   B. Wire up the 7404 and 7406 D-to-A converter and display the stair-step pattern on the scope.
   C. Wire up the op amp comparator and the 7400 NAND gate.

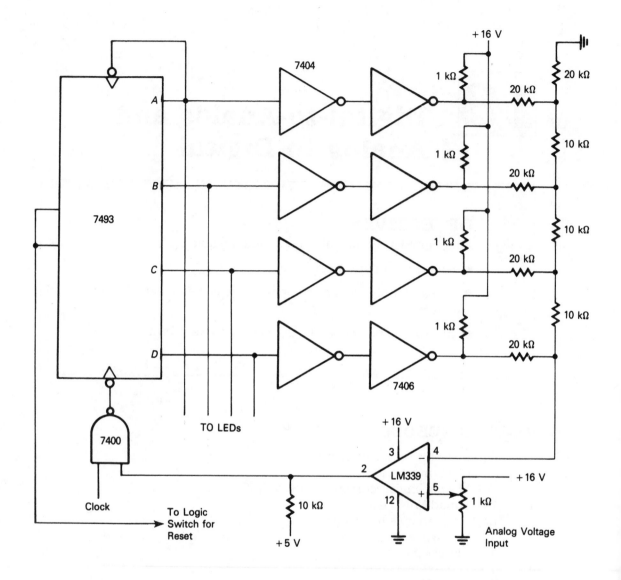

2. Set the clock at the 1 Hz rate, put 5.5 V on the analog input and reset the counter. At which binary number does the counter stop and why?
3. Now raise the voltage to 10.5 V. What is the number now?
4. Now lower the voltage to 6.5 V. What is the number now and why?

If your circuit does not work properly, consider these points:

1. Check the power supply connections to all the components in the circuit.
2. Disconnect pin 2 of the LM339 from the NAND gate input. This will allow the counter to count from 0 to 15.

3. Put channel one of the scope on pin 4 of the LM339. This is the output of the D-to-A converter. You should see the characteristic stair-step waveform from about 0.2 V to 15 V. If the waveform is not in even 1-V increments then check the resistor network for wiring errors.

4. If the stair-step waveform is okay, then put the scope probe on the output of the LM339 comparator. You should be able to vary the duty cycle of the output waveform by varying the input voltage on pin 5 of the LM339. If you cannot do this, then something is wrong with the comparator part of the circuit.

CHAPTER 14

OUTLINE

Decoders, Multiplexers, Demultiplexers and Displays

14.1   DECODERS

Figure 14-1 shows a full 2-bit decoder which will enable one and only one of the four AND gates for each possible binary number input to the 2^0 and 2^1 inputs of

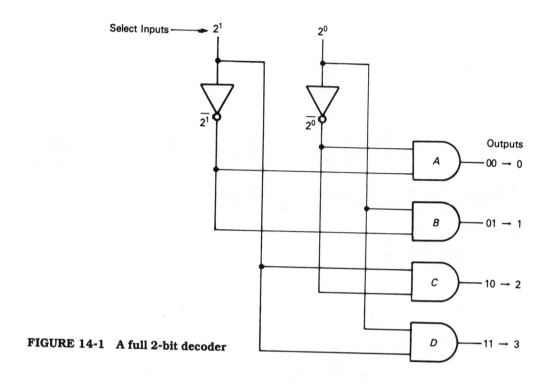

FIGURE 14-1   A full 2-bit decoder

2^1	2^0	$\overline{2^1}$	$\overline{2^0}$	A	B	C	D
0	0	1	1	1	0	0	0
0	1	1	0	0	1	0	0
1	0	0	1	0	0	1	0
1	1	0	0	0	0	0	1

FIGURE 14-2   The truth table for a full 2-bit decoder

the decoder. Figure 14-2 shows the truth table for the full 2-bit decoder shown in Figure 14-1.

The two inverters in the decoder in Figure 14-1 provide $\overline{2^0}$ and $\overline{2^1}$ which is fed along with 2^0 and 2^1 to the AND gates in the proper order to enable each gate when the proper binary number is input. This is called a full decoder because it has an active output line for each possible binary number which can be input to the decoder.

If you were to increase the size of the binary number input to the decoder by 1 bit to 3 bits, then the number of outputs would be 2^3 or 8 if the decoder was to be a full decoder as shown in Figure 14-3. Notice that as the number of bits in the input increases so does the number of inputs of the AND gates used. If the binary number gets very large, a full decoder can become very large. Consider a full 8-bit decoder; it would need 256 AND gates, each with 258 inputs.

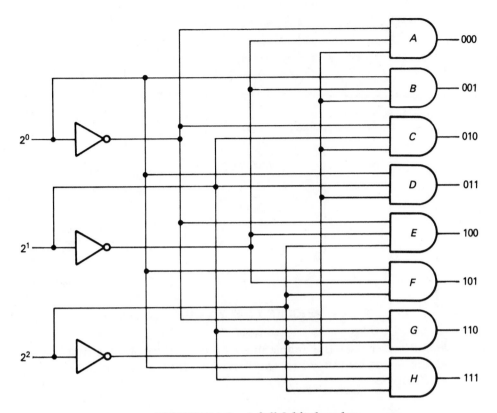

FIGURE 14-3   A full 3-bit decoder

In most cases we do not need to decode every bit of a long binary number. Therefore, a full decoder is not really necessary. An address decoder for a typical Z80 computer output port system may only need unique output for 0, 1, 2 and 3 from an 8-bit binary address. All other possible outputs are not needed. Therefore, the decoder would be constructed as shown in Figure 14-4.

By using a wired NOR gate made from open-collector inverters, if any 1 is put on the upper address bits (bit A_2 to A_7) the output of the wired NOR gate will go LOW causing the four AND gates to be inhibited. The only numbers which can enable any of the AND gates are 0, 1, 2 and 3 because they do not have a 1 in bits A_2 to A_7. This is called a partial decoder and is frequently used in computers.

14.2   DEMULTIPLEXERS

A demultiplexer is a digital switch which allows us to switch one input to one of many possible output lines. The line which we want the input to be connected to is determined by a binary number which is input to the demultiplexers. Figure 14-5 shows a 1-to-4 demultiplexer. It looks very much like a decoder; in fact, the two are almost the same. The only difference is in the use of the enable line of the decoder. A

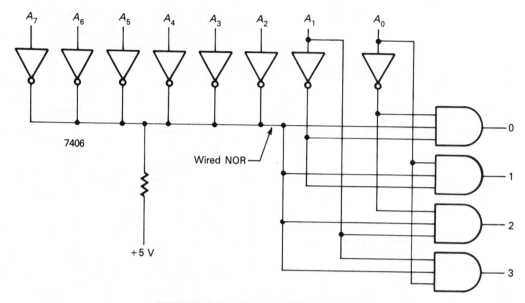

FIGURE 14-4   A partial 8-bit decoder

demultiplexer uses the enable line as a data input. Notice that the data appears on the select output when the corresponding binary number is input to the select inputs.

14.3   MULTIPLEXERS

The multiplexer is the opposite of the demultiplexer. It selects one channel as an input and connects it to a signal output. Figure 14-6 shows a full 4-to-1 multiplexer. The outputs of the AND gates are ORed together to produce a common output. The AND gate, which will control the output, is selected by the binary number input to the select inputs.

14.4   USING A MULTIPLEXER TO REPRODUCE A DESIRED TRUTH TABLE

To build a digital circuit which will conform to the truth table shown in Figure 14-7, use a 4-to-1 multiplexer as shown in Figure 14-7. The select inputs become the truth table input variables and the channel inputs are made LOW or HIGH to reflect the desired output for a given combination of A and B inputs. When the A and B select inputs are sequenced through the values of the truth table, the output of the multiplexer will go LOW or HIGH according to the values which are on the channel inputs, thus reproducing the truth table.

At first glance, this method of using a multiplexer to reproduce a truth table seems to require a multiplexer with at least the same number of select inputs as inputs in the truth table. With a little ingenuity, we can make our 2-select input

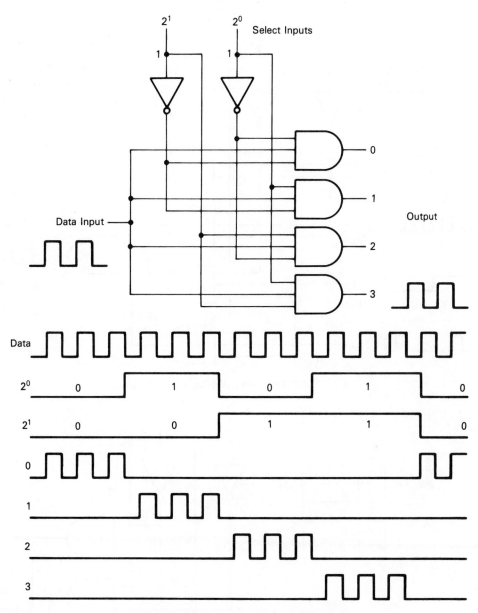

FIGURE 14-5 A 1-to-4 demultiplexer

multiplexer work like a 3-select input multiplexer. Consider the truth table in Figure 14-8 (A). This is the standard method of writing a truth table, starting with 000 and counting in binary to 111 which is the largest binary number that the 3-bit number can express. This gives all the possible combinations for the given 3 inputs A, B and C.

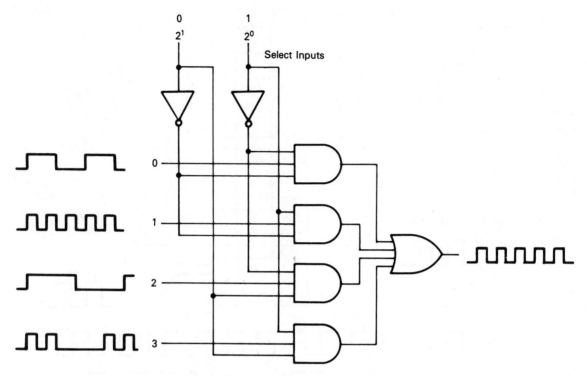

FIGURE 14-6   A 4-to-1 multiplexer

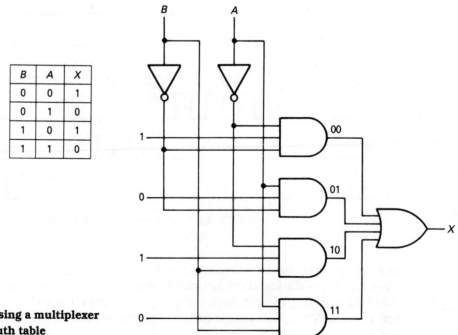

B	A	X
0	0	1
0	1	0
1	0	1
1	1	0

FIGURE 14-7   Using a multiplexer to reproduce a truth table

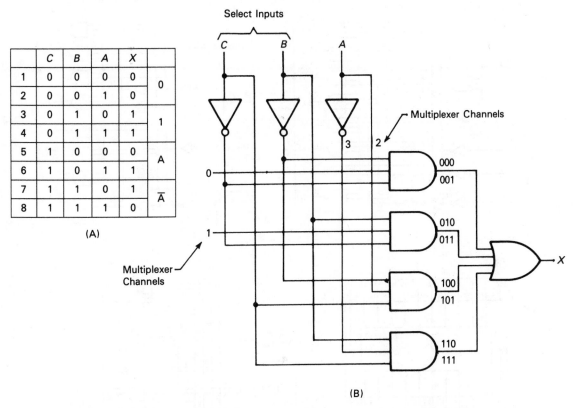

	C	B	A	X	
1	0	0	0	0	0
2	0	0	1	0	
3	0	1	0	1	1
4	0	1	1	1	
5	1	0	0	0	A
6	1	0	1	1	
7	1	1	0	1	$\overline{A}$
8	1	1	1	0	

(A)

(B)

FIGURE 14-8 Using a 4-to-1 multiplexer to reproduce a 3-bit truth table

Notice the two most significant bits in the truth table, C and B, change in value every other line. Therefore, you can group the lines of the truth table into four groups of 2 each in which C and B are the same. In the first group C and B are both 0 and A is 0, then 1, but the outputs for both lines 1 and 2 are 0. If C and B are placed on the select input of a 4-to-1 multiplexer and then 0 on the channel 0 of the multiplexer, the output is 0 when C and B input are 0 no matter what A is. This is shown in Figure 14-8 (B). Lines 3 and 4 of the truth table are similar but in this case, the output X is 1 in both cases. Therefore, we make the 1 channel HIGH, or 1 which gives us a 1 on the multiplexer output no matter what A is. The next two lines of the truth table (lines 5 and 6) do not have the same output value. When A is 1 the output X is 1 and when A is 0, the output X is 0. Therefore, we simply tie input A to the input for channel 2. This will cause the output to follow the A input when C and B are 1 and 0, respectively, thus fulfilling the truth table. The last two lines (lines 7 and 8) do not have the same output either. When A is 0, the X output is 1 and when the A input is 1, the X output is 0 or opposite the value of A. Therefore, we use $\overline{A}$ to

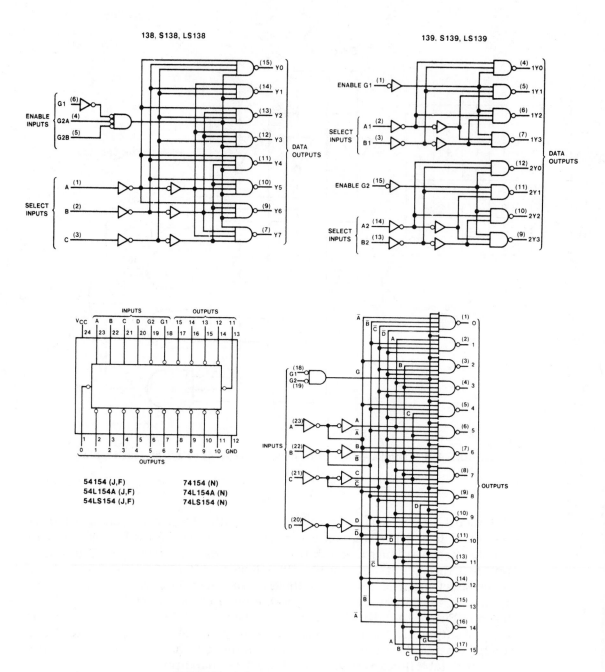

FIGURE 14-9 Decoders and demultiplexers

feed the third input channel of the multiplexer, thus completing the circuit for the truth table. This method of using a multiplexer to produce a pattern of pulses is quite handy for sequencing the operation of a digital machine.

14.5 MULTIPLEXER AND DEMULTIPLEXER ICS

There are many different types of multiplexers and demultiplexers which are put on ICs. Figure 14-9 shows the logic drawing for 3 typical demultiplexer/decoders which are TTL in construction. Notice the 74138 has three enable inputs which can be used as data inputs or enables. The 74154 is a full 4-bit decoder with 16 output lines and two enable lines. Notice all three of these ICs have active LOW outputs. Figure 14-10 shows the 74150 and 74151 multiplexer ICs.

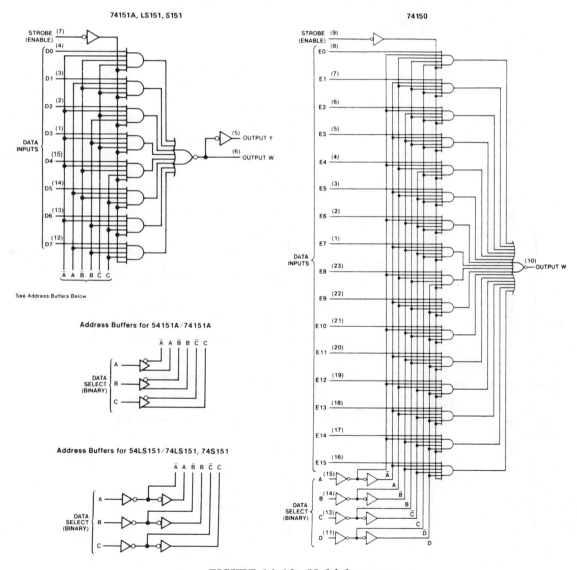

FIGURE 14-10 Multiplexers

In the CMOS family there are several analog multiplexers and demultiplexers such as the 4051, 4052 and 4053 ICs. An analog multiplexer can pass an analog signal from the channel input to the output. These types of ICs could be used to multiplex the inputs to a multitrace scope or multiplex analog phone lines.

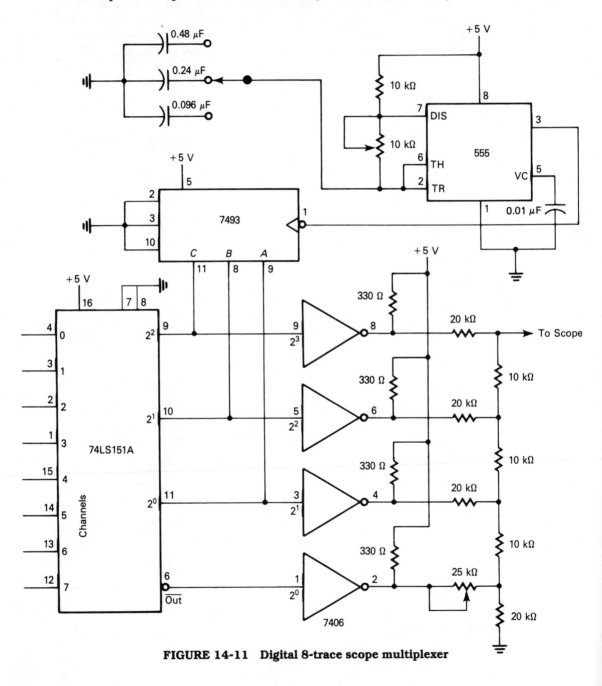

FIGURE 14-11 Digital 8-trace scope multiplexer

14.6   THE 8-TRACE SCOPE MULTIPLEXER

Figure 14-11 shows the circuit for an 8-trace scope multiplexer. This is a very handy gadget when working with digital circuits having several signals which are to be observed at the same time and in relation to each other.

The 7493 counter is used as a divide-by-8 binary counter. Its outputs are fed to the select inputs of an 8-to-1 multiplexer (the 74151A) and the upper 3 bits of a 4-bit D-to-A converter. The least significant bit of the D-to-A converter is driven by the output line of the multiplexer. As the 7493 counts, the analog voltage at the output of the D-to-A converter will increase by increments of 2 because the counter's least significant bit is connected to the 2^1 input of the D-to-A converter.

When the output of the multiplexer changes, the analog voltage changes by 1 increment because the 2^0 input of the D-to-A converter is controlled by the output of the multiplexer. Therefore, the data on channel 0 will be displayed at the 0 to 1 increment of the D-to-A converter. Channel 1 will be displayed at the 2 to 3 increments and channel 2 will be displayed at the 4 to 5 analog voltage level of the D-to-A converter. This will continue until the divide-by-8 counter starts over at 0.

When the counter is run at a speed of at least 10 times less than the sweep frequency, all the 8 inputs will appear on the scope at their separate voltage levels because the counter is running faster than the eye can detect. All the waves will appear on the scope at the same time.

A 555 timer is used to provide a clock for the counter. The rotary switch changes the frequency of the counter to produce a multiplex rate of at least 10 times less than the sweep frequency. An external synch probe connected to the slowest input frequency to the multiplexer is the best way to synch the scope to the wave patterns. The 25 kΩ pot is used to adjust the 0 to 1 logic level of the 8 waves on the scope. Similar schemes can be used in conjunction with an analog multiplexer to produce an analog multitrace scope.

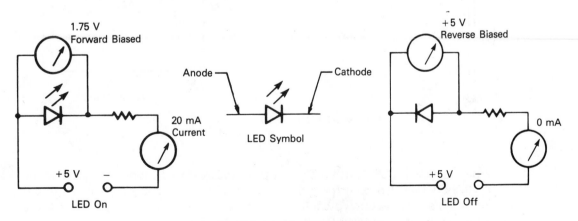

FIGURE 14-12   LED biasing

14.7 THE LIGHT-EMITTING DIODE

When electrons fill a positive hole at the junction of *PN* material, the electron loses some energy which is given off as heat and light. All *PN* junctions do this, but the *PN* junctions made of gallium emit sufficient amounts of light to be used as a visible light source. Depending on the type and amount of crystal doping, the light emitted can be red, green or yellow.

Because the LED is a *PN* junction, it exhibits all the properties of a typical diode. The LED will produce light when the diode is forward biased but not when reverse biased. This is shown in Figure 14-12.

When the LED is forward biased, the voltage drop is about 1.75 V for a typical red LED. The forward biased voltage is higher for yellow and green LEDs. A typical LED needs between 5 mA to 20 mA to be seen well. The resistor in Figure 14-12 is used to limit the current through the diode. If the resistor was not there, the diode would burn out due to high current.

The intensity of light produced by the LED is directly proportional to current flowing through the diode and can be used as a modulated light source. The on/off speed is also quite high, in the order of 10 nanoseconds for a typical red LED. The speed of the LED lends it to such uses as a high speed optocoupler. LEDs come in several packages as shown in Figure 14-13. The cathode of the LED can be found by looking for the flag lead inside the plastic case.

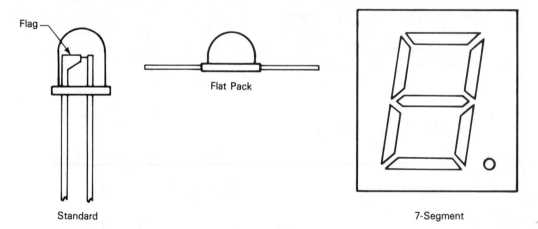

Flag

Flat Pack

Standard

7-Segment

FIGURE 14-13   LED packages

When driving an LED with a TTL output, it is best to design the circuit so the LED will be forward biased (or on) when the TTL output is LOW or at ground potential. This is because a typical TTL output can produce up to 16 mA when at the LOW state without raising the 0 voltage above 0.4 V. This is shown in Figure 14-14.

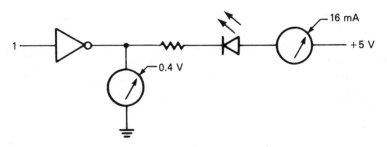

FIGURE 14-14   Driving an LED with a TTL output

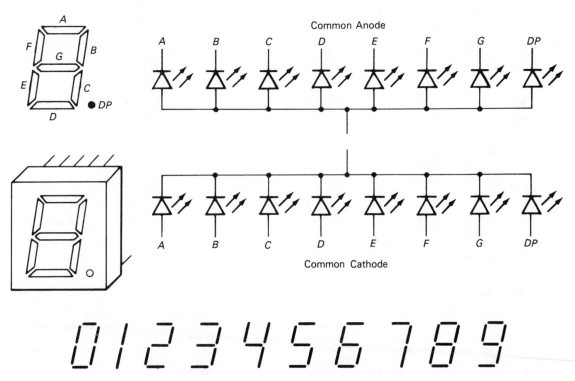

FIGURE 14-15   7-segment displays

14.8   THE 7-SEGMENT DISPLAY

As can be seen in Figure 14-15, the 7-segment display is actually 8 separate LEDs (7 segments and 1 decimal point). The 7-segment display format is used in other types of display and can display any number from 0 to 9. Figure 14-15 shows the typical segments used to display 0 through 9.

There are two types of LED 7-segment displays: common cathode and common anode. As can be seen in Figure 14-15, the common cathode has all the cathodes of the 7 segments connected together and the common anode is the same except that the anodes are all tied together. Also, notice the way the segments are labeled. This is a de facto standard for 7-segment displays and MSI ICs designed to work with 7-segment displays.

Figure 14-16 shows a logic diagram for a 7447 and 7448 TTL decoder driver. These ICs will decode a 4-bit BCD number into the proper output to display the BCD number on the 7-segment display.

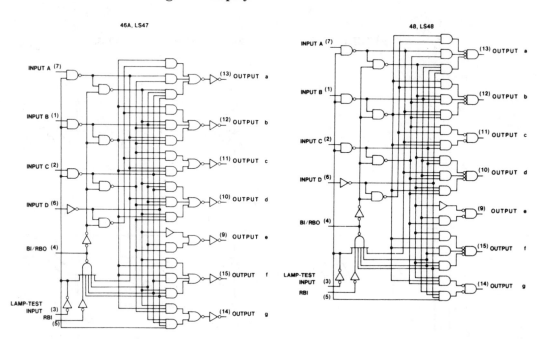

General Description

The 46A, 47A and LS47 feature active-low outputs designed for driving common-anode LED's or incandescent indicators directly; and the 48, LS48 and LS49 feature active-high outputs for driving lamp buffers or common-cathode LED's. All of the circuits except the LS49 have full ripple-blanking input/output controls and a lamp test input. The LS49 features a direct blanking input. Segment identification and resultant displays are shown on a following page. Display patterns for BCD input counts above nine are unique symbols to authenticate input conditions.

All of the circuits except the LS49 incorporate automatic leading and/or trailing-edge, zero-blanking control (RBI and RBO). Lamp test (LT) of these devices may be performed at any time when the BI/RBO node is at a high logic level. All types (including LS49) contain an overriding blanking input (BI) which can be used to control the lamp intensity (by pulsing), or to inhibit the outputs.

FIGURE 14-16   7-segment decoder drivers

14.9   THE LIQUID CRYSTAL DISPLAY

There are two types of LCDs in use today, dynamic and field effect LCDs. The two types use different material for the liquid crystal and work differently. Neither type emits any light and each must have an external light source to be seen.

The dynamic LCD has the liquid crystal material sandwiched between clear pieces of glass. The 7-segment pattern which is etched on the front glass plate is made of a clear electrically conductive material such as indium oxide. The back glass is coated with this clear conductor which corresponds to the 7 segments as shown in Figure 14-17. In this way, only the digit segments will be seen when an electric current is applied to the LCD.

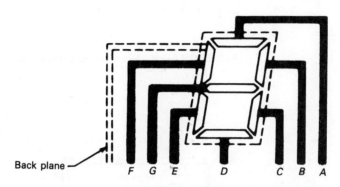

Back plane

F G E D C B A

FIGURE 14-17 Clear LCD segment conductors

When a voltage is applied between the segment pattern and the back conductor on the back sheet of glass, the liquid crystal diffuses the light. This happens because the index of refraction changes randomly causing the light to be refracted randomly as it passes through the liquid crystal material. The scattering action causes the segment to appear milky white in color.

A dc voltage will produce this effect on a dynamic LCD but an ac voltage is used. This is because even a small dc current can cause the segment conductor to be electrically plated with material from the liquid crystal. An ac current will prevent this. The current for a typical dynamic 7-segment LCD is very small, about 25 μA at 30 V_{pp}, 60 Hz. This is the main reason for using LCD displays.

The field effect or twisted nematic LCD is the most commonly used LCD display. This is the type of LCD used by most battery operated calculators, watches and computers. The most common LCD of this type produces a black segment on a reflective background.

To understand how this LCD display operates, you must first understand the operation of a polarized sheet of glass. Figure 14-18 shows a vertically polarized sheet of glass. Notice only light rays that are vertically polarized will pass through the glass. The light which passes through the glass is all vertically polarized and of course less in intensity because some of the light ray could not pass.

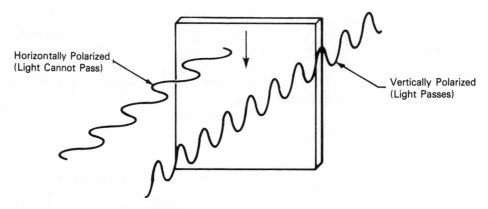

Horizontally Polarized
(Light Cannot Pass)

Vertically Polarized
(Light Passes)

FIGURE 14-18   Vertical polarized glass

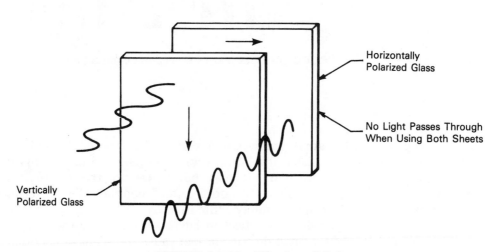

Horizontally
Polarized Glass

No Light Passes Through
When Using Both Sheets

Vertically
Polarized Glass

FIGURE 14-19   Filtering all light

If two polarized sheets of glass are placed at right angles, no light will pass through them. This is because the first polarized glass will stop all light rays which are not vertically polarized and the second polarized glass will only pass horizontally polarized light. Therefore, no light can pass through when using both sheets. This is shown in Figure 14-19.

If it were possible to twist the vertical light rays passing the first vertically polarized glass 90°, then they would pass through the second horizontal polarized glass. This is exactly what the liquid crystal material can do, that is, twist light 90°. By placing the liquid crystal material between the two polarized glass sheets, the vertically polarized light is twisted 90° and passed through the rear horizontally polarized glass, thus passing the light through the two polarized sheets of glass. The twisting of the vertical light rays will continue until an electrical current is passed through the liquid crystal material. When a current passes through the liquid crystal, it stops twisting the light and passes it unaltered to the horizontal polarized glass. This blocks the passage of the light because the light is vertically polarized. This is shown in Figure 14-20. Notice that only the area under the segment conductor will be affected, thus producing a black segment.

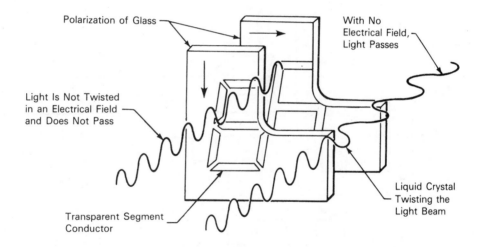

FIGURE 14-20   Twisted nematic LCD

The field effect or twisted nematic LCD typically runs on a 60 Hz ac, 8 V_{pp} voltage with about 300 μA of current. Again, the main advantage of LCDs of this type is low current. The ac voltage used to drive an LCD display can be obtained with some CMOS XOR gates and a 50% duty cycle clock. As can be seen in Figure 14-21, when the input is 0 the voltage differential between the segment conductor and the back plane is 0. When the input is 1, the voltage potential is an ac voltage.

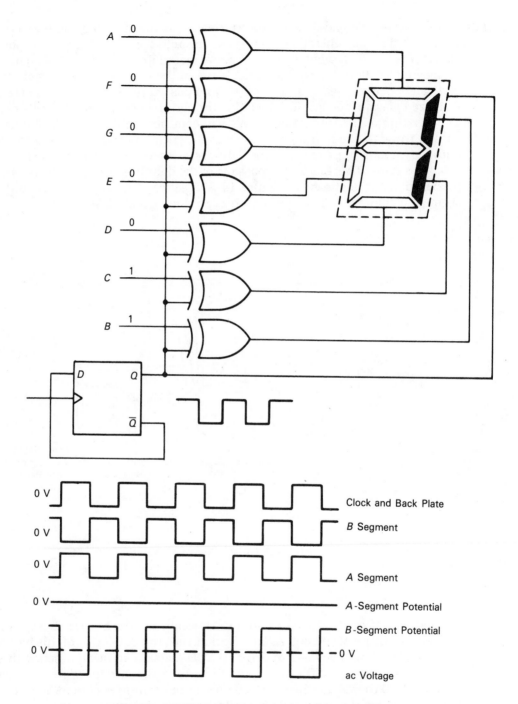

FIGURE 14-21 Driving an LCD with an ac voltage

Exercise 14

1. Draw the logic diagram for a 3-bit full decoder.
2. Draw the logic diagram for a partial decoder which will give active LOW output for FB, FA, FC, and FF.
3. Use a 74150 multiplexer to reproduce the following truth table. Draw the logic diagram.

Inputs					Output
E	D	C	B	A	X
0	0	0	0	0	0
0	0	0	0	1	1
0	0	0	1	0	1
0	0	0	1	1	1
0	0	1	0	0	0
0	0	1	0	1	0
0	0	1	1	0	1
0	0	1	1	1	0
0	1	0	0	0	1
0	1	0	0	1	0
0	1	0	1	0	0
0	1	0	1	1	1
0	1	1	0	0	1
0	1	1	0	1	1
0	1	1	1	0	0
0	1	1	1	1	0

Inputs					Output
E	D	C	B	A	X
1	0	0	0	0	0
1	0	0	0	1	0
1	0	0	1	0	0
1	0	0	1	1	0
1	0	1	0	0	1
1	0	1	0	1	1
1	0	1	1	0	0
1	0	1	1	1	1
1	1	0	0	0	1
1	1	0	0	1	0
1	1	0	1	0	0
1	1	0	1	1	1
1	1	1	0	0	1
1	1	1	0	1	1
1	1	1	1	0	1
1	1	1	1	1	0

4. What is the typical forward biased voltage for a red LED?
5. Draw and label the segments of a typical 7-segment LED.
6. Using a data book, draw the logic diagram for a 3-digit, 7-segment LED display using 7447 decoders and implement the leading 0 suppression.
7. What are the two types of LCD displays?
8. Why are LCD displays driven with an ac voltage?
9. Which type of display is faster, the LCD or the LED?
10. Could a twisted nematic LCD display be made to have a black background and white segments?
11. Draw the logic diagram for a partial decoder which will decode the first 8 binary numbers of an 8-bit number. Use 74138 and 7406 ICs. Show pinouts.

12. Redesign the 8-trace scope multiplexer in Figure 14-11 to be a 16-trace scope multiplexer.
13. Make a list of CMOS analog multiplexers and show the pinouts.
14. Look up the 4511 CMOS IC and describe its operation.
15. Look up the 74C945 CMOS IC and describe its operation.

LAB **14** # Multiplexers, LEDs and 7-Segment Displays

OBJECTIVES

After completing this lab, you should be able to:

- use a 74150 to construct a circuit to reproduce a 5-bit input truth table.

- construct a one-digit 7-segment LED display.

- test the operation of a red LED.

COMPONENTS NEEDED

1	FND-510 7-segment LED common anode
1	7447 BCD-to-common-anode 7-segment LED driver
8	330-Ω, ¼-W resistors
1	red LED
1	50-Ω, 1-W resistor
1	1-kΩ pot, 1 W
1	74150 multiplexer

PROCEDURE

1. Determine the pinout of the 7-segment LED by the following method:
   A. Connect pin one to +5 V.
   B. Use a 330-Ω resistor connected to ground as a probe and test all the other pins to see if you can light a segment.
   C. If no segment lights, move the 5 V connector to the next pin and repeat the probing with the 330-Ω resistor connected to ground. When a segment lights, you have found the common anode.
   D. After the first segment lights, leave the +5 V connection on that pin and use the 330-Ω resistor to determine which pin is the A, B, C, D, E, F, G and decimal point segments.

+5 V

FND 510
7-Segment LED

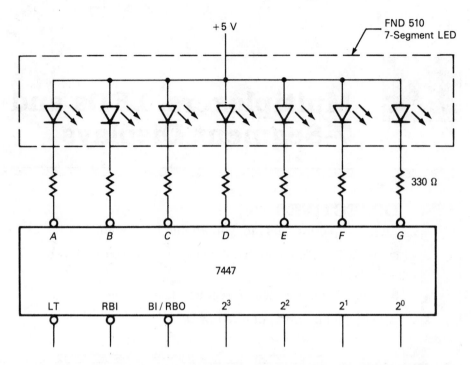

330 Ω

| A | B | C | D | E | F | G |

7447

| LT | RBI | BI / RBO | 2^3 | 2^2 | 2^1 | 2^0 |

2. Use your data book to determine the pinout for the 7447 decoder driver and connect the 7-segment LED to it as shown on the previous page. Have your instructor check its operation.

3. Construct the simple circuit shown. Complete the table and draw the graph for current versus voltage.

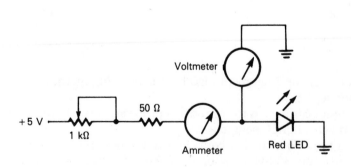

Current (mA)	Forward Biased Voltage
0	
0.5	
1	
2	
3	
4	
5	
10	
20	
30	
40	
50	
60	

4. Implement a 4-input expression using a 74150 multiplexer as follows:
   A. Write a 4-input truth table that you wish to design.

Inputs				Output
D	C	B	A	X
0	0	0	0	1
0	0	0	1	0

Etc.

   B. Place each value of X on the corresponding data input pin.
   C. Sequence the 74150 through the truth table by using a 7493 and the pulser switch on the trainer.
   D. Observe the outputs on pin 10. (You may want to invert the output.)
5. Implement a 5-input as follows:
   A. Write a 5-input truth table that you wish to design.

Inputs					Output
E	D	C	B	A	X
0	0	0	0	0	1
0	0	0	0	1	0
0	0	0	1	0	1
0	0	0	1	1	1

Etc.

   B. For the first two lines of the truth table place the appropriate value on data Channel 0.

Inputs					Output	
E	D	C	B	A	X	
0	0	0	0	0	0	Place 0 V on Channel 0
0	0	0	0	1	0	
0	0	0	1	0	1	Place 5 V on Channel 1
0	0	0	1	1	1	
0	0	1	0	0	0	Place E on Channel 2
0	0	1	0	1	1	
0	0	1	1	0	1	Place E on Channel 3
0	0	1	1	1	0	

Etc.

CHAPTER # 15

Tri-State Gates and Interfacing to High Current

OBJECTIVES

After completing this chapter, you should be able to:

- explain the operation of a tri-state gate.

- interface logic gates to transistors for higher current control.

- construct a circuit using tri-state gates that will multiplex two or more signals onto two or more displays.

- describe the use of tri-state gates with computer buses.

- use relays and optocouplers to isolate circuits.

15.1 TRI-STATE GATES

In the gates with totem-pole outputs studied earlier in this text, either the top transistor was on, or the bottom transistor was on, and the gate output was a 1 or a 0. In tri-state gates, both transistors can be turned off and the output is not pulled up to V_{CC} or pulled down to ground. The gate assumes a third state or high impedance state. In this HiZ state, the gate has no effect on the gates to which it is connected. If the outputs of several gates are connected together, only one of the ICs can be active at a time. The remaining gates must be in their HiZ state.

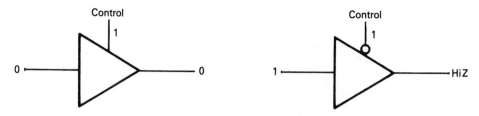

FIGURE 15-1   Tri-state gates

Figure 15-1 shows two tri-state buffers. In the first buffer, the control is not bubbled. A 1 on the control enables the IC and the output is either HIGH or LOW as determined by the input. A 0 on the control inhibits the gate, and the gate enters the high impedance state. The second buffer in Figure 15-1 has a bubble on the control line. This implies that a 0 enables the gate and the output assumes a 1 or 0 state. When the control line goes HIGH, the output enters the HiZ state and another tri-state gate can control the output.

Figure 15-2 shows three tri-state buffers connected together with the common output driving an OR gate. Since the control input is active HIGH (no bubble) only one control input can be HIGH at a time. Table 15-1 shows the wide variety of tri-state devices that are available.

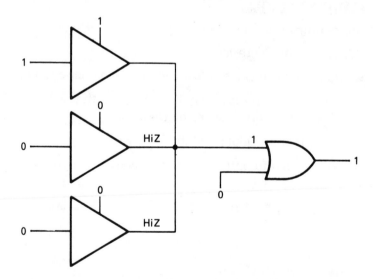

FIGURE 15-2   Only one control input can be HIGH at a time

15.2   TRI-STATE INVERTERS AND BUFFERS

Buffers are single input circuits that do not alter the signal; 1 in, 1 out. Several of the ICs listed in Table 15-1 are octal buffers, such as the 74LS240, 241, 242, 243, and 244. They vary in their configurations. For example, Figure 15-3 shows the

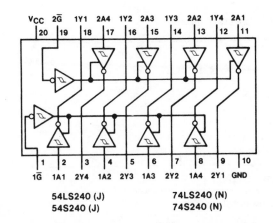

FIGURE 15-3 The 74LS240 tri-state inverter

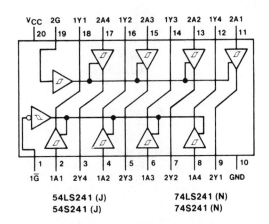

FIGURE 15-4 The 74LS241

pinout for a 74LS240. There are two groups of four tri-state inverters. One group is controlled by $2\overline{G}$ on pin 19. When $2\overline{G}$ goes LOW, the top set of inverters, 2A1, 2A2, 2A3, and 2A4 are enabled and data can pass inverted from 2A1 to 2Y1, 2A2 to 2Y2, 2A3 to 2Y3, and 2A4 to 2Y4. When $2\overline{G}$ goes HIGH, 2Y1, 2Y2, 2Y3, and 2Y4 enter a high impedance state. Other tri-state gates could control those output lines. $1\overline{G}$ on pin 1 controls the other set of inverters. Each group is convenient for handling four bits in parallel.

The 74LS241, Figure 15-4, is designed in a similar manner except that the gates do not invert, and the top group of four gates are enabled when 2G goes HIGH. Even though the gates do not alter the signal, they do buffer the output from the input. Buffers are used to provide increased current drive capabilities. Often a source cannot provide the current drive required by the following circuitry. In those cases, the source drives a buffer and the buffer drives the following circuitry. Figure 15-5 shows the 74LS242 quad-bus transceiver.

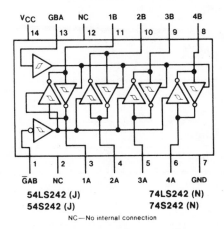

FIGURE 15-5 The 74LS242 quad bus transceiver

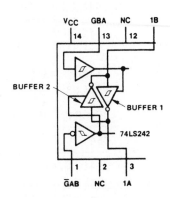

FIGURE 15-6 Only one of the buffers can be enabled at a time

TABLE 15-1 Available tri-state gates

DEVICE NUMBER	DESCRIPTION
54175/74125 54LS125/74LS125	Tri-state quad buffer
54126/74126 54LS126/74LS126	Tri-state quad buffer
54S134/74S134	Tri-state 12-input NAND
54LS240/74LS240 54S240/74S240	Tri-state inverting octal buffer
54LS241/75LS241 54S241/75S241	Tri-state octal buffer
54LS242/74LS242 54S242/75S242	Tri-state octal buffer
54LS243/75LS243 54S243/75S243	Tri-state octal buffer
54LS244/75LS244 54S244/75S244	Tri-state octal buffer
54S244/74S244	Tri-state octal buffer
54LS245/75LS245	Tri-state octal buffer
54365/74365 54LS365/75LS365	Tri-state hex buffer
54366/74366 54LS366/74LS366	Tri-state hex buffer
54367/74367 54LS367/74LS367	Tri-state hex buffer
54368/74368 54LS368/75LS368	Tri-state hex buffer
545940/745940	Tri-state octal buffer
545941/745941	Tri-state octal buffer
54173/74173 54LS173/74LS173	Tri-state quad-*D* registers

TABLE 15-1 continued

54251/74251 54LS251/74LS251 545251/745251	Tri-state data selector/multiplexer
54LS254/74LS253 54S253/74S253	Tri-state data selector/multiplexer
54LS257/74LS257 54S257/74S257	Tri-state quad 2-data selector/multiplexer
54LS258/74LS258 54S258/74S258	Tri-state quad 2-data selector/multiplexer
54S299/74S299	Tri-state 8-bit universal shift/storage registers
54LS353/74LS353	Tri-state data selector/multiplexer
54LS373/74LS373 54S373/74S373	Tri-state octal latch
54LS374/74LS374 54S374/74S374	Tri-state octal latch
54LS670/74LS670	Tri-state 4 × 4 register file
54C173/74C173	Tri-state quad-D flip-flop
54C373/74C373	Octal latch with tri-state outputs
54C374/74C374	Octal D-type flip-flop with tri-state outputs
4034BM/4034BC	8-stage tri-state bidirectional parallel/serial input/output bus register
4043BM/4043BC	Quad tri-state NOR rs latches
4048BM/4048BC	Tri-state expandable 8-function 8-input gate
4076BM/4076BC	Tri-state quad-D flip-flop
4503BM/4503BC	Hex noninverting tri-state buffer

The 74LS242 shown in Figure 15-5 has eight buffers but they are arranged in pairs. One pair is shown in Figure 15-6. When GBA is HIGH, $\overline{\text{GAB}}$ must be HIGH also. Data can pass inverted from 1B to 1A through inverter 1. The output of inverter 2 is in the high impedance state and does not compete with the signal at 1B. Data passes from line or bus 1B to bus 1A. When $\overline{\text{GAB}}$ is LOW, inverter 2 is enabled. GBA must be LOW to put inverter 1 into its high impedance state so that its output does not interfere with the signal on 1A. By controlling these two tri-state inverters, data can flow in either direction between 1A and 1B. This combination is called a bidirectional bus driver or a bus transceiver since it can transmit or receive data. The 74LS242 is a quad-bus transceiver.

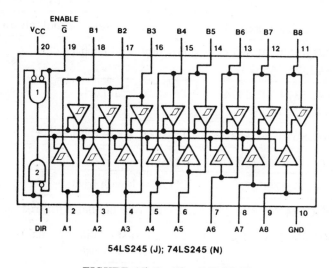

54LS245 (J); 74LS245 (N)

FIGURE 15-7   The 74LS245

Digital systems are often configured in parallel, with each bit having its own data line. Microprocessor-controlled systems often work with 8 or 16 data lines called a bus. ICs like the 74LS245 shown in Figure 15-7 are ideal for controlling the bus of microprocessors. When $\overline{\text{G}}$ is LOW, gates 1 and 2 are both enabled. When direction control DIR goes HIGH, gate 2 produces a 1 out which enables all 8 buffers that pass data from A1 through A8 to B1 through B8. The output of gate 1 is 0 and the buffers that pass data from B1 through B8 to A1 through A8 enter the HiZ state. When direction control goes LOW, gate 1, a NOR gate, produces a 1 out and the eight buffers that pass data from B1 through B8 to A1 through A8 enter the HiZ state. Only one set is active at a time and no conflict occurs.

15.3 COMPUTER BUSES AND THE TRI-STATE GATE

A *bus* is a group of conductors used to transfer digital binary numbers from one device to another. Microcomputers use three types of buses (address, control, and data buses). As shown in Figure 15-8, the address bus is used by the central processor to select the memory location or input/output device to be accessed. This bus is unidirectional and transfers the address or data, only in one direction, from the CPU to the devices connected to the bus. Figure 15-8 shows a typical input/output port wired to the bus of an 8080A CPU. The address bus is buffered with 74LS244 tri-state gates. This gives them good current driving capability, plus they can be put into HiZ by a control signal called $\overline{\text{HOLD}}$. This ability to disconnect the CPU from its bus allows other devices, such as a direct memory access device, to gain control of the bus and use any of the memory or input/output devices on the bus.

The data bus is used to transfer data in the form of binary numbers to and from the CPU and memory or input/output devices. This bus is bidirectional because data flows both ways. To buffer this bus, a bidirectional bus driver such as the 74LS245 must be used. The 74LS245 IC gives good current driving ability and can be switched to control the direction of data flow. The data bus can have many devices connected to it, each of which draws current from it. This means that a typical NMOS IC, such as a CPU, cannot supply the current needed to produce a good logic 0 or 1, therefore the need for current buffering.

Tri-state gates are also used to input data to a data bus. As shown in Figure 15-8, the input device is a 74LS244 which is controlled by the combination of an address decoder and the control signal $\overline{\text{I/O RD}}$. When the CPU wishes to receive data from this input port it places 00000000 on the lower 8 bits of the address bus which enables the 0 select line of the address decoder 74LS154. The CPU then makes the $\overline{\text{I/O RD}}$ signal LOW which causes the tri-state gates of the input port to turn on and control the bus. When the CPU has latched the data into a register inside the CPU it raises the $\overline{\text{I/O RD}}$ to a 1, thus inhibiting the tri-state gates and putting them in HiZ state. In this way the CPU can cause 1 and only 1 device to have control of the bus at a time.

15.4 BUFFERING TO HIGH CURRENT AND HIGH VOLTAGE

The digital engineer and technician would like to turn the world on and off with a 5-V signal but the world runs on such voltages as 120 V ac, 440-V 3-phase ac and currents of milliamperes to megamperes. This has not deterred them though; instead they design ways to control these large voltages and currents with their 5-V digital signal.

Control of dc voltages and currents, in excess of that which a digital IC can supply, can be accomplished by use of a buffer IC or transistor. The ac voltages can be controlled by triacs if they are not extremely large or too high in current. If the control does not need to be extremely fast, a relay can be used to buffer the digital signal to extremely high currents and voltage. Relays offer high current capability

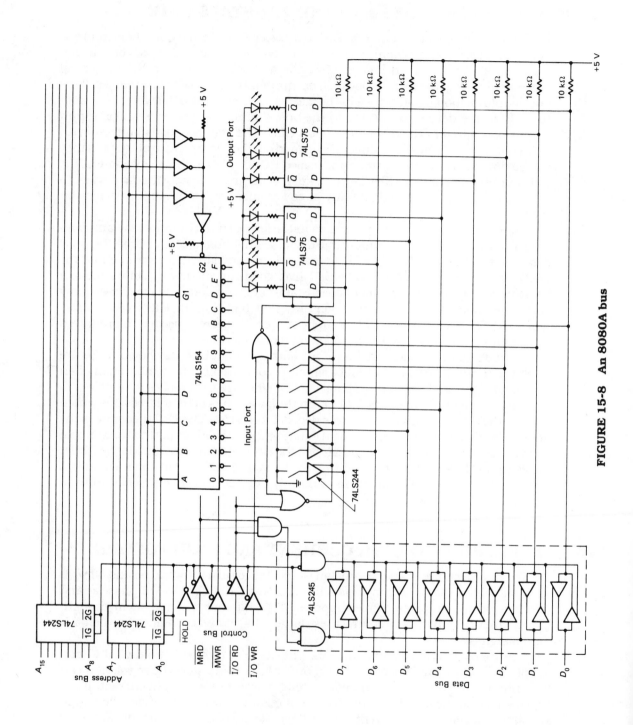

FIGURE 15-8   An 8080A bus

and complete circuit isolation from the high current and high voltage. Complete circuit isolation can also be obtained with optocouplers. Each of these options will be examined more closely in this chapter.

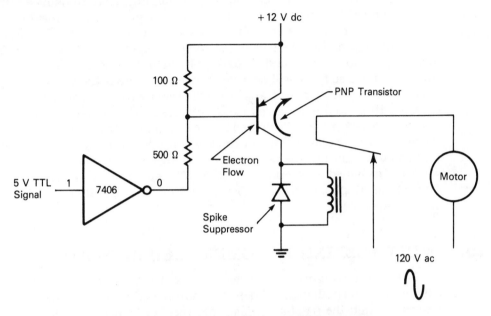

FIGURE 15-9   Buffering to high current

Figure 15-9 shows the use of a transistor to supply enough current to energize a relay which will turn on a pump motor. The 7406 is an open-collector inverter which is designed to be used for buffering to high voltage (up to 30 V) and high currents (up to 40 mA). When a TTL logic 1 is input to the 7406 input, the inverter's output goes LOW causing a current of about 23 mA to flow through the emitter-base junction of the transistor. This saturates the transistor and turns it on hard, producing a large collector current to energize the relay. The current through the transistor's collector is limited only by the resistance of the relay's coil, 50 Ω, and therefore can be quite large. The collector current in Figure 15-9 is 12 V divided by 50 Ω or 240 mA.

When the TTL logic signal goes to 0, the inverter's output goes to HiZ and no base current can flow, thus shutting off the transistor and stopping the current flow in the relay coil. This causes the magnetic field of the coil to fall in on its own coil windings, inducing a back emf or voltage opposite of the voltage which produced the magnetic field. This reverse voltage spike can be quite large and damaging to the transistor or other components in the circuit. To prevent this spike from being too large, a diode is placed across the relay's coil in a reverse bias mode to the circuit's voltage. The diode will be forward biased by the back emf generated by the relay's coil. This diode prevents the back voltage from going any larger than -0.7 V which is the forward biased voltage of the diode.

There are two things to consider when designing a circuit using a transistor to turn on or off a higher current. The first is to use a power transistor which has an I_C or collector current large enough to handle the current of the device being controlled. In Figure 15-9 the transistor must be able to handle at least 0.25 A or it would burn up. The second is to supply enough base current to turn on the transistor hard so there will be little voltage drop across the transistor. If the base current is too little then the transistor will decrease the current flow from emitter to collector, thus producing a voltage drop across the transistor. This voltage drop causes more power to be dissipated by the transistor and it may burn up from heat. Remember power equals current times voltage. When the voltage drop across the transistor is almost 0, the power dissipated by the transistor decreases.

An IC designed to interface CMOS or TTL logic levels to a relay is the 74C908. This IC can supply a current of 250 mA at 30 V. This is sufficient to energize most small relays.

15.5   MULTIPLEXING 7-SEGMENT LED DISPLAYS

The circuit in Figure 15-10 uses a 74LS241 octal buffer to multiplex two digits onto two 7-segment displays. When the Q output of the 7476 is LOW, the BCD digit 1 passes through the tri-state buffers into the 7447 where it is decoded to drive a 7-segment display. The displays are common anode. The common anode must be connected to a positive supply for the displays to be lit. The $\overline{Q}$ output is inverted by inverter number 1 of the 7406. The LOW level turns on $Q1$, and 7-segment display number 1 is connected to $+12$ V through the turned on transistor.

During this time outputs 2Y1, 2Y2, 2Y3, and 2Y4 of the 74241 are in their high impedance state and do not interfere with BCD digit number 1. The 7406 is open collector. When Q goes LOW and gets inverted by the 7406 inverter number 2, the output is pulled HIGH by the 470-Ω resistor. The emitter-base junction of $Q2$ is not forward biased and $Q2$ turns off. Only 7-segment display number 1 is on to display the BCD digit number 1.

When the 7476 changes states, Q is HIGH, the BCD digit number 2 passes through to the 7447 and is decoded. The 7406 inverter number 2 goes LOW and $Q2$ turns on. The BCD digit number 2 is displayed on the 7-segment display number 2. Display number 1 is off.

This process alternates with the output of the 7476. The 7476 is wired to toggle. Its output has an even duty cycle so that each set of tri-state buffers is enabled for an equal amount of time, and the brightness of each 7-segment display should be about the same. Each display is on for one-half each clock cycle. They appear to be about one-half as intense as they would be if they were on continuously. The current limiting resistors could be reduced in value to allow more current to flow through the LEDs during the time they are on. If this is done and the clock stops, the LEDs can be destroyed from excessive current.

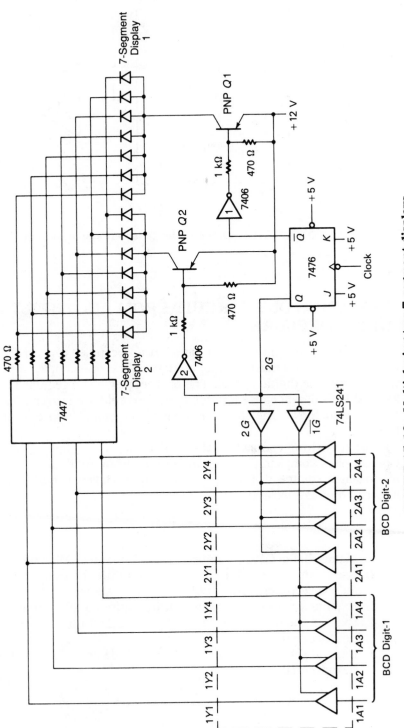

FIGURE 15-10 Multiplexing two 7-segment displays

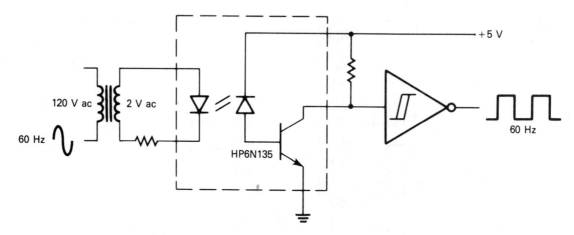

FIGURE 15-11   Optocoupler to isolate the ac voltage

15.6   ISOLATING ONE CIRCUIT FROM ANOTHER WITH OPTOCOUPLERS

The circuit in Figure 15-11 uses an optocoupler to transfer the 60-Hz line frequency to a standard TTL logic signal. The optocoupler also isolates the ac voltage from the digital signal. The optocoupler has a regular LED which, when forward biased, will cause a photodiode to conduct a current. This current then turns on an NPN transistor to supply a TTL signal to a Schmitt-trigger input inverter.

The only thing connecting the two circuits is the light given off by the LED. Optocouplers completely isolate one circuit from another, just as relays do, but much faster. The speed of the optocoupler makes it very useful for such applications as serial interface isolation.

Exercise 15

1. Make a list of CMOS tri-state gates. (Use your CMOS data book.)
2. Use Figure 15-9 as the circuit and compute the values for the base resistor for a collector current of 300 mA. (The transistor HFE is 15.)
3. Expand the multiplexing circuit in Figure 15-10 to multiplex 4 displays.
4. Expand the bus in Figure 15-8 to include an input port at address 000E hexadecimal.
5. Expand the bus in Figure 15-8 to include an output port at address 00F0 hexadecimal.

LAB **15** **Tri-State Gates**

OBJECTIVES

After completing this lab, you should be able to:

- use a transistor to control higher current.
- multiplex 7-segment displays.
- use tri-state gates to control a bus.

COMPONENTS NEEDED

1	7447 BCD to common anode 7-segment LED driver IC
1	74LS241 tri-state gate buffer IC
1	7476 dual *JK* flip-flop IC
1	7406 hex open-collector output inverter IC
2	FND-510 common anode 7-segment LEDs
2	small-signal PNP transistors
9	470-Ω, ¼-W resistors
2	1-kΩ, ¼-W resistors

PROCEDURE

1. Construct the circuit in Figure 15-10.
2. Pick the transistor to do the current buffering.
3. Have the instructor check the operation.
4. Run the displays at frequencies of 1000 Hz and 1 Hz. At which speed does the display appear brightest and why?

CHAPTER 16

Memories and Introduction to Microcomputers

OBJECTIVES

After completing this chapter, you should be able to:

- use and understand semiconductor memories.

- understand the basic microcomputer structure and operation.

- have some basic knowledge of the Z-80 CPU and its operation.

16.1 THE MICROCOMPUTER AND ITS PARTS

It is in the world of the computer and its hardware that much of what you have learned in this textbook is used. The computer is a digital machine with four main parts: central processor, memory, input/output, and program. The central processor is the digital brain behind the computer. The CPU (central processing unit) controls all other parts of the computer and is responsible for interpreting the program stored as binary numbers in the memory.

Memory is a large set of storage registers that can be accessed by the CPU. It is used to store the program and data. We will discuss several types of memory later in this chapter.

The program is stored in the memory as a set of binary numbers. This set of sequential binary numbers is a set of instructions used by the CPU to accomplish a task that the programmer wishes done by the computer. This part of the computer is not a set of gates or semiconductors, but is a set of instructions created by the mind of the computer programmer. For this reason the program is called *software*, in contrast to hardware, which is the electronics of the computer. The hardware of the computer would be of little use if the program were not there to run it. You might say the computer program is the fuel that the computer hardware uses to accomplish a job. As many computer hardware manufacturers have learned the hard way, you can have a very well designed, efficient piece of hardware, but if no one has developed good software for it, it just will not sell.

The last basic part of a computer is the input/output to the computer. This is the human interface to the computer. The computer can do many different things very fast, but the human cannot pry off the lid to the CPU and look at what is going on inside. The computer must output its answer to some device that the human can see and understand. Also the computer must be able to get input from the outside world. This input/output is not always from a human; it can come from other machinery or devices that the computer is to control.

We will look at the four main parts of a computer and how they work together.

16.2   THE CENTRAL PROCESSING UNIT

Before the advent of single-chip CPUs, central processors were made from many chips and placed on printed circuit boards. They were large and power-hungry. Large-scale integration of the CPU has produced a great number of single-chip central processors, such as the 6800, 6502, 8080, 8085, and Z-80. Because all of these CPUs are in a single 40-pin package, they are called *microprocessors*. The term "micro" seems to infer that the computing power is somewhat less than a minicomputer or mainframe CPU, and that is true for most of the 8-bit CPUs. An 8-bit CPU has an 8-bit data bus. Today's new 16-bit and 32-bit single-chip CPUs, such as the Z8000, 8086, 68000, 68020, and 80386, are quite powerful and can no longer be thought of as being small. Because the Z-80 is one of the most powerful and widely used 8-bit CPUs in production, we will use it as an example. This chapter is not designed to be a complete documentation on microprocessors, so we will not go into detail on many topics. The subject of microprocessors would take a whole textbook.

Figure 16-1 shows a Z80-based CPU that is part of a simple computer trainer. The Z-80 CPU has 16 address lines labeled A_0 to A_{15}. It uses these lines to supply the binary address of the memory or I/O with which the CPU wishes to communicate. This bus is a unidirectional bus and is buffered for more current drive by a set of 74LS245 ICs. This buffering is required because the Z-80 can only supply a little over 1 TTL load on any of its outputs. A 1-TTL output load is typical of most NMOS LSI CPUs.

The data bus is 8 bits wide, bidirectional, and used for data transfer to and from the CPU. The direction of data flow through the 74LS245 is controlled by a Z-80 control signal called $\overline{RD}$. This control signal will go LOW any time the CPU wishes to

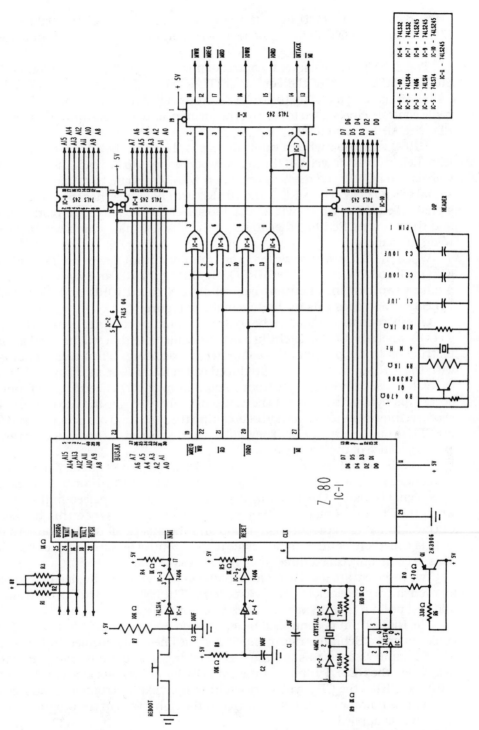

FIGURE 16-1   Z-80-based CPU for simple computer trainer

bring data into the CPU from the outside. When pin 1 of the data bus buffer (74LS245) goes LOW, data will be transferred from pin 18 to pin 2, from pin 17 to pin 3, and so on.

The Z-80 generates several control signals to control the operation of the data bus and other parts of the computer. These control signals are buffered with a 74LS245 also and make up the third Z-80 bus called the control bus. There are four basic control signals used to control the memory and input/output of the computer: $\overline{RD}$, $\overline{WR}$, $\overline{MREQ}$, and $\overline{IORQ}$.

$\overline{MREQ}$ will go low when a valid address is placed on the address bus by the Z-80 for a memory read or write. If the memory operation is to be read, then the $\overline{RD}$ signal will also go LOW. If the memory operation is to be a write to memory, then the $\overline{WR}$ signal will go LOW. The $\overline{MREQ}$ and the $\overline{RD}$ are ORed together to produce a control signal called $\overline{MRD}$, which will be used to select the memory for a read. The $\overline{MREQ}$ and the $\overline{WR}$ are ORed together to make the $\overline{MWR}$ signal, which will be LOW for a write to memory.

When the CPU reads or writes to an input or output device on the data bus, the $\overline{IORQ}$ will go LOW but not the $\overline{MREQ}$. By ORing the $\overline{RD}$ and $\overline{WR}$ to $\overline{IORQ}$, we can produce two new input/output control signals called $\overline{IORD}$ and $\overline{IOWR}$. IC6 is a 74LS32 which does this in the schematic of Figure 16-1.

One more control signal on the control bus of this computer is called $\overline{M1}$. $\overline{M1}$ will go LOW when the CPU is fetching a byte from memory that will be used as the next instruction in the program the computer is executing. When this signal is ORed with the $\overline{IORQ}$, a new signal is produced called $\overline{INTACK}$ or interrupt acknowledge. $\overline{INTACK}$ is used to read a byte from a special input port called an interrupt vector port. This byte is used to tell the computer where to get the next instruction to execute. Interrupts are a method of breaking the execution of a program and forcing the CPU to do something else. They will not be discussed in this chapter to any great extent because of their complexity.

You will probably notice that the chip select (CS) pins of all the 74LS245 buffers are connected to a signal through an inverter called $\overline{BUSAK}$ or bus acknowledge. This signal is made LOW, causing all the bus drivers to be placed in their HiZ state when the $\overline{BUSRQ}$ is made LOW by some outside device. When this happens, the CPU will finish the current instruction and then place all the bus buffers in HiZ. This is a method for an outside device to gain control of the computer bus system. This is commonly called *direct memory access* or DMA.

Two inverters are used to make a 4-MHz clock, which is divided by 2 to produce an even duty cycle clock for the Z-80 at pin 6. The PNP transistor is used as a hard-and-fast pull-up to +5 V for the clock to maintain a fast rise time.

The $\overline{RESET}$ pin is controlled by a power-on one-shot made from an *RC* timing constant and a Schmitt trigger, much like the one-shots studied in this text. The $\overline{RESET}$ input is used to start the Z-80 fetching instructions from address 0000 hex. When the $\overline{RESET}$ input is brought LOW for at least six clock cycles and then returned HIGH, the CPU will start fetching the next instruction at address 0000 hex. This is how the computer is started at the right place in the program each time the power is applied.

The $\overline{\text{NMI}}$ or nonmaskable interrupt is used to reboot the Z-80 after the computer has had power applied. This input causes the program to begin from 0066 hex when brought LOW. The *RC* one-shot is the same as the $\overline{\text{RESET}}$ except that a button has been placed in the circuit to retrigger the one-shot.

The $\overline{\text{RFSH}}$ signal is an output used by the Z-80 to refresh dynamic memory. This signal will go LOW when the address contains the address for refreshing the dynamic memory. The Z-80 has a hidden refresh cycle for dynamic memory, which makes it very easy to use with dynamic memory. We will take a look at dynamic memory later.

The $\overline{\text{HALT}}$ output indicates that the CPU has executed the halt instruction and is halted. Only an $\overline{\text{NMI}}$ or $\overline{\text{RESET}}$ can make the CPU continue after the halt instruction has been executed.

The CPU may be interrupted by bringing the $\overline{\text{INT}}$ pin LOW. This interrupt can be masked by the CPU through programming, and there are three different modes for interrupts in the Z-80.

The $\overline{\text{WAIT}}$ input is used to make the CPU stop and wait for slow memory, which takes longer to get the data on the data bus than the normal bus cycle.

16.3 COMPUTER MEMORY

A semiconductor memory is an integrated circuit capable of storing a binary number and recalling it when addressed or selected by a computer or other digital device. A simple latch made from *D* flip-flops, such as in Figure 7-20, can be considered memory because it can store a binary number.

There are two major types of memory, ROM and RAM. ROM stands for read-only memory. This type of memory has a preset value set in its memory cells that cannot be easily changed. RAM is random-access memory. RAM is read/write memory. This means that the value stored in its memory cells can be changed to a new value easily and quickly. The name random-access memory is not a good name because it implies that memory cells can be accessed randomly or in any order, not that they are read/write memory cells. The fact is that most ROM and RAM can be accessed randomly. Nevertheless, the acronym RAM has come to mean read/write semiconductor memory.

16.4 ROM

Figure 16-2 shows how a ROM can be constructed by using a decoder, four tri-state gates, and some diodes. This ROM can store eight 4-bit numbers or 32 bits of information. Each 4-bit word, or *nibble*, can be read or placed on the output line by supplying the correct address to the inputs of the decoder and enabling the tri-state gates with a logic LOW on the IC chip select ($\overline{\text{CS}}$) input. If address 1 1 1 is placed on the address inputs, then output number 7 of the decoder will go LOW or 0 V. This will forward bias the diode between the output number 7 and D_0 output, thus pulling the D_0 output to a logic LOW. All of the other diodes connected to the D_0 output are reverse biased because the other decoder outputs are 1 or positive voltage. Because D_1, D_2, and D_3 do not have any diodes to pull them LOW, their output value is 1. Each time the address to the ROM is changed, the D_0 to D_3 outputs reflect the value stored in the ROM. That value is determined by the placement of diodes between the decoder output and the D_0 to D_3 outputs.

Memories are measured by the number of memory addresses and the number of bits each address can store. The memory in Figure 16-2 is an 8 by 4 ROM. This means that it has eight address locations of 4 bits each, or 32 bits of total storage. ROMs of the type in Figure 16-2 are manufactured as single IC semiconductors with a predefined bit pattern stored in them. They are used for computer memories, character generators, and code converters.

16.5 PROM

The problem with ROM is that once the IC is made, the bit pattern in it cannot change, and to have a new IC manufactured is quite expensive. The PROM was

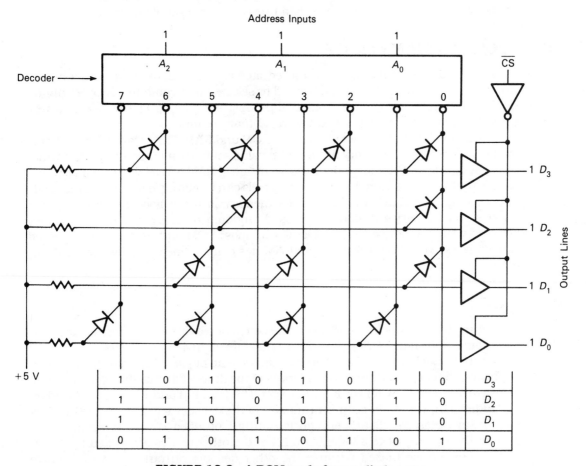

FIGURE 16-2 A ROM made from a diode array

introduced to help alleviate this problem. PROM stands for programmable read-only memory; its bit pattern can be set by the user. The programming is done by blowing a small semiconductor fuse in the memory cell in which you wish to make a 1. This is shown in Figure 16-3.

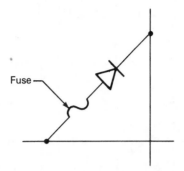

FIGURE 16-3   A PROM memory cell

To program a typical PROM, place the address of the memory location on the address lines of the PROM, place the data to be stored on the data output lines, hold $\overline{CS}$ HIGH, and pulse the program input PM HIGH for a few milliseconds. This will blow the fuse on each diode where a 1 is placed, but not on the diode where a 0 was placed. When a fuse is blown, the memory cell will, from that time on, read as a 1 because the diode has effectively been disconnected. Figure 16-4 shows the circuit for a 4 by 4 PROM that is programmed as just described.

16.6   EPROM

Once a PROM has been programmed by blowing the fuses at the bit location where a 1 was needed, it cannot be reprogrammed. Once a fuse is blown, it cannot be put back together again. The EPROM solved the problem by allowing the IC to be cleared and then programmed to a new bit pattern. EPROM stands for erasable programmable read-only memory.

EPROMs use a light-sensitive memory cell that, when exposed to ultraviolet light, will return to a 1 value. Therefore, most EPROMs have all 1s in their memory cells after they have been cleared by exposing them to ultraviolet light for about 20 minutes. See Figure 16-5.

As shown in Figure 16-6 the memory cell of an EPROM has a floating gate to the field-effect transistor that can be charged by placing a high voltage of about 12.5 to 25 volts on it. The exact programming voltage depends on the type of EPROM being programmed. You should check the IC specifications for the exact programming voltage to be used. Charging this floating gate causes the memory cell to become a 0. Because the electrons are forced across a very thin barrier of silicon dioxide (an insulator) to charge the floating gate, they will not cross back unless their energy level is increased artificially. The ultraviolet light will cause the floating gate to lose

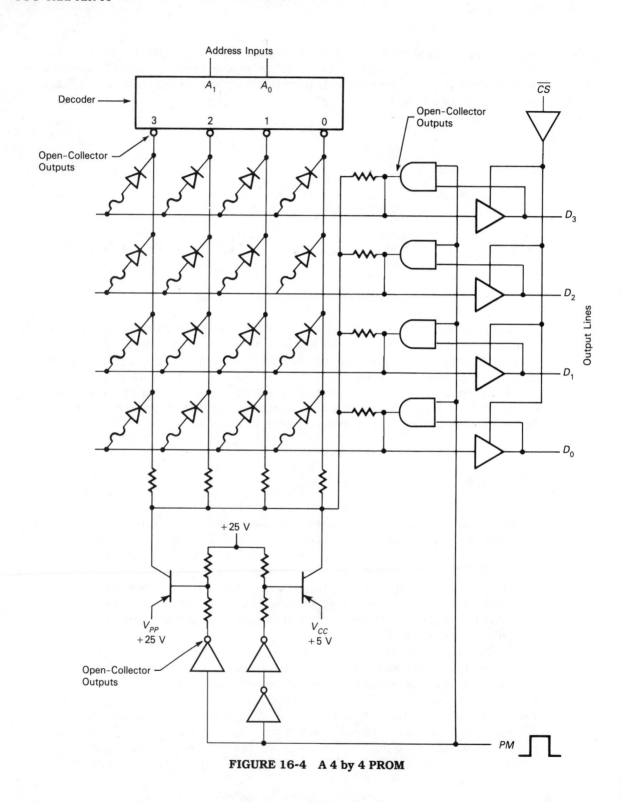

FIGURE 16-4   A 4 by 4 PROM

FIGURE 16-5 Typical EPROM chips

its charge, thus making the memory cell regain its 1 value. Most EPROMs will be completely erased by exposure to ultraviolet light with a wavelength of 2500 Å or less and an intensity of 15 W-s/cm² for 15 to 20 minutes. A standard 15-W fluorescent germicidal lamp (barbers put their cutting instruments under such lamps) works quite well for erasing EPROMs. The EPROMs should be placed about 1 inch below the light. The light should be enclosed so that your eye is not exposed to it for long periods. Actual erasing begins at a wavelength of about 4000 Å. This means that a standard fluorescent light used for visual lighting can erase an EPROM in about three to four years. Direct sunlight can do the job in as little as one week. Therefore, the EPROM should have the optical window in the chip covered to prevent external light from entering the chip.

The 2716 is a 2K by 8 EPROM, which is typical of the EPROMs used today. We will study its operation and programming methods because it is typical. It should be noted here that the 2708, which was a very early EPROM, is somewhat different in its programming methods. It is still found in older equipment.

Figure 16-7 shows the pinout of the 2716 EPROM. There are 11 address lines (A_0 to A_{10}) that select the memory location to read or program and eight data lines (D_0 to D_7), which are used to output data or place data into memory. The IC has one +5-V power supply and +25 V on the V_{PP} input for programming of memory. The $\overline{OE}$ (output enable) controls the internal tri-state gates on the D_0 to D_7 output pins. The $\overline{CE}$ (chip enable) also controls the output tri-state gates. The difference in the two is that when $\overline{CE}$ returns to the inactive state (HIGH) the 2716 goes into its standby mode, which causes it to draw about 75% less power.

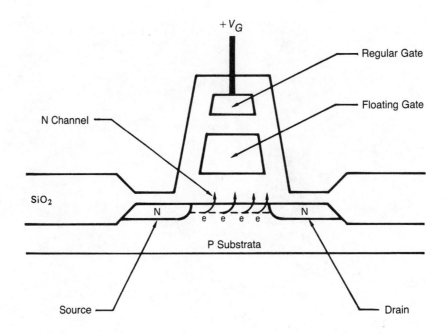

FIGURE 16-6   The EPROM FET transistor

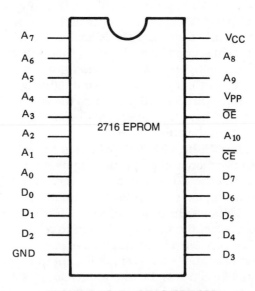

FIGURE 16-7   2716 EPROM

To program the 2716 EPROM, the V_{PP} pin is brought up to $+25$ V, the $\overline{OE}$ is made HIGH, and the $\overline{CE}$ is used to control programming. The byte to store in the memory is placed on the output lines (D_0 to D_7), and the $\overline{CE}$ pin is pulsed HIGH from a LOW state for 50 ms. You can randomly program the memory locations or do them sequentially. Once programmed, a 0 will remain a 0 until it is erased by exposure to ultraviolet light.

EPROMs have quickly become the major IC used for storage of boot programs and operating systems in computers today. Some typical chips used today are listed in Figure 16-8. The 2716 and the 2732 are 24-pin chips, and the rest are 28-pin chips. They have a similar pin configuration, which allows the designer to make provision for EPROM upgrade in the design with some simple jumper on the board.

16.7 EEPROM

EEPROM stands for electrically erasable programmable read-only memory. This type of memory will retain its stored bit pattern when power is removed. It can have the bit pattern programmed, and it can be changed by applying an electric field to the memory cell. The chief advantage to this type of memory is the ease with which it can be changed.

EPROMs cannot be selectively cleared, nor can they be cleared very fast. The EEPROM is an improvement over the basic EPROM technology. Figure 16-9 shows the basic EEPROM memory transistor and the floating gate. In the EPROM the electrons were forced across the silicon dioxide insulation by placing a high voltage across the P substrata and the normal gate. The electrons then collected on the floating gate and were trapped, thus charging the gate. When the gate is charged, the field-effect transistor will not conduct.

In the EEPROM the floating gate and normal gate have a protruding portion that comes very close to the drain of the transistor. Electrons are forced onto the floating gate by producing a high voltage of $-$ to $+$ from drain to normal gate. Then, just as in the EPROM transistor, the electrons collect on the floating gate, thus charging it negatively. Reversing the voltage polarity removes electrons and reverses the charge. This gives the EEPROM the ability to be cleared and then reprogrammed quickly, with a voltage of about 21 V.

EEPROM technology has not produced the ultimate read/write memory yet. The number of storages is not unlimited. It stands at about 100,000 at present, and it takes much more time to write than typical RAM. Because of these limitations, the EEPROM will not be used as a RAM. The EEPROM is being used today to store the configuration information for such devices as computer terminals, printers, and modems. The operator of the equipment can clear the EEPROM and reprogram it with a new configuration for a piece of computer equipment without removing the chip or using any special light. The configuration can be easily changed and will remain unchanged even when the power is turned off.

27256	2764	2732A	2716
V_{PP}	V_{PP}		
A_{12}	A_{12}		
A_7	A_7	A_7	A_7
A_6	A_6	A_6	A_6
A_5	A_5	A_5	A_5
A_4	A_4	A_4	A_4
A_3	A_3	A_3	A_3
A_2	A_2	A_2	A_2
A_1	A_1	A_1	A_1
A_0	A_0	A_0	A_0
O_0	O_0	O_0	O_0
O_1	O_1	O_1	O_1
O_2	O_2	O_2	O_2
Gnd	Gnd	Gnd	Gnd

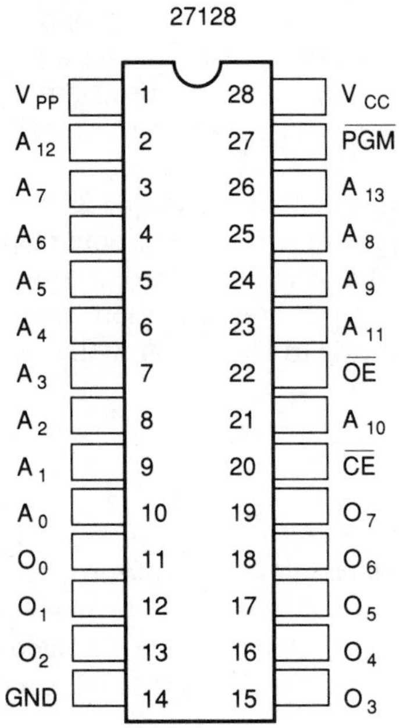

27128

Pin	Left	Pin	Right
1	V_{PP}	28	V_{CC}
2	A_{12}	27	$\overline{PGM}$
3	A_7	26	A_{13}
4	A_6	25	A_8
5	A_5	24	A_9
6	A_4	23	A_{11}
7	A_3	22	$\overline{OE}$
8	A_2	21	A_{10}
9	A_1	20	$\overline{CE}$
10	A_0	19	O_7
11	O_0	18	O_6
12	O_1	17	O_5
13	O_2	16	O_4
14	GND	15	O_3

2176	2732A	2764	27256
		V_{CC}	V_{CC}
		$\overline{PGM}$	A_{14}
V_{CC}	V_{CC}	N.C.	A_{13}
A_8	A_8	A_8	A_8
A_9	A_9	A_9	A_9
V_{PP}	A_{11}	A_{11}	A_{11}
$\overline{OE}$	$\overline{OE}/V_{PP}$	$\overline{OE}$	$\overline{OE}$
A_{10}	A_{10}	A_{10}	A_{10}
$\overline{CE}$	$\overline{CE}$	$\overline{CE}$	$\overline{CE}$
O_7	O_7	O_7	O_7
O_6	O_6	O_6	O_6
O_5	O_5	O_5	O_5
O_4	O_4	O_4	O_4
O_3	O_3	O_3	O_3

PIN NAMES

A_0-A_{13}	ADDRESSES
$\overline{CE}$	CHIP ENABLE
$\overline{OE}$	OUTPUT ENABLE
O_0-O_7	OUTPUTS
$\overline{PGM}$	PROGRAM
N.C.	NO CONNECT

2716 — (2K BY 8)
2732 — (4K BY 8)
2764 — (8K BY 8)
27128 — (16K BY 8)
27256 — (32K BY 8)

FIGURE 16-8 Typical EPROMs

16.8 STATIC RAM

The static RAM IC uses a simple set/reset flip-flop made from cross-coupled transistors as a memory cell. This memory cell can be either set or reset and will hold the set or reset value until the power is turned off. Typical read/write speed for MOS (metal-oxide semiconductor) RAMs ranges from 55 to 450 ns. This is quite

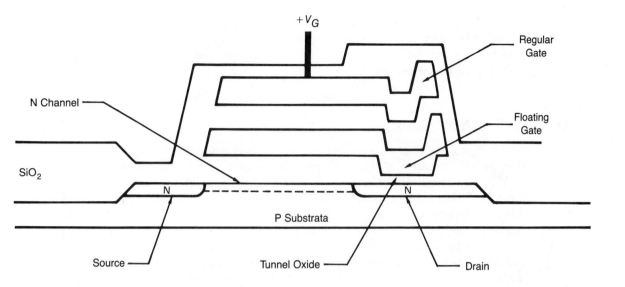

FIGURE 16-9   EEPROM transistor

fast enough for almost all computer operations today. All static RAM ICs are volatile, which means they lose their memory patterns when electrical power is lost. The old magnetic core memory is nonvolatile and static. Magnetic core memory used an array of magnetic doughnut-shaped circles to store bits. This memory was the major type used in computers before the advent of good semiconductor memories. The term *core memory*, which is often used today to mean the central RAM of a computer, came into use because in the past the RAM was made from magnetic core memory.

Static RAM is made with four basic technologies: MOS (metal-oxide semiconductor), CMOS (complementary metal-oxide semiconductor), TTL (transistor-transistor logic), and ECL (emitter-coupled logic). TTL RAM is much faster than MOS RAM but is not as dense. A typical TTL RAM, like Motorola's MCM93415, is a 1K by 1 RAM with an access time of 45 ns. CMOS RAM is slower than TTL RAM, somewhat denser, and uses much less power. When a CMOS RAM is not being read from or written to, it usually uses no power at all. This means that a small long-life battery, such as a silver oxide cell, can be used as a backup power supply for the RAM when the main power is turned off. This makes the CMOS RAM appear to be nonvolatile, and it is used in many applications, such as portable computers. A typical CMOS RAM is Motorola's MCM61L16P20, which is a 2K by 8 CMOS RAM with an access time of 200 ns.

ECL RAM is the fastest of the four basic technologies, but it uses more power. A typical ECL RAM, such as Motorola's MCM10474-15, is a 1K by 4 RAM with a 15-ns access time. The access time for a memory IC is the time it takes for the number stored in the memory to become stable on the data lines after the address and CS have become active. Figure 16-10 shows some typical static RAMs used in many of today's computer designs.

16.9 DYNAMIC RAM

Dynamic RAM uses a single transistor and a small capacitor for a memory storage cell. Because of the few components used per memory cell, the dynamic RAM is very dense. At present, the dynamic RAM can hold more memory cells per IC than any other type of IC memory. Because of the high density and relatively low power consumption, the dynamic RAM has quickly become the major computer memory used today. The only problem with dynamic RAM is that it must have every memory cell read or written to every 2 ms or the small capacitor in the memory cell will discharge and the RAM will lose the stored bit pattern. This refresh operation takes computer time and other circuits to accomplish. The Z-80 CPU has a built-in transparent refreshing system for dynamic RAM that does not waste any CPU time. This makes Z-80 CPU very attractive to designers.

The 4164 dynamic RAM is a 64K by 1 NMOS (N-channel metal-oxide semiconductor) made by many companies. It has a single +5-V power supply, and its outputs can handle two TTL loads. Figure 16-11 shows the pinout of the RAM.

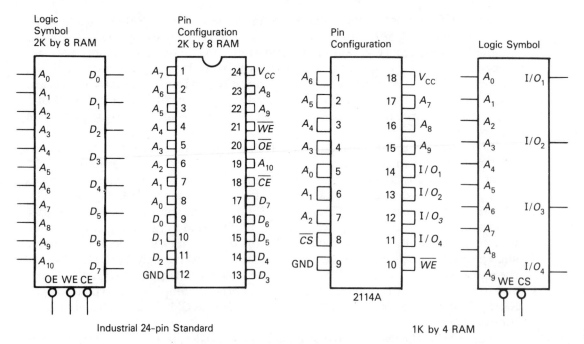

FIGURE 16-10 Typical static RAM

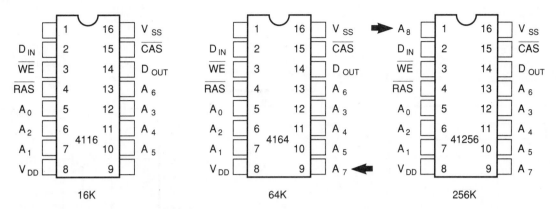

FIGURE 16-11   The industrial standard 16-pin package pinout for dynamic RAMs

The RAM comes in a 16-pin package. To address 64K of memory, you need 16 address pins (A_0 to A_{15}). To get all 16 address inputs, one data input, V_{CC}, ground, and the control inputs into one 16-pin package, the address inputs were cut in half and multiplexed into the chip. This means that only eight pins are used to supply the RAM IC with 16 address inputs. To do this, two new control lines were added: $\overline{RAS}$ (row address strobe) and $\overline{CAS}$ (column address strobe). These two strobes latch address information into the row and column of the memory cell matrix used by the RAM. Figure 16-12 shows the memory cell array and the row and column address latches. As can be seen in Figure 16-11, the industrial standard pin package for the three dynamic RAMs is the same, except for the addition of one address input for each larger IC memory. This means that a computer design could be made that would allow an upgrade in memory size by simply changing memory ICs.

To read or write to the RAM, the least significant byte (LSB) of the address is placed on the A_0 to A_7 address input, and then the $\overline{RAS}$ (row address strobe) is brought LOW, which latches the LSB of the address into the row latches of the memory array. Then the most significant byte (MSB) of the address is placed on the A_0 to A_7 address inputs and latched into the column latches of the memory array when $\overline{CAS}$ (column address strobe) goes LOW. Shortly after the LOW-going CAS signal, the data is put on the D_{out} pin, or the data on the D_{in} pin is stored in the RAM. Reading or writing is controlled by the $\overline{WE}$ input to the IC.

To refresh the memory cells in the array, you only need to latch in a new row address, because each time a new row is selected all the memory cells on that row are refreshed. Notice that the row address is actually only 7 bits wide, and the column address is 9 bits wide. Therefore, only a 7-bit refresh address is needed to refresh the entire dynamic memory.

During the M1 machine cycle or instruction fetch machine cycle of the Z-80 CPU, the 7-bit internal refresh register is incremented and placed on the lower 7 bits of the address bus. Then the $\overline{MREQ}$ and $\overline{RFSH}$ go LOW. By logically ORing these two signals you can produce a dynamic refresh signal.

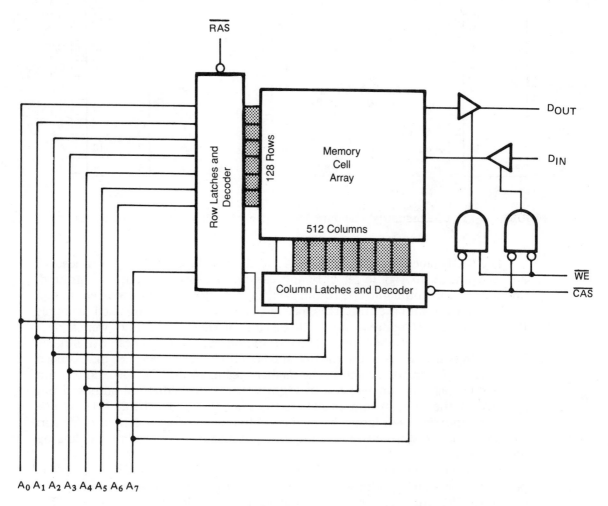

FIGURE 16-12   Typical DRAM configuration

The schematic shown in Figure 16-13 is the memory decoders and ROM/RAM for the computer trainer shown in Figure 16-1 (page 391). The ROM is a 2716 2K by 8 EPROM, and the RAM is two 2114 1K by 4 static RAM chips. The address decoder is a 74LS138 octal decoder. This decoder was studied in this text, so its operation should be clear. The decoder will allow the ROM to be accessed from address 0000 hex to 07FF hex and the RAM from 0800 hex to 0BFF hex. The decoder supplies a LOW output for each of the first eight 1K blocks of memory of the possible 64K the CPU can address.

The 74LS245 is used to add drive current to the data bus because the ROM and RAM ICs can only supply about one TTL load to an output. The direction of the data flow is determined by the $\overline{MRD}$ control signal from the CPU, and the bus driver is

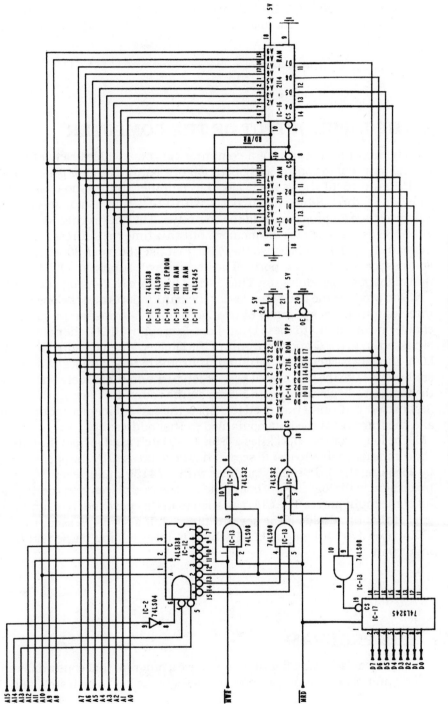

FIGURE 16-13   Trainer RAM and ROM

turned on and off with the address decoder. The bus driver will only be on if an address that is in the memory area of the ROM and RAM is present on the address bus.

16.10   THE INPUT/OUTPUT OF THE COMPUTER

The input and output of a computer can be a wide variety of things, from a typical terminal to control and operation of a running motor generator on a power grid. Figure 16-14 shows the I/O for the computer trainer shown in Figure 16-1. It consists of three 8-bit output ports with LEDs and a hex keypad for input of machine code for programming the computer. The address decoder is the same old 74LS138 as was used for the memory, but this time the address is only 8 bits wide instead of 16. This is because the Z-80, just like the older 8080, has only an 8-bit address for input/output ports. The decoder will supply a LOW select line for the first eight possible addresses. These select lines are NORed together with $\overline{IORD}$ and $\overline{IOWR}$ to produce the latch enables for the output ports and the one input port. The three output ports are made of two 7475 quad D transparent latches. When the enable or clock input to these D flip-flops is brought HIGH, the data on the data bus is passed on to the Q and $\overline{Q}$ outputs of the latches. When the enable or clock returns LOW, the data is latched on the Q and $\overline{Q}$ outputs of the D latches. When a 1 is present on the data line, the $\overline{Q}$ is made LOW, which turns on the associated LED. In this way the CPU can display 3 bytes in binary to the computer user. In this computer the first two ports (0 and 1) are used to display the current memory address being used, and the third (2) is the data contained in that address.

The fourth output port (3) is a 4-bit 7475 latch used to hold the current key that the computer wishes to test to see if it has been pressed by the operator. This is done by decoding the 4-bit binary number with a 74150 multiplexer. The 74150 multiplexer selects the logic level of one of the 16 keys and places it on the most significant bit of the input port 0. The CPU can then read this port and, by examining the logic level of that bit, decide if the key was pressed. If a key is pressed, the input to the multiplexer (74150) will be LOW, which is inverted and passed on the computer input port.

16.11   THE PROGRAM

As you can see, the computer must have a program that continually changes the binary number on the output port that selects the key to be tested, and then reads

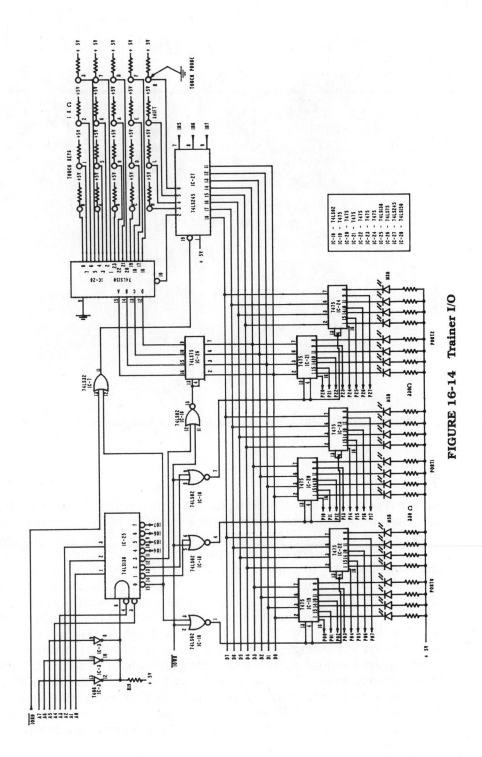

FIGURE 16-14 Trainer I/O

FIGURE 16-15   Z-80 motor generator control computer (Courtesy of Precise Power Co.)

the input port to see if it has been pressed. This program is stored in the ROM so that it will be present each time the power is applied to the computer.

To create a program for a microprocessor, you must first understand the microprocessor's internal architecture and its instruction set. Each microprocessor has a set of internal registers that can be used in different ways to manipulate binary numbers. The internal register set for the Z-80 is shown in Figure 16-17. The A register is the accumulator, which is an 8-bit storage register where the answer to arithmetic instructions is stored. The F or flag register is a collection of 1-bit flip-flops that indicate the status of the last instruction that has just occurred. For example, suppose the binary number 70 hex that was stored in the B register was subtracted from the A register, which contains 70 hex also. The answer 00 hex would be stored in the A register. The zero flag in the flag register would be set to 1, indicating that the result of the previous subtraction was zero. The microprocessor has other instructions that can test the state of the zero flag and do one of two things based on its value. The Z-80 has the following flags in its flag register.

FIGURE 16-16 **Control and input/output board for the Z-80 motor generator control computer** (Courtesy of Precise Power Co.)

S — Sign flag, used in signed two's complement arithmetic

Z — Zero flag, indicates a zero answer

H — Half-carry, indicates a carry from D_3 to D_4, used for binary-to-BCD conversion

P/V — Parity and overflow flag, used to indicate even or odd parity or an overflow in two's complement

N — Negative flag, used to indicate a subtraction

C — Carry flag, used to indicate a carry-out or borrow from the MSB in an arithmetic instruction

Main Reg Set	
Accumulator A	Flags F
B	C
D	E
H	L

General-Purpose Registers

Alternate Reg Set	
Accumulator A′	Flags F′
B′	C′
D′	E′
H′	L′

General-Purpose Registers

Interrupt Vector I	Memory Refresh R
Index Register IX	
Index Register IY	
Stack Pointer SP	
Program Counter PC	

Special-Purpose Registers

FIGURE 16-17   Internal register set for the Z-80

The BC, DE, and HL registers are used to store binary numbers or to hold addresses for memory locations where other binary numbers can be stored. They can be used as 8-bit registers or 16-bit registers when needed.

The IX and IY registers are called index registers. They are used to hold address locations for tables of binary numbers stored in memory.

The SP is the stack pointer. It holds the address of a last-in–first-out storage stack in memory for temporary storage of the CPU registers.

The PC is the program counter, which contains the address of the next instruction to be executed. This register keeps track of the location of the program in memory and the location of the next instruction.

The R or refresh register is used to produce an incremental address for the hidden refresh of dynamic memory if it is used for the computer's memory.

The I or interrupt register is used in the vectored interrupt mode of the Z-80. Interrupts are a method for an outside device to interrupt the flow of the Z-80 program and cause it to jump to a new program, which will survive the device that produced the interrupt. The Z-80 has three interrupt modes which are beyond this discussion.

The AF', BC', DE', and HL' registers are an alternative set of registers that can be exchanged with the regular set at any time with an instruction in the program.

The instructions for a microprocessor program are stored as binary numbers in the memory and called *op codes.* Op codes are read into the CPU and decoded to determine which instruction should be executed. Each op code will work on or affect another binary number, such as the number stored in the A register. The binary number operated on by the instruction is called the *operand.* The operand can be another register or binary number stored in the memory.

To help in writing programs, each of the major types of instructions have been given short alpha codes to help the programmer remember them. These alpha codes are called *mnemonics.* An instruction that will load the A register with the B register is

Op	Mnemonic	Operand	Comment
78	LD	A,B	;Load register A from register B

Programs can be written using the mnemonics of the instruction set alone and then processed by a program called an assembler to produce the actual op codes. This is a much easier way to produce a program than by looking up the op codes by hand and placing them in memory by hand.

There is much more to programming a microprocessor than is discussed here, but you must learn programming if you wish to completely understand the operation of a microcomputer.

Exercise 16

1. What is the frequency of the clock fed to pin 6 of the Z-80 in Figure 16-1?
2. What are the names of the three busses in the Z-80 computer in Figure 16-1?
3. What will happen if $\overline{\text{BUSRQ}}$ is brought LOW?
4. What is the CPU doing if the $\overline{\text{M1}}$ control signal is LOW in Figure 16-1?
5. What would be the address of the 2716 ROM in Figure 16-9 if the inverter on address line 15, which drives pin 6 of the 74LS138, was not there?

6. Why is the 74LS245 buffer IC used in the computer in Figure 16-1?
7. How many output ports can be added to the computer in Figure 16-10 if the address decoder is not changed?
8. Why were 74LS02 NOR gates used in Figure 16-10 instead of 74LS32 OR gates?
9. Draw the standard pinout for a 16-pin 64K dynamic RAM.
10. Redraw the ROM in Figure 16-2 to be 8 by 8.

16 RAM

OBJECTIVES

After completing this lab, you should be able to:

- understand the read and write operation of a static RAM.
- understand the use of a memory as a code translator.

COMPONENTS NEEDED

2	8-pin dip switches
14	10-kΩ, ¼-W resistors
8	330-Ω, ¼-W resistors
1	FND-510 common anode seven-segment LED
2	2114 1K by 4 static RAM
2	7406 quad open-collector inverter ICs

PROCEDURE

This circuit uses a pair of 2114 RAM ICs to convert a standard hex digit expressed in 4 bits of binary to the 8-bit binary code necessary to display the hex digit on a seven-segment LED. Code conversion is quite often done with ROM instead of RAM. A typical example of this is the character generator ROM used in a CRT controller circuit. This ROM generates the correct code to be shifted out across the computer monitor screen from the ASCII code input to the address pins of the ROM.

Follow the procedure below to program the RAM with the correct code for each hex number. After you have programmed the RAM with the correct code for each hex number, the seven-segment LED will display the corresponding 4-bit binary input.

1. Wire the circuit shown and open all the switches before connecting the power. This will place the RAM outputs in the HiZ state', and SW1 cannot pull down an output that might be 1. If this happens, the RAM could be damaged.

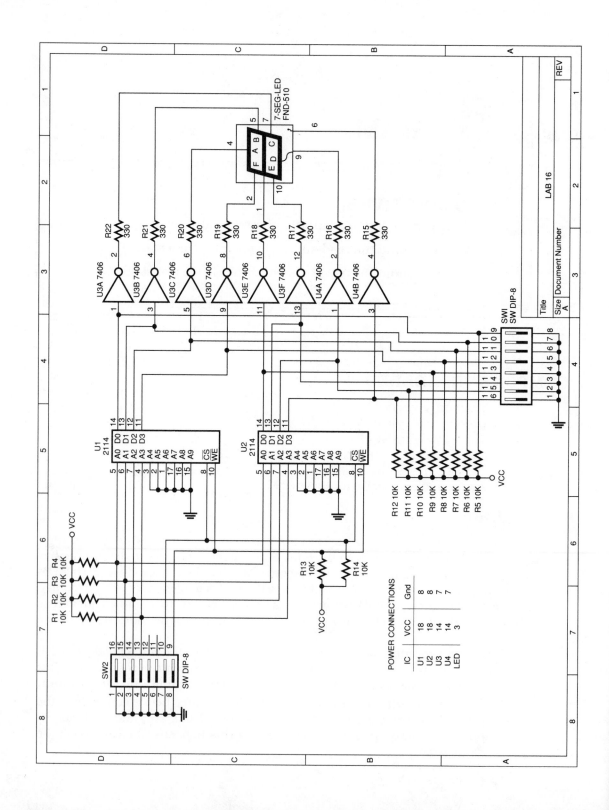

2. Now set the four RAM address inputs A_0, A_1, A_2, and A_3 to ground or the 0 value.
3. Use SW1 to display a 0 on the seven-segment display.
4. Store the 0 code in the RAM by first making $\overline{\text{WE}}$ LOW, and then bringing $\overline{\text{CS}}$ LOW and then back HIGH. This procedure will write the binary value on the RAM outputs in to memory location 0.
5. Next change the 4-bit memory address to binary 1 and repeat steps 3 and 4 to program the next RAM location with the correct code for binary number one.
6. Repeat this procedure for all 16 numbers of the hex number system. Use lowercase b and d for binary 1011 and 1101.
7. After you have programmed the RAM, place all the switches in SW1 to the open position and bring $\overline{\text{WE}}$ HIGH.
8. Now place a binary number on the 4-bit address of the RAMs, and the corresponding hex number should be displayed on the seven-segment LED.
9. Connect A_4 of the RAMs to SW2 pin 12 and make it HIGH. Now reprogram the seven-segment LED with the same codes as before, but make the hex point light each time. This will make the A_4 input the hex point of 1 of the next hex number in a two-number hex value.
10. Use the remaining switch on SW2 for A_5 of the RAMs and program two more sets of different codes for the seven-segment LED.

Appendix A

LAB TRAINER PLANS

The labs in this text are designed to be constructed on a solderless breadboard or protoboard and require an external power supply, clock or frequency generator and some buffered LEDs for logic indicators. These can be purchased separately or as a complete trainer which has everything in one package.

Another and much more desirable option is to build the equipment needed yourself. Schematics are shown in the following figures: Figure A-1 shows the debounce switch; Figure A-2 shows 8 buffered LED logic indicators; Figure A-3 shows 8 logic switches; Figure A-4 shows the clock generator; and Figure A-5 shows a power supply to operate these items and your lab circuit.

These parts are easily obtained and the trainer can be constructed in many different ways. The parts list for the digital trainer follows.

Quantity	Part Description
2	IC 1-2, 7406 open-collector inverters
1	IC 3, 7408 quad NAND gate
2	IC 4-5, 4050 CMOS to TTL buffers
1	IC 6, 555 timer
1	7805 +5 V voltage regulator
1	LM317 positive variable voltage regulator
8	1 kΩ, 0.5 W resistors
4	1 kΩ, 0.25 W resistors
1	240 Ω, 0.25 W resistor
1	5 kΩ potentiometer
1	20 kΩ potentiometer
8	Red LEDs
10	SPDT switches
1	5 position rotary switch
2	0.01 μF capacitors
1	0.1 μF capacitor
1	1 μF capacitor
3	10 μF capacitors
1	100 μF capacitor
2	4000 μF capacitors, 25 V dc
1	4 A bridge rectifier
1	2 A, 18 V center tap transformer
8	10 kΩ, 0.25 W resistors

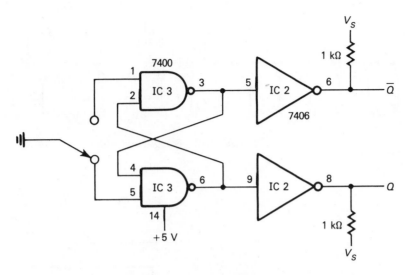

FIGURE A-1   Debounce switch

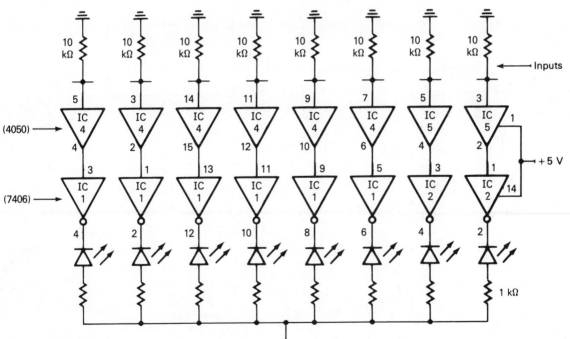

Note: 4050 V_{DD} is pin 1.

4050 GND is pin 8.

7406 V_{CC} is pin 14.

7406 GND is pin 7.

FIGURE A-2   8 buffered LEDs

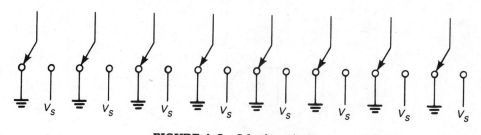

FIGURE A-3   8 logic switches

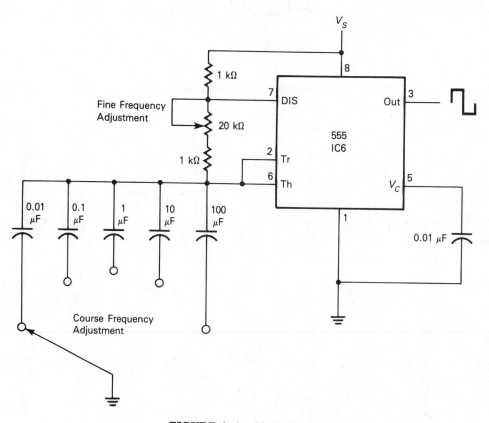

FIGURE A-4   Clock generator

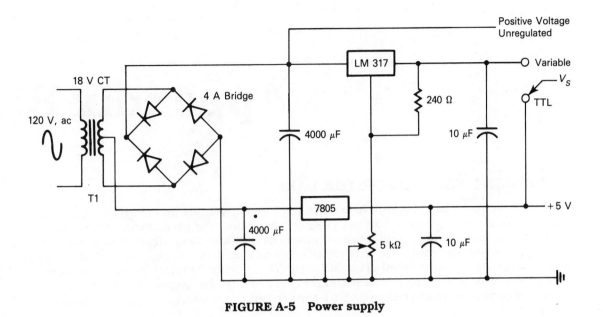

FIGURE A-5 Power supply

Appendix B

EQUIPMENT NEEDED FOR THE LABS

Most electronics labs in schools have all the needed equipment shown in the following list and more, but some items such as oscilloscopes may be few in number or nonexistent. In this case, the instructor can omit portions of a lab or use other methods to explain the lab in a class demonstration.

The labs were designed to be constructed on protoboards such as explained in Lab 1. This protoboard can be a separate board all by itself or part of a complete trainer which has its own power supply, clock, debounce switch, etc. Such a trainer is very useful in the completion of these labs. There are several companies which produce such a trainer or you can construct your own from the plan in Appendix A.

The equipment list needed to do the labs follows.

Quantity	Description	Labs Used In
1	Digital or analog multimeter	all labs
1	10 MHz dual trace scope	8, 9, 10, 11, 12, 13
1	0 to 20 V power supply	13
1	An ac signal generator	9, 11
1	Digital trainer with protoboard or 1 protoboard, 5 V power supply, TTL signal generator and 1 debounce switch	all labs
3	7400 ICs	2, 3, 7, 8, 9, 13
1	7402 IC	2, 3, 8, 9
1	7404 IC	2, 3, 6, 8, 13, 14
2	7406 ICs	12, 13, 14, 15, 16
1	7408 IC	2, 3, 5, 7, 8, 9, 10, 13
1	7410 IC	9
1	7411 IC	2, 3, 8
1	7414 IC	9, 11, 12
1	7420 IC	9
3	7432 ICs	2, 3, 5, 8
1	7447 IC	14, 15
2	7475 ICs	9
2	7476 ICs	8, 9, 10, 15
2	7483 ICs	1, 5
2	7486 ICs	4, 5
1	7490 IC	10, 15
1	7493 IC	10, 13, 14
3	7495 ICs	9

Quantity	Description	Labs Used In
1	74121 IC	12
1	74150 IC	14
1	74LS241	15
1	74C14 IC	6, 12
1	4069	2
1	4071	2
1	4081	2
1	4011	2
1	4070	4
1	4012	4
1	4009	4
2	74180 ICs	4
1	555 timer	11
1	LM339 IC	13
1	4001 IC	2, 6, 11
1	1 kΩ resistor array	6, 9, 11, 12, 13, 15
2	10 kΩ resistor arrays	13, 16
1	20 kΩ resistor array	13
1	330 Ω resistor array	all labs
1	10 MΩ resistor	11
1	22 kΩ resistor	11
2	100 Ω resistors	5, 6, 12
8	Red LEDs	all labs
2	FND-510 7 segment display	14, 15, 16
2	0.01 μF capacitors	11, 12
1	0.68 μF capacitor	11
2	20 pF capacitors	11
2	crystals of various frequencies	11
2	PNP power transistors	15
1	1 kΩ pot	6, 13, 14
1	914 Diode	12
10	470 Ω resistors	15
2	8-pin DIP switches	16
2	2114 1K by 4 static RAM	16

Appendix C

PINOUTS OF THE ICS USED IN THE LABS (TTL)

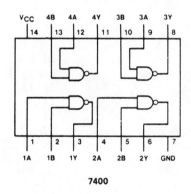

7400

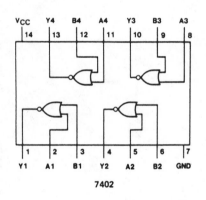

7402

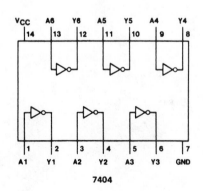

7404

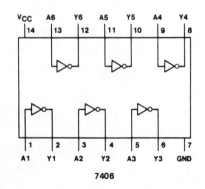

7406

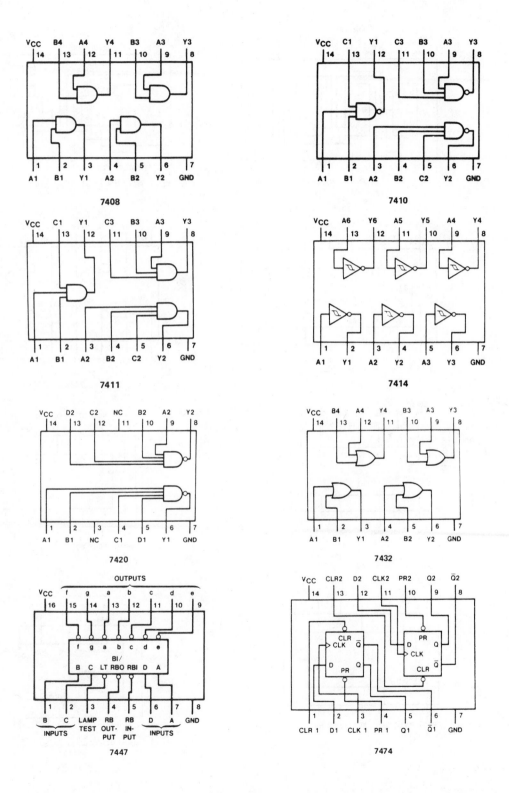

7408

7410

7411

7414

7420

7432

7447

7474

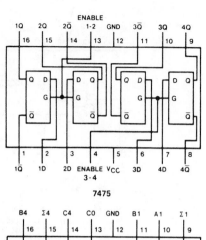

7475

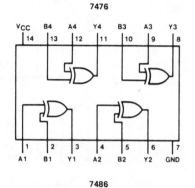

7476

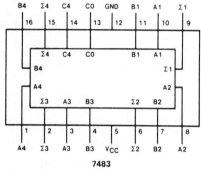

7483

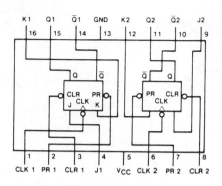

7486

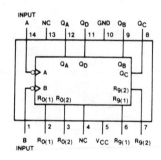

7490

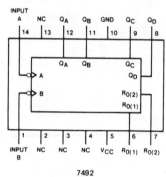

7492

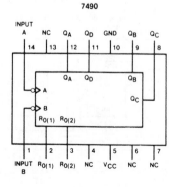

7493

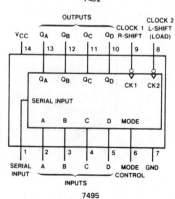

7495

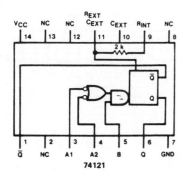

74121

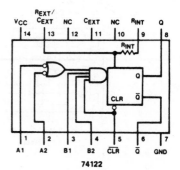

74122

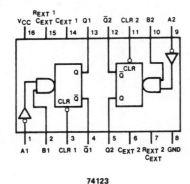

74123

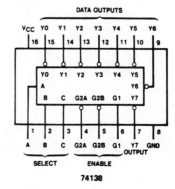

74138

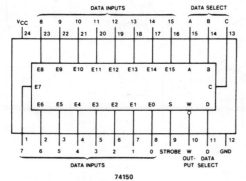

74150

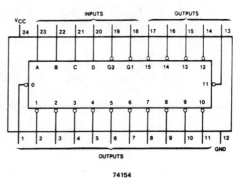

74154

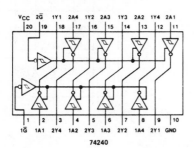

74240

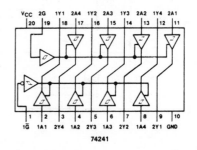

74241

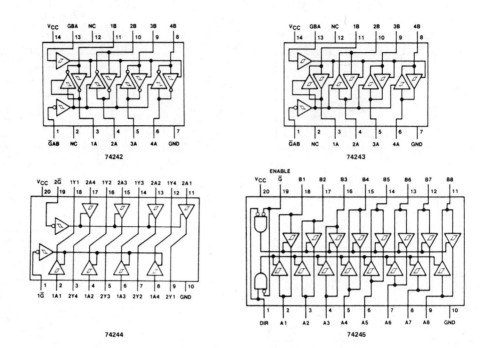

74242

74243

74244

74245

PINOUTS OF THE ICS USED IN THE LABS (CMOS)

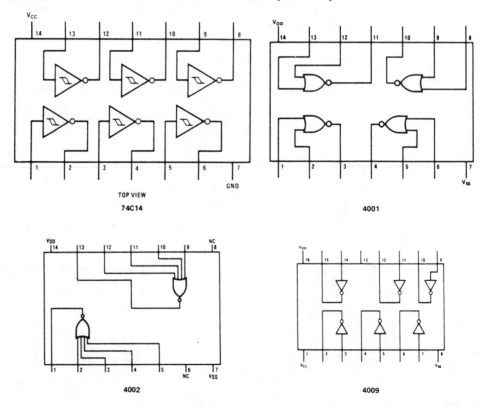

TOP VIEW

74C14

4001

4002

4009

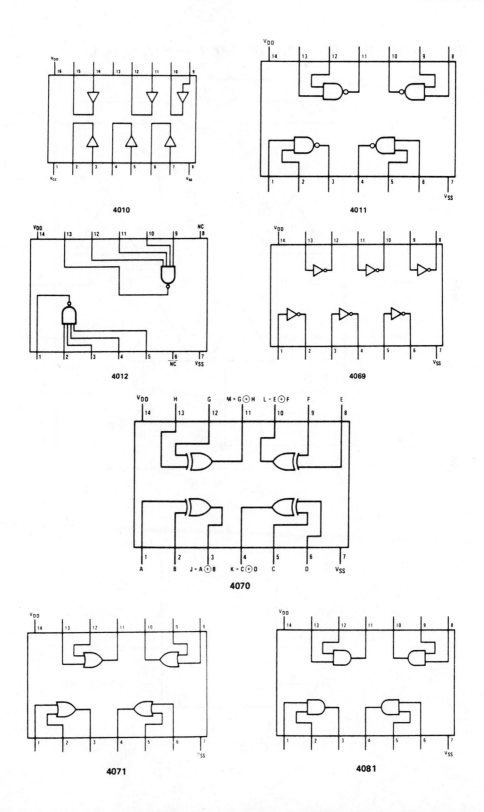

4010

4011

4012

4069

4070

4071

4081

PINOUTS OF THE ICS USED IN THE LABS (ANALOG)

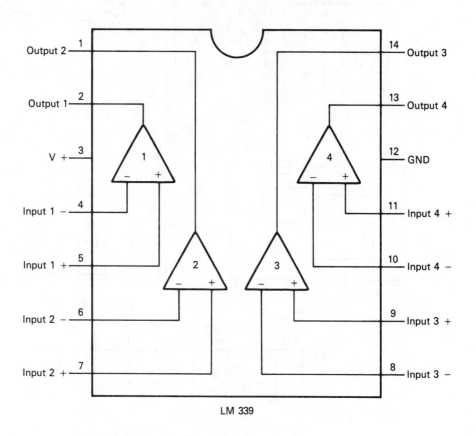

LM 339

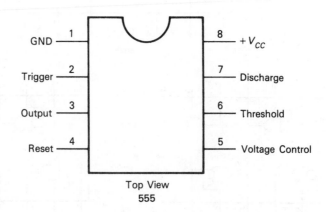

Top View
555

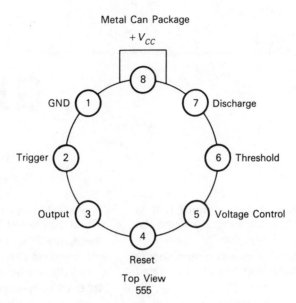

Metal Can Package

$+ V_{CC}$

GND 1

Trigger 2

Output 3

Reset

4

5 Voltage Control

6 Threshold

7 Discharge

8

Top View
555

Glossary

Active HIGH input. A circuit input that is "looking for" or "waiting for" a 1 to cause the circuit to activate or function.

Active HIGH output. A circuit output that is normally 0 and switches to a 1 when activated by the circuit.

Active LOW input. A circuit input that is "looking for" or "waiting for" a 0 to cause the circuit to activate or function.

Active LOW output. A circuit output that is normally 1 and switches to a 0 when activated by the circuit.

Analog. Pertaining to data that is continuously variable and not broken into discrete units. An example of an analog device is an ordinary car speedometer.

Analog-to-digital. Converting a continuous or analog quantity to a signal with a proportional value, usually a binary number.

AND gate. A circuit that combines ones and zeros according to the rule, "all 1s in, 1 out or any 0 in, 0 out."

Anode. One lead of a diode or LED which is connected toward the positive supply terminal to forward bias the diode.

ASCII (American Standard Code for Information Interchange). A 7-bit code that represents the decimal digits, letters of the alphabet, symbols, and control characters.

Astable multivibrator. A free-running clock or oscillator.

Asynchronous serial data transmission. A system whereby data is transmitted one bit at a time on a single data line at a predetermined baud rate. Asynchronous implies that there is no specific time between the start of one word and the start of the next.

Baud rate. The number of signal transitions per second that are transmitted or received, usually bits per second.

BCD (Binary-coded decimal). The code in which each decimal digit is represented by four bits.

BCD counter. A 4-bit binary counter that counts from 0000 to 1001 and then resets to 0000. It advances one number for each pulse received.

Binary. Base-2 number system which uses 2 digits, 0 and 1.

Binary ladder. A resistive network based on resistor values which increase in value by a power of 2. The resistor network will produce a proportional voltage for a given binary number input to the network.

Bit. A contraction for binary digit. Each place in a binary number is a bit; for example 1011 is a 4-bit number.

Boolean algebra. An algebra used to express the output of a base-2 digital circuit in terms of its inputs and used to reduce the output to lowest terms.

Boolean expression. Terms written in Boolean algebra that express the output of a circuit in terms of the input.

Broadside load. A parallel load in which all bits of a data word are loaded into a register on a single clock pulse.

Bubble. The small circle used on inputs and outputs of logic symbols to indicate the complement operation.

Byte. A binary number 8 bits long.

Carry-in. The carry into the first stage of an adder from a previous addition. It is sometimes called C_0.

Carry-out. The carry from the last stage of an adder.

Cathode. One lead of a diode or LED which is connected toward the negative supply to forward bias the diode.

Channel. The path of the current flow in an MOS transistor.

Clear. To reset or turn off a flip-flop and cause the Q output of a flip-flop to assume a 0 level.

Clock. A continuous rectangular waveform used for timing.

Collector-follower stage. Configuration in which the input signal is on the emitter and output is on the collector. It is the input stage of TTL circuits.

Combinational logic. Use of more than one gate to produce the required output.

Comparator. Digital circuit that compares two binary numbers and outputs whether they are equal.

Complement. (1) Invert; (2) A number which when added to a given number yields a constant. For example, the 9's complement of 7 is 2.

Conventional current. Indicates the current flow is from positive to negative.

CPU. Central processing unit. The part of a computer that interprets the instruction fetched from the memory and executes them.

Data bus. Circuits that generate, store, use, input, or output data are connected to the data bus. One line is provided for each bit of data.

Data separator. A circuit that can separate multiplexed data into its component parts.

Decimal. A base-10 number system which uses the digits 0 through 9.

Decoder. A circuit that converts a number from another number system or code back into the decimal system.

Delayed clock. A nonoverlapping clock or a dual clock system. The two rectangular waves are offset so that only one is HIGH at a time.

De Morgan's theorems. Two theorems of Boolean algebra where $\overline{A \cdot B} = \overline{A} + \overline{B}$ and $\overline{A + B} = \overline{A} \cdot \overline{B}$.

Demultiplexer. A logic circuit which will switch the digital or analog data coming in on one input to one of many possible output lines. The output line selected to receive the input data is chosen by a binary number input to the logic circuit.

Digital. Pertaining to data that is noncontinuous in nature. It changes in discrete steps. The data is represented by zeros and ones.

Digital-to-analog. Converting a number (usually binary) to a proportional continuous analog quantity.

Diode. A semiconductor device that conducts in one direction but not in the other. It has one anode lead and one cathode lead.

DIP (Dual in-line package). One style of integrated circuit package which has two rows of leads.

Drain. An element of an MOS transistor. Analogous to the collector of a bipolar transistor.

Dynamic RAM. Read-write memory that must have all the memory locations read every 2 ms to maintain the bit pattern stored in the memory.

EAC (End-around carry). The process in which the overflow in a 1's complement subtraction problem is added to the least significant column (rightmost).

Edge-triggered flip-flop. A flip-flop in which the data on the inputs is clocked into the flip-flop and appears at the output all on the same edge of the clock, unlike the master-slave flip-flop.

EEPROM. Electrically erasable programmable read-only memory. Nonvolatile memory that can be programmed and erased by electrical means.

Electron current flow. The actual flow of the electrons in the conductor from negative to positive.

Enable. To supply a control signal for one of the basic gates that will allow data to pass through the gate.

Encoder. A circuit that converts a number from decimal into another number system or code.

EPROM. Erasable programmable read-only memory. Nonvolatile memory that can be programmed and erased with an ultraviolet light.

Even parity. A system used to detect binary data transmission errors that uses a parity bit to make the total number of ones in a word even.

Exclusive-NOR gate. A two-input gate that produces a 1 out when its inputs are alike. Also called nonexclusive OR.

Exclusive-OR gate. A two-input gate that produces a 1 out when its inputs are different.

Expand. To use additional gates to increase the number of inputs to a gate.

Fan-out. A measure of how many loads a circuit can drive.

Fast carry. A carry signal is developed at the same time that the other signals are being developed. Carry does not have to ripple through other gates. A look-ahead carry.

Flip-flop. A bistable multivibrator. A circuit that is capable of assuming one of two conditions, on or off, in response to input signals, and maintaining that condition until signaled to change by the input.

Frequency. The number of cycles that a waveform completes in one second. It is measured in hertz.

Full adder. A circuit that adds three inputs and outputs a sum and a carry.

Full decoder. A circuit which will make one output line active for a given binary number input to the circuit. Each possible binary number has its corresponding output line.

Functional logic symbol. An alternate symbol used to represent the operation of one of the basic gates.

Gates. A circuit used to combine ones and zeros in specific ways. The basic gates are AND, NAND, OR, NOR.

H (High speed). Subfamily of TTL. 74HXX, 54HXX.

Half adder. A circuit that adds two inputs and outputs a sum and a carry.

Hexadecimal. A base-16 number system which uses the 16 digits, 0 through 9 and A through F.

HiZ (High impedance). The term used to indicate very high impedance in the order of 10 MΩ to 20 MΩ or higher.

Inhibit. To supply a control signal for one of the basic gates that will keep data from passing through the gate.

Inverted logic symbol. An alternate symbol used to represent the operation of one of the basic gates. A functional logic symbol.

Inverter. A circuit with one input and one output which functions according to the rule, "1 in, 0 out," or "0 in, 1 out."

JK flip-flop. A flip-flop that can be turned on, turned off, toggled, or left the same according to the control signals on the J and K inputs.

L (Low power). Subfamily of TTL. 74LXX, 54LXX.

Latch. A data flip-flop. A circuit that assumes an on state or an off state according to the input signal.

LCD (Liquid Crystal Display). A non-light-emitting method of displaying information.

Leading edge. The first transition of pulse, be it LOW-to-HIGH or HIGH-to-LOW.

Leading zeros. Zeros to the left of the most significant nonzero digit.

LED (Light-emitting diode). Emits light when forward biased.

LED display. A light emitting diode used to make a letter or number.

Logic diagram. A schematic showing the gates, flip-flops, and other modules used in the circuit.

Look-ahead carry. A carry signal is developed at the same time the other outputs are being generated. Carry does not have to "ripple through" other stages. A fast carry.

LS (Low-power Schottky). Subfamily of TTL. 74LSXX, 54LSXX.

LSB (Least significant bit). The right-most bit in a binary number.

LSI (Large scale integration). An IC that contains circuitry equivalent to 100 gates or more.

Master-slave *D* flip-flop. A flip-flop in which the input data is latched into the master section of the flip-flop on the leading edge of the clock and into the slave on the trailing edge.

Military specifications. Military standard for IC construction.

Minuend. A first or top number in a subtraction problem.

Mnemonic. The alpha code for machine level instructions used in assembly language.

MSB (Most significant bit). The leftmost bit in a binary number.

MSI (Medium scale integration). An IC that contains circuitry equivalent to more than 11 and less than 100 gates.

Multiplexer. A logic circuit which will switch one of the many inputs to one output. The input which is connected to the output is selected by a binary number input to the logic circuit.

NAND gate. A circuit that combines ones and zeros according to the rule, "all 1s in, 0 out or any 0 in, 1 out."

Negative edge. The transition of a signal from HIGH to LOW.

Nine's complement. The decimal number that is formed by subtracting a decimal number from all nines.

Nine's complement subtraction. Method in which 9's complement of the subtrahend is added to the minuend.

Noise immunity. One method of expressing the tolerance of a family of ICs to noise. It measures the range of acceptable input levels from the positive supply or ground.

Noise margin. One method of expressing the tolerance of a family of ICs to noise. Measures the voltage difference between an acceptable input level and the corresponding acceptable output level.

Nonexclusive-OR gate. A two-input gate that produces a 1 out when its inputs are alike. Also called an exclusive NOR.

Nonoverlapping clock. Delayed clock or a dual clock system. The two rectangular waves are offset so that only one is HIGH at a time.

Nonretriggerable one-shot. A logic device which when triggered by the rising or falling edge of a digital signal will produce an output pulse of a predetermined length. It cannot be triggered again until the output pulse has timed out.

NOR gates. A circuit that combines ones and zeros according to the rule, "any 1 in, 0 out or all 0s in, 1 out."

Octal. Base-8 number system which uses the 8 digits, 0 through 7.

Odd parity. A system used to detect binary data transmission errors. It uses a parity bit to make the total number of ones in a word odd.

One's complement. The binary number formed by inverting each bit of a binary number.

One's complement subtraction. Method of subtraction in which 1's complement of the subtrahend is added to the minuend.

One-shot. A monostable multivibrator. A logic device which when triggered by the rising or falling edge of the input pulse will produce a pulse of a predetermined output pulse width.

Open collector. Circuit in which the output has no internal path to the power supply. An external pull-up resistor is usually added.

Operational amplifier. A high gain amplifier with an inverting and noninverting input.

OR gate. A circuit that combines ones and zeros according to the rule, "any 1 in, 1 out or all 0's in, 0 out."

Oscillator. A circuit that switches its output from 0 to 1 and back according to its internal circuitry. The three types include the astable or free running (generates a clock), the monostable or one-shot (generates a single pulse), and the bistable or flip-flop.

Output port. A register that latches data to be moved from the system to the outside world.

Overflow. The carry from the most significant column (leftmost) in an addition problem.

Parallel data. Each bit has its own data line. The entire word is transmitted on the same clock pulse.

Parity. A system used to detect errors in binary data transmission.

Parity bit. An extra bit used with data bits to make the total number of ones even or odd.

Parity checker. Circuit that can determine whether the total number of ones in a binary word is even or odd.

Parity generator. Circuit that can generate the proper parity bit for an even- or odd-parity system.

Partial decoder. A logic circuit which will give an active signal on one output line for a given binary number input to the circuit. The circuit does not have a unique output line for each possible binary number input to it as a full decoder has.

Period. Time required for a signal to complete one cycle.

Positive edge. The transition of a signal from LOW to HIGH.

Preset. To set or turn on a flip-flop and cause the output, Q, of a flip-flop to assume a 1 level.

Presettable counter. A counter that can be loaded with the starting number. It then advances one count for each pulse received.

PROM. Programmable read-only memory. Nonvolatile memory that can have the bit pattern programmed once.

Propagation delay. The measure of time between change in input and corresponding change in output.

Pull-up resistor. A resistor that provides a connection to the power supply external to the IC. It is used with open-collector devices.

RAM. Random access memory. The read-write memory in a computer.

RC time constant. Resistance in ohms times capacitance in farads yields seconds. The time for the capacitor to charge to 63.2% of applied voltage.

Reset. To clear or turn off a flip-flop and cause the output, Q, of a flip-flop to assume a 0 level.

Retriggerable one-shot. A logic device which when triggered will produce an output pulse of predetermined pulse width. The time

width of the pulse or time-out period will start again each time the circuit is triggered even if the output has not yet timed out.

Ring counter. A shift register in which the output of the last flip-flop is fed back to the inputs of the first flip-flop.

Ripple counter. A counter designed so that each flip-flop generates the clock pulse for the flip-flop that follows. A ripple delay results.

ROM. Read-only memory. The nonvolatile memory in a computer.

S (Schottky). Subfamily of TTL. 74SXX, 54SXX.

Schmitt-trigger input. An input to a digital device which will change the output logic level at a fixed upper and lower threshold voltage of the input.

Serial data. A single data line exists and data is transmitted or received one bit at a time.

SET. To preset or turn on a flip-flop and cause the output, Q, of a flip-flop to assume a 1 level.

SET-RESET flip-flop. A flip-flop that can be turned on by a signal on the SET input and turned off by a signal on the RESET input.

Seven-segment display. An alpha numeric display made up of seven segments.

Shift counter. Also called a Johnson Counter. It produces output waveforms that are offset timewise and is used for producing control waveforms.

Signed two's complement. System in which a sign bit signifies whether a number is positive or negative and the remaining bits specify the magnitude. Negative numbers are represented in two's complement form.

Sinks. To provide a path for conventional current to flow to ground.

Source. Element of a field-effect transistor. Analogous to the emitter of bipolar transistor.

Sources. To provide a path for conventional current to flow from the power supply.

SSI (Small-scale integration). An IC that contains circuitry equivalent to less than 12 gates.

Static RAM. Read-write memory that does not need to be read or refreshed to maintain the memory's bit pattern.

Substrate. The silicon materials, P type or N type, upon which a transistor is fabricated.

Subtrahend. The second or bottom number in a subtraction problem.

Successive approximation. A method of using a digital-to-analog converter and a voltage comparator to produce a binary number which is proportional to the given analog input voltage.

Successive division. Method for converting a decimal number to a binary number.

Switch bounce. The making and breaking of switch contacts during a single switch closure.

Synchronous counter. A counter designed so that each flip-flop receives the clock pulse at the same time.

Ten's complement. The decimal number that is found by subtracting a decimal number from all nines and then adding one.

Ten's complement subtraction. Method in which the 10's complement of the subtrahend is added to the minuend.

Time-out period. Time required for a one-shot multivibrator to turn itself back off after being turned on.

Toggle. To change state. To switch from 1 to 0 or 0 to 1.

Totem pole. Circuit in which output has internal paths to the power supply and ground.

Trailing edge. The second transition of a pulse as it returns to its original level, be it HIGH-to-LOW or LOW-to-HIGH.

Transparent D flip-flop. A flip-flop that allows the input data to pass through to the output unaltered during one phase of the clock and latches the input data as the clock changes levels.

True magnitude. Actual value as opposed to a complemented value.

Truth table. A table listing all possible inputs to a circuit and the corresponding outputs.

TTL (Transistor-transistor logic). One of the popular families of digital integrated circuits.

2R ladder. Sometimes called a 2R binary ladder. A resistive network based on two resistor values which will produce an output voltage proportional to the binary number input to the network.

Two's complement. The binary number formed by inverting each bit of a binary number and adding 1.

Two's complement subtraction. Method of subtraction in which the 2's complement of the subtrahend is added to the minuend.

Unique state. The output state of one of the basic gates that occurs for only one combination of inputs.

Up-down counter. Counter that can increment or decrement according to a control signal.

V_{CC}. Positive supply voltage in a TTL IC (5 V). Sometimes used to designate the power supply voltage for a CMOS IC.

V_{DD}. Positive supply voltage in a CMOS IC. ($+ 3$ V to $+ 18$ V).

Voltage comparator. A circuit that compares the relative amplitudes of two input signals. The output is HIGH when the voltage on the noninverting input exceeds that on the inverting input.

Waveforms. A graphical representation of a signal. It is the plot of the amplitude as a function of time.

Zener diode. A diode that conducts in the reverse direction at a definite voltage level.

Answers to the Odd-Numbered Problems

CHAPTER 1

1. 100_2
    101_2
    110_2
    111_2
    1000_2

3.

66_8	72_8	76_8	102_8	106_8
67_8	73_8	77_8	103_8	107_8
70_8	74_8	100_8	104_8	110_8
71_8	75_8	101_8	105_8	

5.

DD_{16}	E6	EF	F8
DE	E7	F0	F9
DF	E8	F1	FA
E0	E9	F2	FB
E1	EA	F3	FC
E2	EB	F4	FD
E3	EC	F5	FE
E4	ED	F6	FF
E5	EE	F7	100
			101_{16}

7.

10001001_{BCD}	10010100_{BCD}	10011001_{BCD}
10010000_{BCD}	10010101_{BCD}	100000000_{BCD}
10010001_{BCD}	10010110_{BCD}	100000001_{BCD}
10010010_{BCD}	10010111_{BCD}	
10010011_{BCD}	10011000_{BCD}	

9. a. 15   b. 16

11. a. 65, 535   b. 65, 636

13.

Octal	Hexadecimal	Binary	Decimal	BCD
36	1E	11110	30	110000
251	A9	10101001	169	101101001
22	12	10010	18	11000
143	63	1100011	99	10011001
103	43	1000011	67	1100111

15. a. 10110   b. 10101   c. 111011

17. a. 111   b. 1111   c. −110   d. −1101

19. a. 1001   b. −11

21. a. 11100   b. −11

23. a. 11001110   c. 11111110   e. 10001000
     b. 00101011   d. 00001000   f. 01010011

25. a. 00010011   (Answer correct)
     b. 00110000   (Answer correct)
     c. 00110111   (Answer incorrect; no carry from column 7, but overflow)
     d. 11111110   (Answer incorrect; carry from column 7, but no overflow)

27. a. 01110011   (Answer correct)
     b. 00010000   (Answer correct)
     c. 01010010   (Answer correct)
     d. 00010110   (Answer incorrect; no carry from column 7 to 8, but overflow)

29. A one as the most significant bit indicates that the number is negative and is represented in two's complement form. A zero as the most significant bit indicates that the number is positive and is represented in true magnitude form.

31. Binary numbers are easy to represent electronically (on, off). Hex numbers are used to represent binary numbers.

CHAPTER 2

1. a.

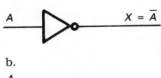

b.

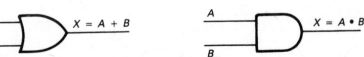

c.

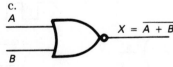

d.

e.

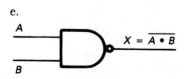

$X = \overline{A \cdot B}$

3.

AND		OR		NAND		NOR	
AB	X	AB	X	AB	X	AB	X
00	0	00	0	00	1	00	1
01	0	01	1	01	1	01	0
10	0	10	1	10	1	10	0
11	1	11	1	11	0	11	0

5. a.

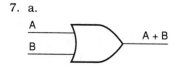

AB

b.

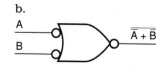

$\overline{\overline{A} + \overline{B}}$

c.

AB	X
00	0
01	0
10	0
11	1	Unique state

7. a.

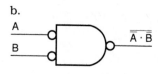

A + B

b.

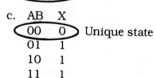

$\overline{\overline{A} \cdot \overline{B}}$

c.

AB	X
00	0	Unique state
01	1
10	1
11	1

9. 1,   0,   0,   1

11. 0,   1,   1,   0

13. 0,   1,   0,   0

15. Enables

17. Put a 1 on the control input.

19. 0

21. Inverted

23. Put a 1 on the control input.

25. Locked up in the 1 state.

27. Unaltered

29.

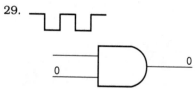

31.

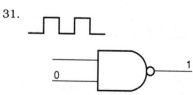

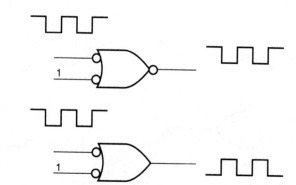

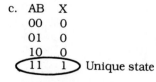

33.

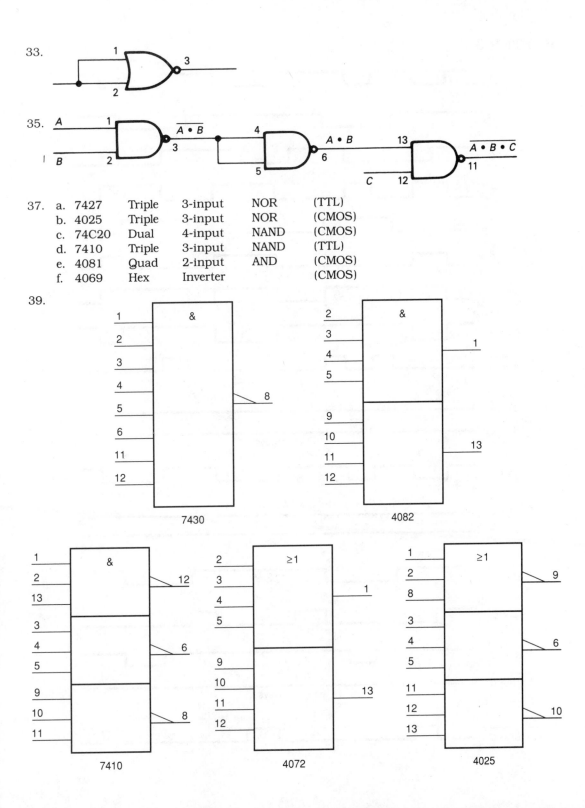

35.

37. | | | | | | |
|---|---|---|---|---|---|
| a. | 7427 | Triple | 3-input | NOR | (TTL) |
| b. | 4025 | Triple | 3-input | NOR | (CMOS) |
| c. | 74C20 | Dual | 4-input | NAND | (CMOS) |
| d. | 7410 | Triple | 3-input | NAND | (TTL) |
| e. | 4081 | Quad | 2-input | AND | (CMOS) |
| f. | 4069 | Hex | Inverter | | (CMOS) |

39.

CHAPTER 3

1.

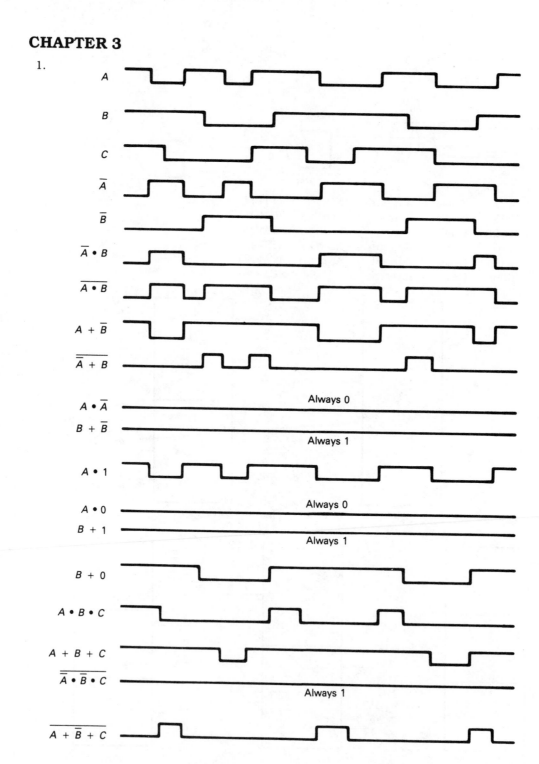

3. a. $\bar{A} + B$ ^{1 2}⊓⊔   b. $\bar{A} + B$ ^{1 2}⊓⊔   c. $C_P' \cdot \bar{B} \cdot A$ ⌐1'⌐

      $\bar{B} + \bar{C}$ ^{3 5}⊓⊔ $\bar{A} + C$ ^{1 3}⊓⊔ $C_P' \cdot B \cdot C$ ⌐3'⌐⌐4'⌐

      $A + \bar{C}$ ^{4 6}⊓⊔ $B + C$ ^{2 3}⊓⊔ $C_P' \cdot \bar{A} \cdot \bar{C}$ ⌐6'⌐

   d. $\overline{A \cdot B}$ ^{1 2}⊓⊔   e. $C \cdot A$ ^{3 4}⊔⊓   f. $C_P' \cdot A \cdot \bar{C}$ ⊔1'⊔2'⊔

      $\overline{AC}$ ^{3 4}⊓⊔ $B \cdot \bar{C}$ ^{2 3}⊔⊓ $C_P' \cdot B \cdot A$ ⊔2'⊔3'⊔

      $\overline{\overline{B}\overline{C}}$ ^{2 3}⊓⊔ $A \cdot \bar{C}$ ^{1 3}⊔⊓ $C_P' \cdot \bar{A} \cdot C$ ⊔4'⊔5'⊔

   g. $C_P' \cdot \bar{A} \cdot \bar{B}$ ²⊔2'⊔3'⊔⁴

      $C_P' \cdot A \cdot C$ ⁶⊔6'⊔¹

      $C_P' \cdot B \cdot \bar{C}$ ⁵⊔5'⊔⁶

5.

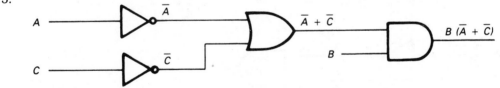

A $\bar{A}$ $\bar{A} + \bar{C}$ $B(\bar{A} + \bar{C})$
C $\bar{C}$ B

7.

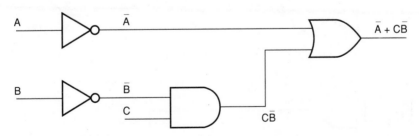

A $\bar{A}$ $\bar{A} + C\bar{B}$
B $\bar{B}$ C $C\bar{B}$

9. a. $\bar{A}(B + C)$ b. A c. $A + D$ d. 0 e. $\bar{A} + B$

11. a. $A + \bar{B}$ b. $A + C$ c. $A(\bar{C} + B)$ d. $A(\bar{C} + \bar{B})$

13. a. $\bar{B} + \bar{A}C$ b. $AB + \bar{C}B + \bar{A}\bar{C}$

15. $AC + \overline{\overline{AC}}$

CHAPTER 4

1.

Inputs		Output
A	B	X
0	0	0
0	1	1
1	0	1
1	1	0

3. Inverted

5.

7.

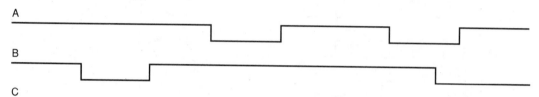

9.

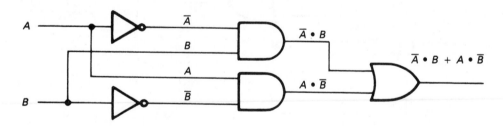

Alternate Solution

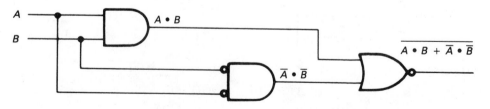

11. a. EVEN: $\underline{0}$ 101101   b. ODD: $\underline{1}$ 110000   c. EVEN: $\underline{0}$ 000011   d. ODD: $\underline{0}$ 110010

13.

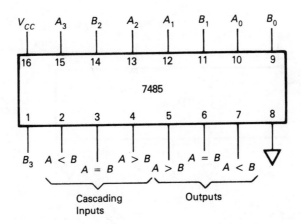

15.

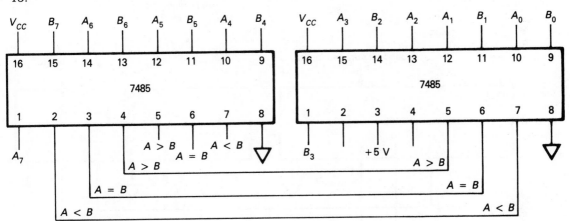

Note: Pin 3 of 1st 7485 tied HIGH for cascading.

17.

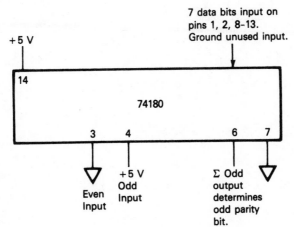

19.

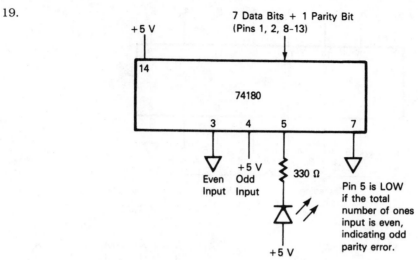

21.

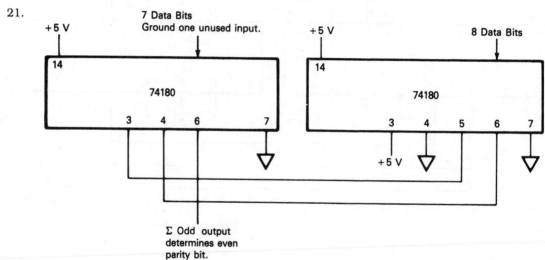

23.

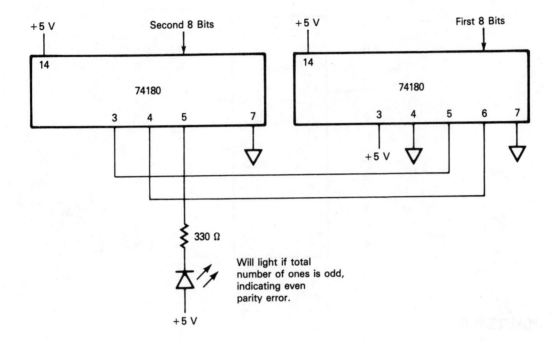

+5 V   Second 8 Bits   +5 V   First 8 Bits

14   14

74180   74180

3   4   5   7   3   4   5   6   7

+5 V

330 Ω

Will light if total
number of ones is odd,
indicating even
parity error.

+5 V

25.

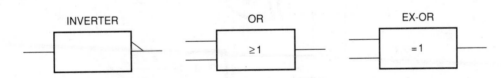

INVERTER   OR   EX-OR

≥1   =1

27.

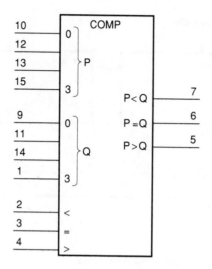

CHAPTER 5

1.

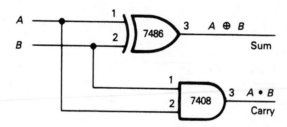

Inputs		Outputs	
A	B	Sum	Carry
0	0	0	0
0	1	1	0
1	0	1	0
1	1	0	1

3. a. 1111

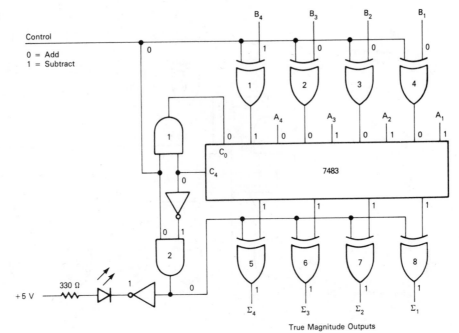

ON = Negative Answer to Subtraction

b. 0011

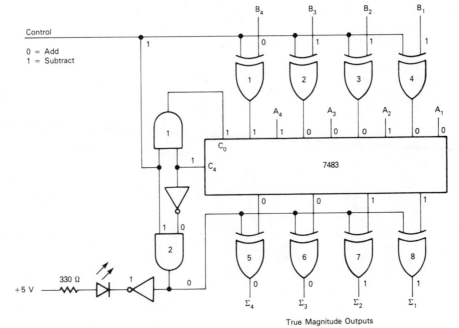

ON = Negative Answer to Subtraction

c. -0101

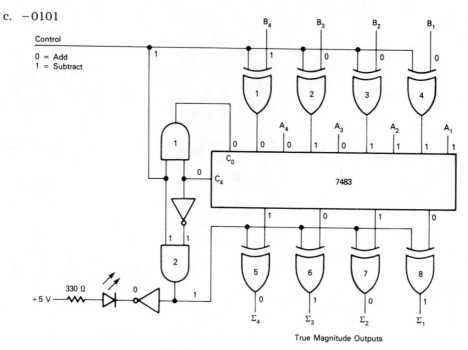

Control

0 = Add
1 = Subtract

True Magnitude Outputs

ON = Negative Answer to Subtraction

5. a. 1101

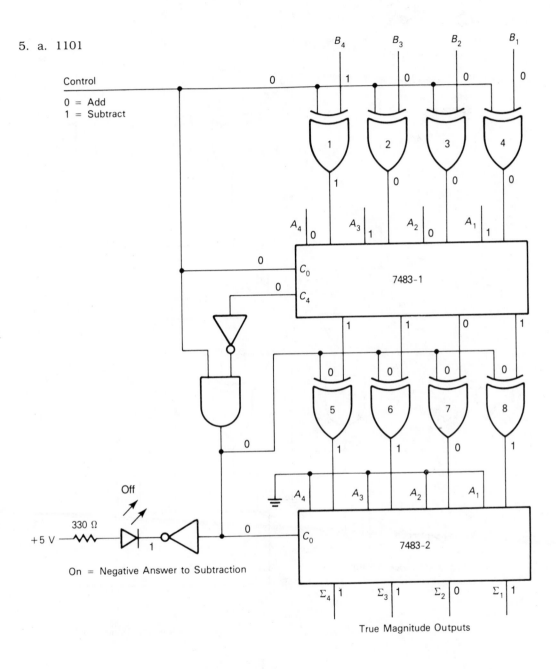

True Magnitude Outputs

b. 0010

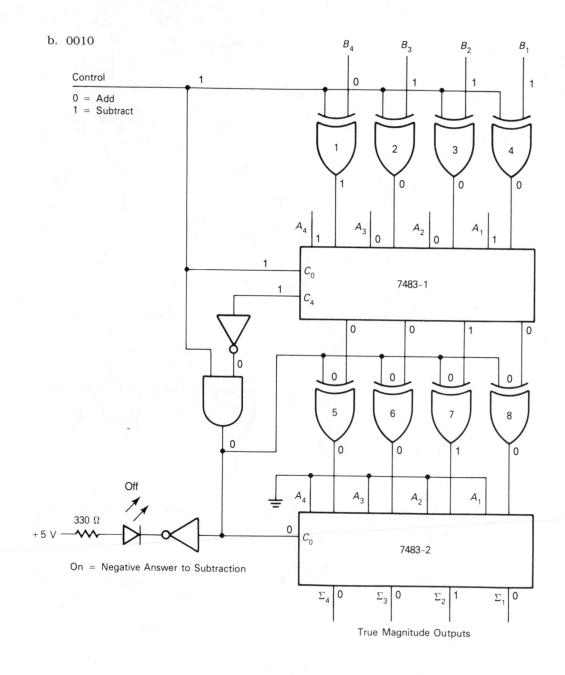

True Magnitude Outputs

c. -0101

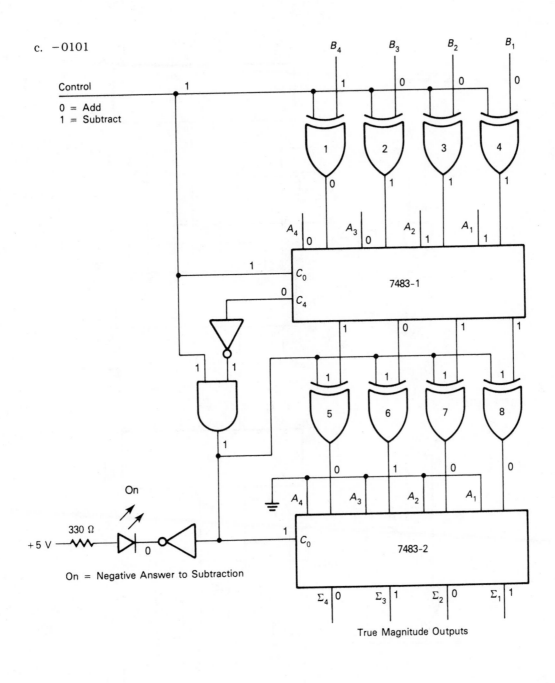

True Magnitude Outputs

7. a. 1001

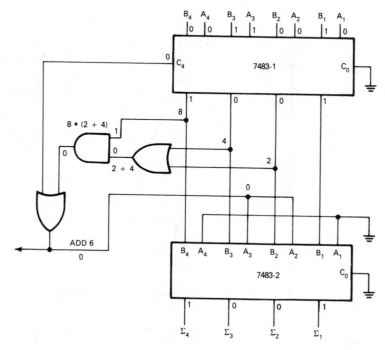

b. 1 0101

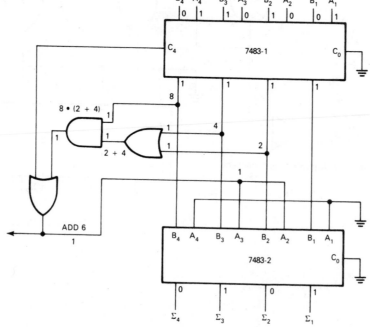

c. 0110

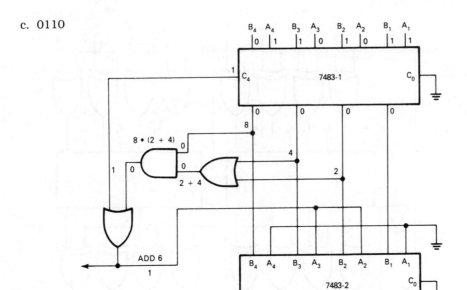

9.

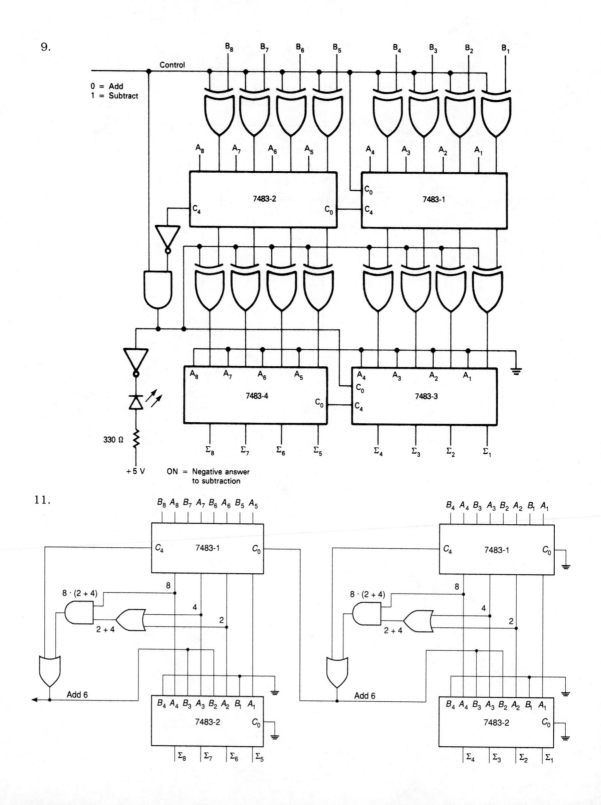

11.

13. a. 10100110
    b. 01011111
    c. 10110101

15. a. 11001011

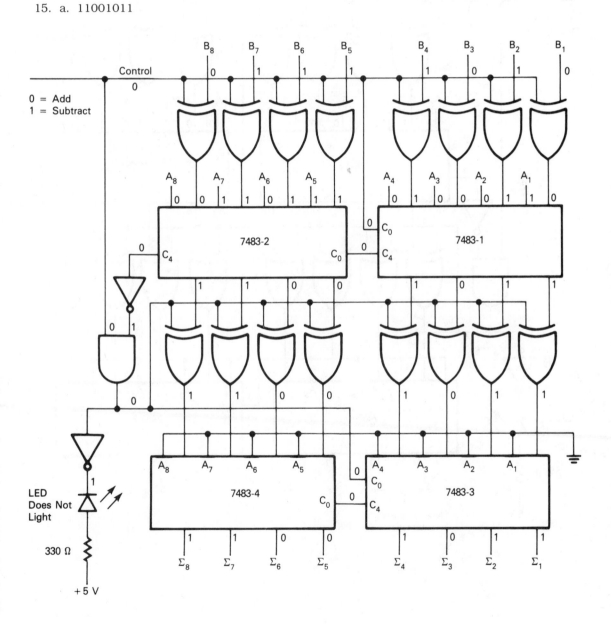

ON = Negative answer
     to subtraction

b. 1 10100110

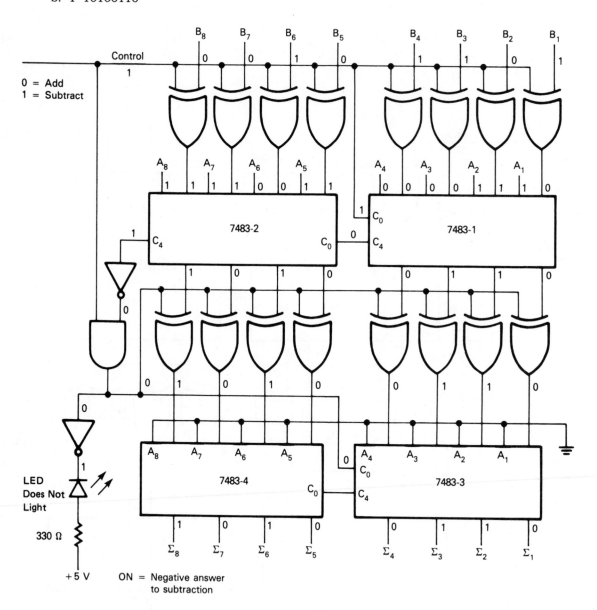

c. -00000001

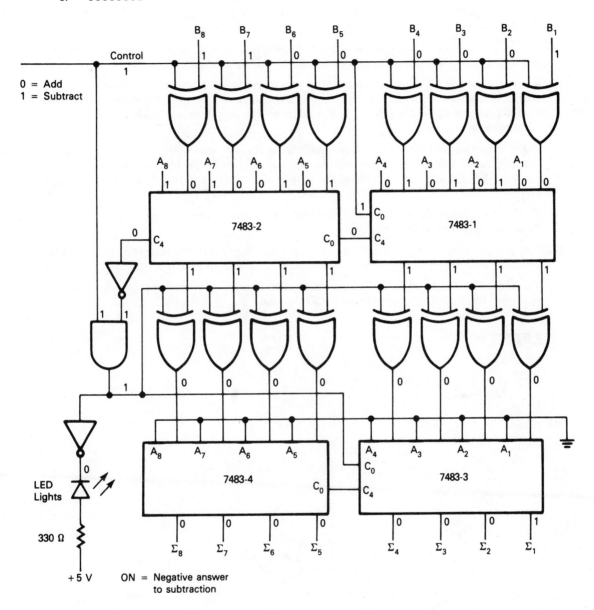

17. a. 10001001
    b. −0010 0001 (Circuit for problem 11 will not handle a subtraction problem.)
    c. 10100 0110

19. A 1 on the control line (subtract) enables AND gate 1. If $C_4 = 1$ (overflow) the output of AND gate 1 is 1. The 1 is fed into C_0 to accomplish the end-around-carry.

21. A 1 on the control line (subtract) and $C_4 = 0$ (no overflow) requires that the sum be complemented to obtain the true magnitude of the answer. AND gate 2 provides a 1 in this case to cause exclusive OR gates 5, 6, 7, 8 to invert the output of the 7483. The true magnitude appears at $\Sigma_4 \Sigma_3 \Sigma_2 \Sigma_1$.

23. In a subtraction problem if there is no overflow the 2's complement must be taken to obtain the true magnitude of the answer. In this case the AND gate provides a 1 which causes exclusive-OR gates 5, 6, 7, 8 to invert the output of the 7483-1 to begin the 2's complement process.

25. The function of the 7483-2 is to add 1 during the 2's complement process. The 1 is added via C_0 so $A_4 A_3 A_2 A_1$ are grounded.

27. The output of the AND gate is a 1 during a subtraction problem when there is no overflow. A 1 indicates that the 2's complement must be taken to obtain the true magnitude of the answer.

29. The sum of 2 BCD numbers can be a legitimate BCD number even though overflow has occurred. In that case C_4 provides the ADD 6 signal.

CHAPTER 6

1. 2.7 V

3. 5

5. 18 ns

7. 4.95 V

9. 4.6 mA

11.

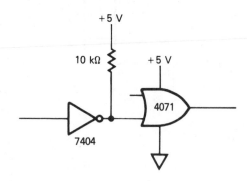

13.

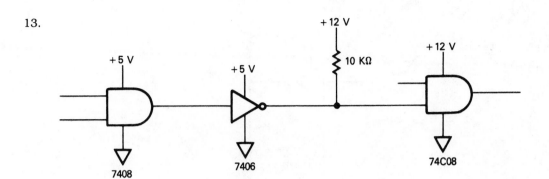

15.

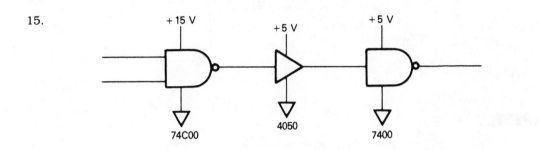

17.

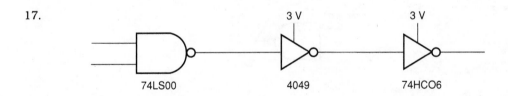

19. 4.5 to 5.5 V.

21. Needs a pull-up resistor

23. 10

25. FAST, S, ALS, LS, TTL, CMOS HC, CMOS

27. 0.001 Microamps

29. 0.05 V

31.

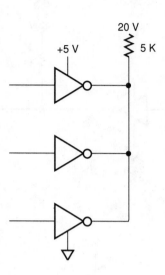

CHAPTER 7

1.

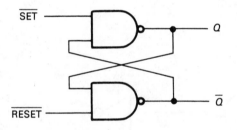

3.

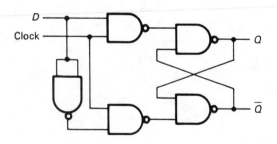 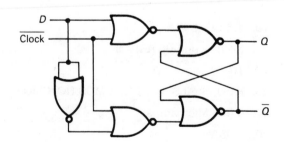

5.

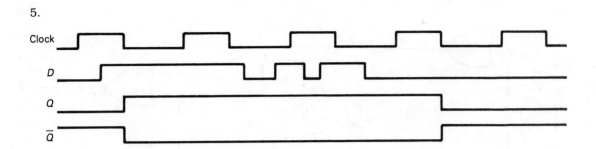

7.

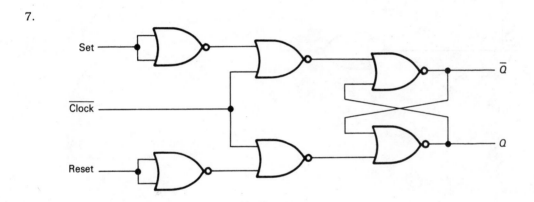

9.

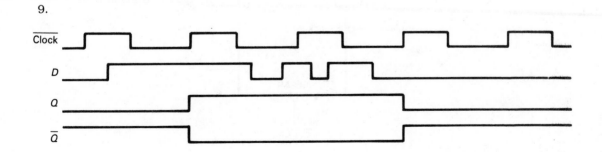

11.

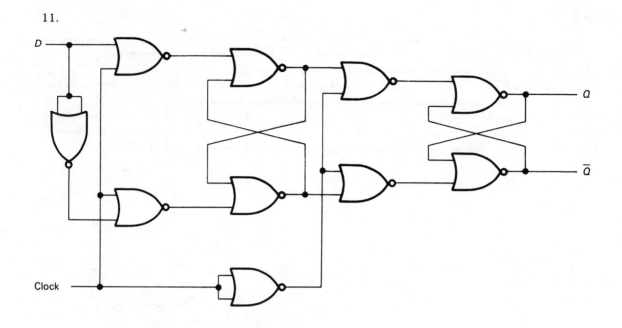

13.

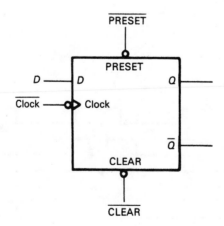

15.

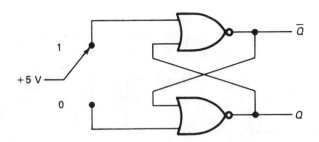

CHAPTER 8

1.

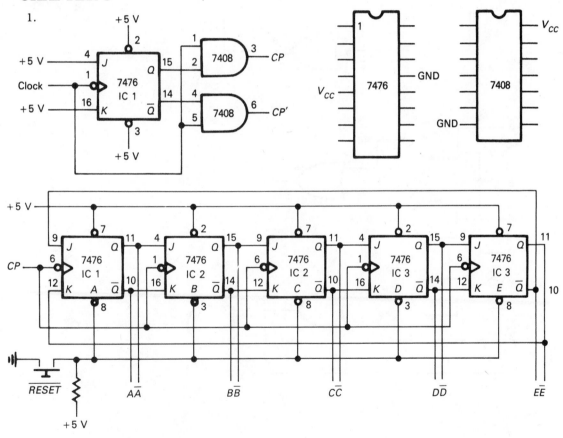

3.

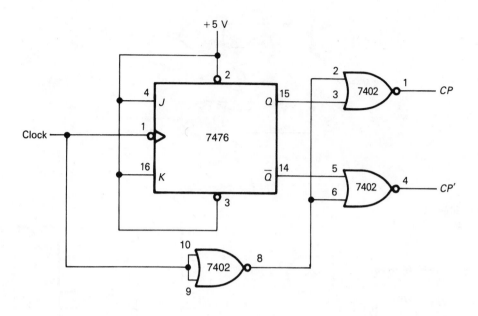

5.

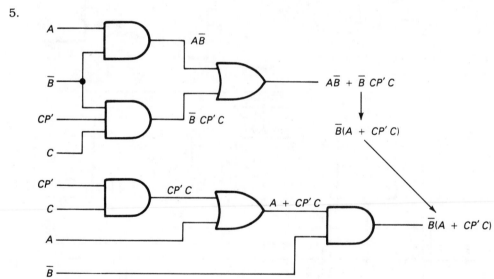

7.

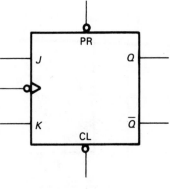

9.

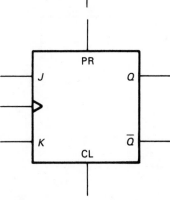

11.

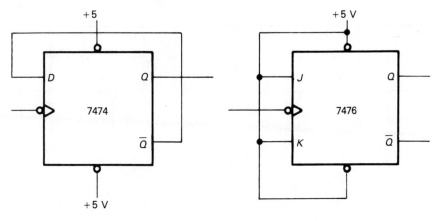

13.

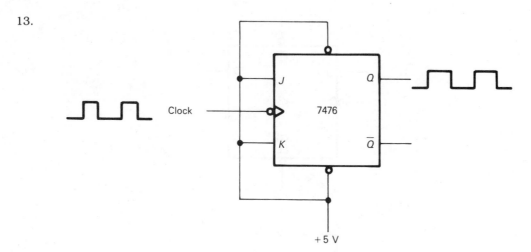

15. The same as the frequency of *CP*.

CHAPTER 9

1.

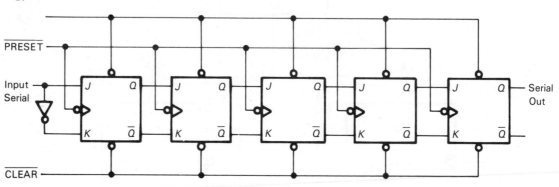

3.

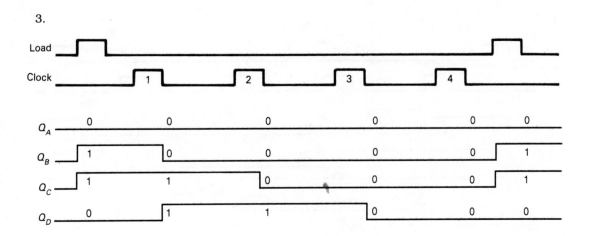

5.

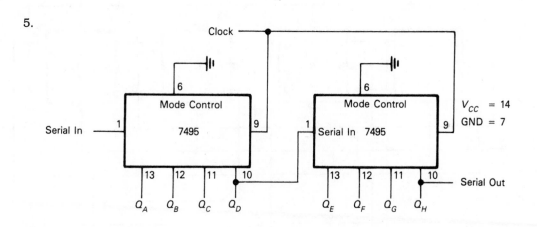

7.

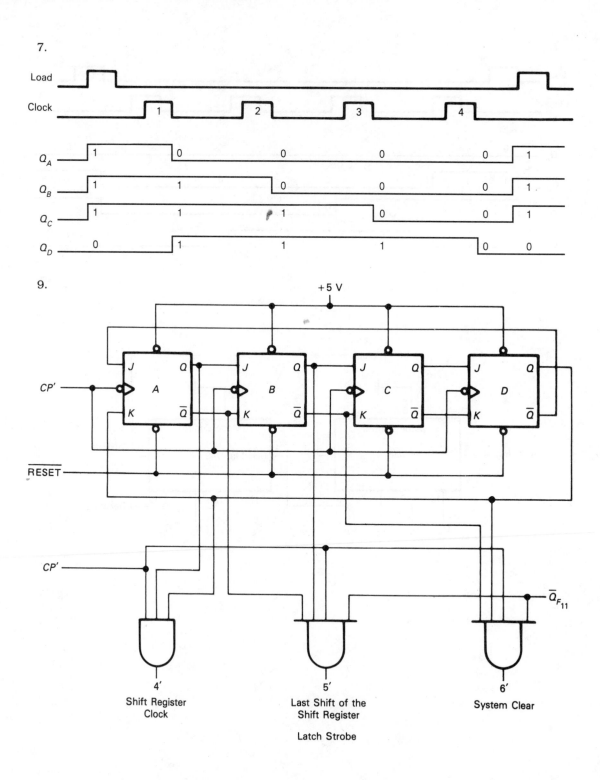

9.

4'
Shift Register
Clock

5'
Last Shift of the
Shift Register

Latch Strobe

6'
System Clear

11.

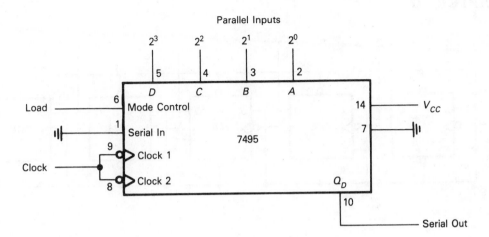

13.

D	I	G	I	T	A	L
44	49	47	49	54	41	4C

E	L	E	C	T	R	O	N	I	C	S
45	4C	45	43	54	52	4F	4E	49	43	53

15.

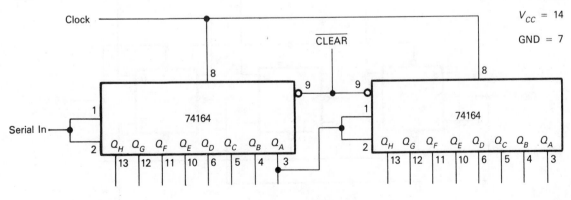

CHAPTER 10

1.

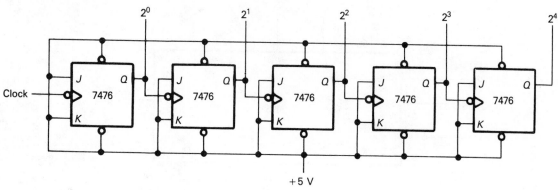

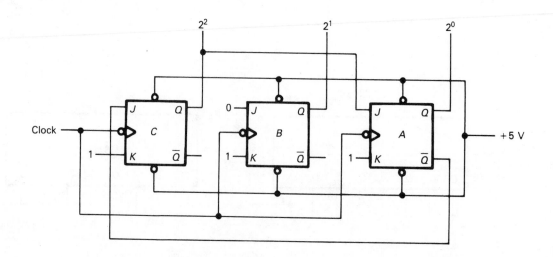

3.

Before Clock	After Clock	Before Clock	
Q	Q	J	K
0	0	0	X
0	1	1	X
1	0	X	1
1	1	X	0

Negative-Edge *JK* Flip-Flop

X = 1 or 0

Before Clock			After Clock			Before Clock					
Q			Q			C		B		A	
C	B	A	C	B	A	J_C	K_C	J_B	K_B	J_A	K_A
0	0	0	1	0	0	1	X	0	X	0	X
1	0	0	0	0	1	X	1	0	X	1	X
0	0	1	0	0	0	0	X	0	X	X	1

$$J_C = \overline{A} \qquad J_B = 0 \qquad J_A = C$$
$$K_C = 1 \qquad K_B = 1 \qquad K_A = 1$$

Clock Count			Outputs			
			2^2	2^1	2^0	
0	0	0	0	0	0	
0	0	1	1	0	0	Cycle 1
0	1	0	0	0	1	
0	1	1	0	0	0	
1	0	0	1	0	0	Cycle 2
1	0	1	0	0	1	
1	1	0	0	0	0	

Note: The *B* Flip-Flop could be eliminated.

5.

CMOS	TTL
74C74	7474
74C174	74174
74C175	74175
74C374	74374
4013	
4027	
4042	
40174	
4723	

7. The answer is the same as problem 3.

9.

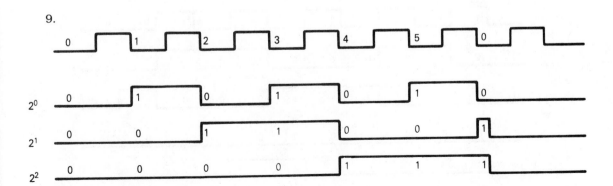

2^0

2^1

2^2

11.

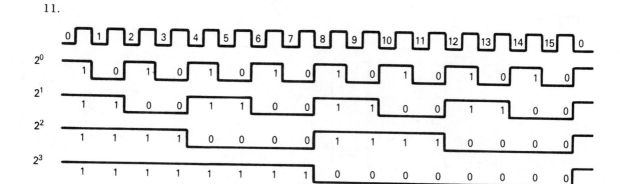

13.

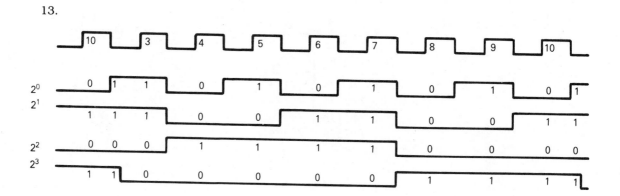

15.

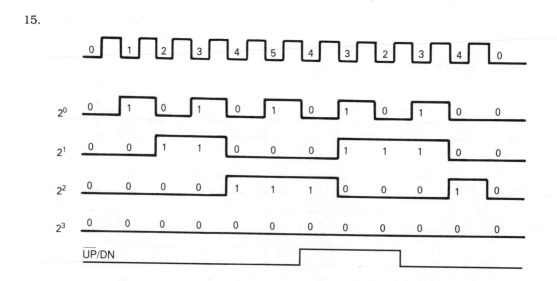

CHAPTER 11

1.

	Min.	Typ.	Max.
Upper Threshold	6.0 V	6.8 V	8.6 V
Lower Threshold	1.4 V	3.2 V	4.0 V

3.

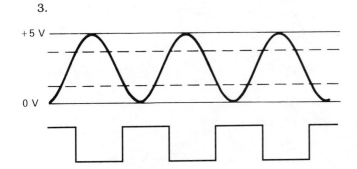

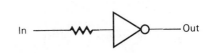

5. $F = \dfrac{0.663}{RC}$

7. When the capacitor charges, it charges through the resistor and the input to the Schmitt-trigger inverter. When the capacitor discharges, it discharges through only the resistor and therefore takes longer.

9.

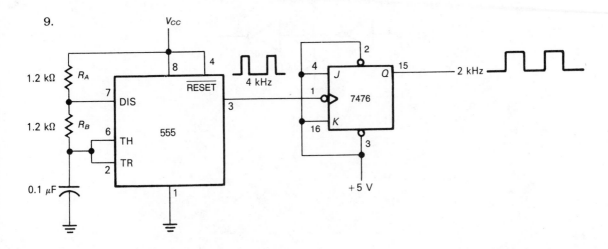

11. $F = \dfrac{3.45}{C(2R_B + R_A)}$

13.

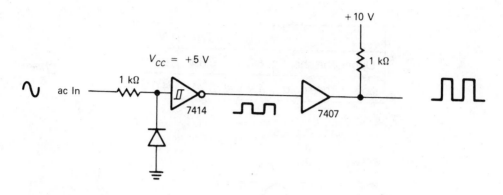

15. $F = \dfrac{0.529}{RC}$

CHAPTER 12

1. 34.184 µs

3. 78.74 kHz

5.

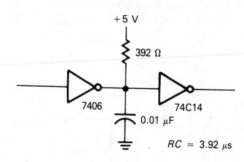

7.

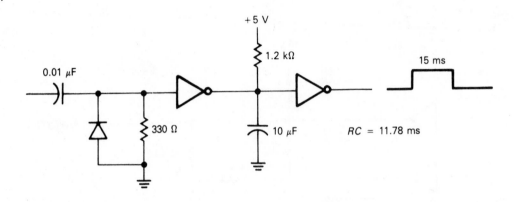

9.

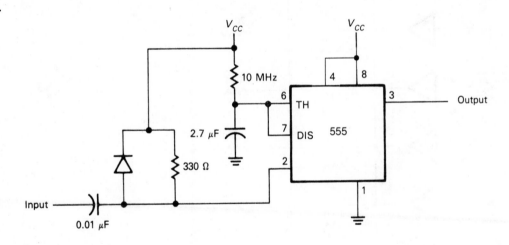

11. $22.\overline{2}$ kHz

13. 74C221, 4528, 4521, 4047

CHAPTER 13

1.

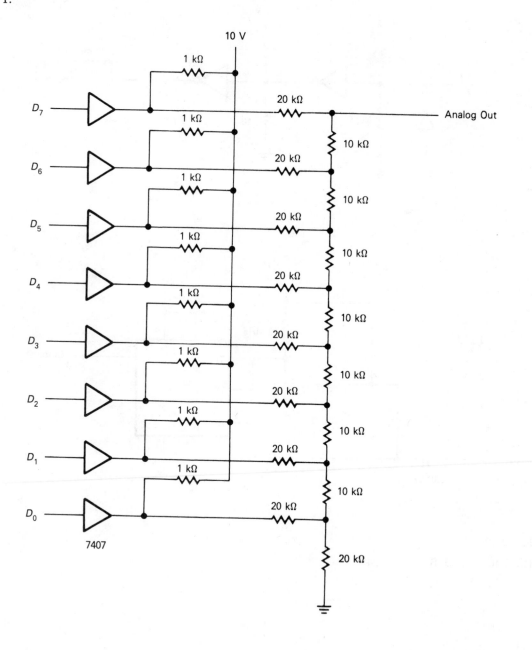

3. If V_S = 5V, the voltage increment is 0.3125 V.
   If V_S = 10 V, the voltage increment is 0.625 V.
   If V_S = 32 V, the voltage increment is 2.0 V.

5.

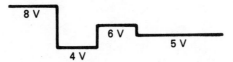

7. The purpose of the op amp is to prevent the 2R D-to-A converter from being loaded and distorting its output.

9.

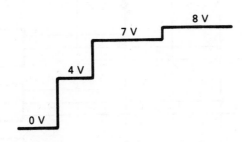

11. 500 Ω

13. V_S = 22.5 V

15.

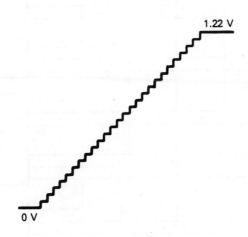

CHAPTER 14

1.

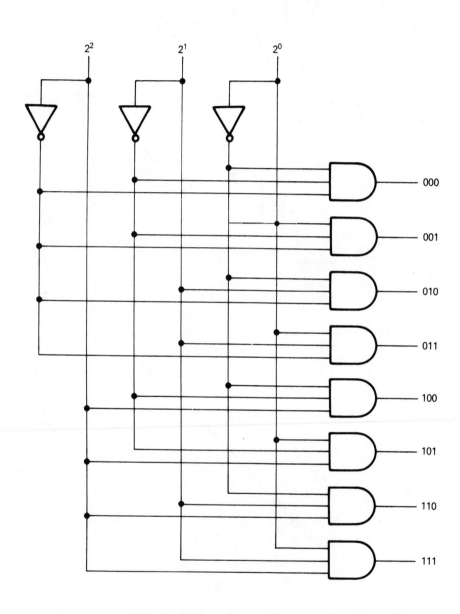

3.

	E	D	C	B	A	X	Input	Value
0	0	0	0	0	0	0	0	A
1	0	0	0	0	1	1		
2	0	0	0	1	0	1	1	1
3	0	0	0	1	1	1		
4	0	0	1	0	0	0	2	0
5	0	0	1	0	1	0		
6	0	0	1	1	0	1	3	$\overline{A}$
7	0	0	1	1	1	0		
8	0	1	0	0	0	1	4	$\overline{A}$
9	0	1	0	0	1	0		
10	0	1	0	1	0	0	5	A
11	0	1	0	1	1	1		
12	0	1	1	0	0	1	6	1
13	0	1	1	0	1	1		
14	0	1	1	1	0	0	7	0
15	0	1	1	1	1	0		

	E	D	C	B	A	X	Input	Value
16	1	0	0	0	0	0	8	0
17	1	0	0	0	1	0		
18	1	0	0	1	0	0	9	0
19	1	0	0	1	1	0		
20	1	0	1	0	0	1	10	1
21	1	0	1	0	1	1		
22	1	0	1	1	0	0	11	A
23	1	0	1	1	1	1		
24	1	1	0	0	0	1	12	$\overline{A}$
25	1	1	0	0	1	0		
26	1	1	0	1	0	0	13	A
27	1	1	0	1	1	1		
28	1	1	1	0	0	1	14	1
29	1	1	1	0	1	1		
30	1	1	1	1	0	1	15	$\overline{A}$
31	1	1	1	1	1	0		

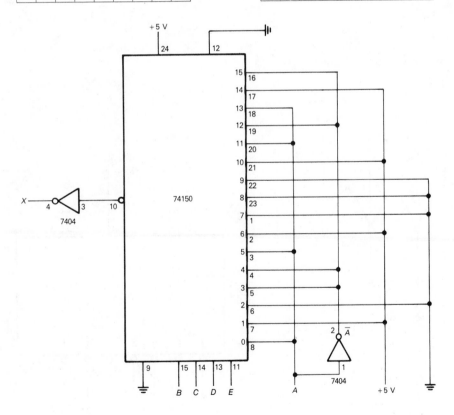

5.

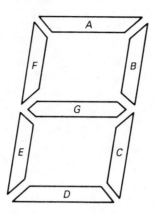

7. Dynamic LCD and Field-Effect LCD

9. LED

11.

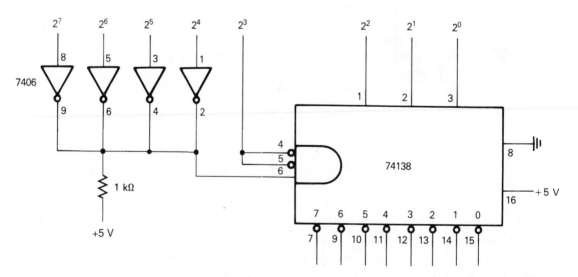

13.

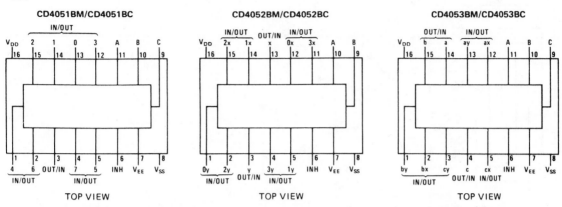

15. The 74C945 is a 4-digit counter for directly driving LCD displays. It contains a 4-decade up/down counter, output latches, counter/latch, select multiplexer and 7-segment decoders, backplane oscillator/driver, segment drivers and display blanking circuitry.

CHAPTER 15

1.

70C95/80C95 70C96/80C96 70C97/80C97 70C98/80C98	Tri-State Hex Buffers
4503	Hex Noninverting Tri-State Buffer
54C240/74C240 54C244/74C244 54C941/74C941	Octal Tri-State Buffers
4076	Tri-State Quad *D* Flip-Flop
74C374	Octal *D* Flip-Flop with Tri-State Outputs
74C373	Octal Latch with Tri-State Outputs
4043	Quad Tri-State NOR R/S Latches
4044	Quad Tri-State NAND R/S Latches
4048	Tri-State Expandable 8-Function, 8-Input Gate
4094	8-Bit Shift/Store Tri-State Register

3.

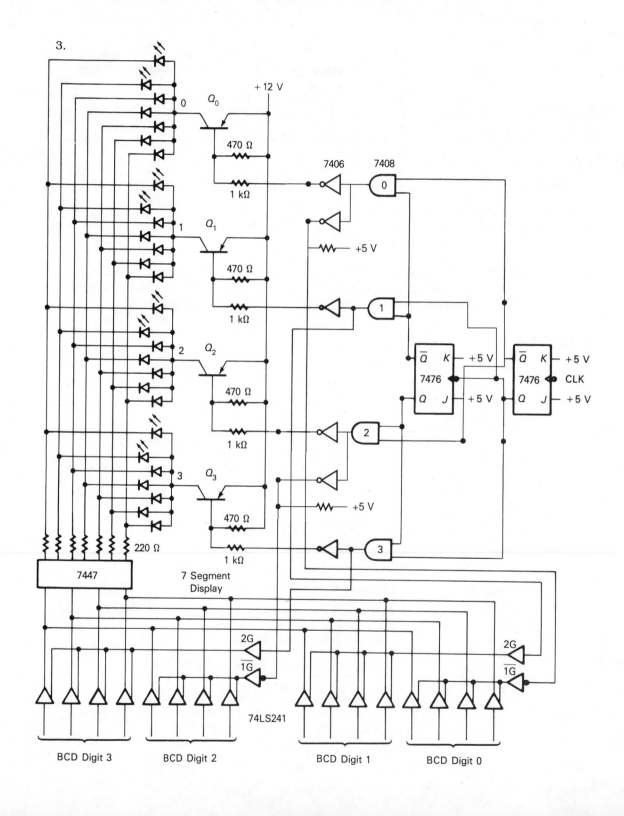

5.

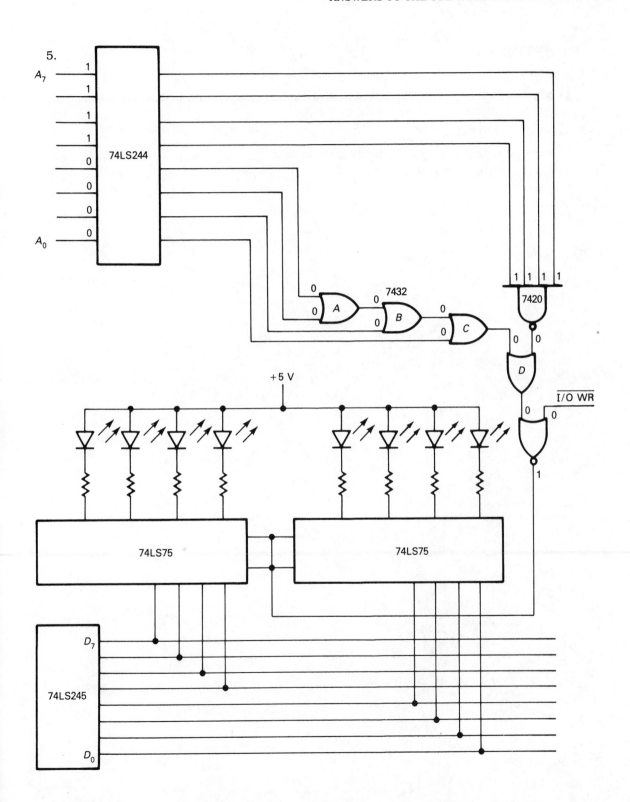

CHAPTER 16

1. 2 MHz

3. The Z-80 CPU will finish the current instruction and then place the address bus, data bus, and the control signals in Hi-Z state.

5. 8000 H to 87FF H

7. There are four present and four more can be added.

9. See Figure 16–14, page 407.

Index